Phased Array-Based Systems and Applications

WILEY SERIES IN MICROWAVE AND OPTICAL ENGINEERING

KAI CHANG, Editor
Texas A & M University

FIBER-OPTIC COMMUNICATION SYSTEMS · *Govind P. Agrawal*

COHERENT OPTICAL COMMUNICATIONS SYSTEMS · *Silvello Betti, Giancarlo De Marchis, and Eugenio Iannone*

HIGH-FREQUENCY ELECTROMAGNETIC TECHNIQUES: RECENT ADVANCES AND APPLICATIONS · *Asoke K. Bhattacharyya*

COMPUTATIONAL METHODS FOR ELECTROMAGNETICS AND MICROWAVES · *Richard C. Booton, Jr.*

MICROWAVE RING CIRCUITS AND ANTENNAS · *Kai Chang*

MICROWAVE SOLID-STATE CIRCUITS AND APPLICATIONS · *Kai Chang*

DIODE LASERS AND PHOTONIC INTEGRATED CIRCUITS · *Larry A. Coldren and Scott W. Corzine*

MULTICONDUCTOR TRANSMISSION-LINE STRUCTURES: MODAL ANALYSIS TECHNIQUES · *J. A. Brandão Faria*

PHASED ARRAY-BASED SYSTEMS AND APPLICATIONS · *Nicholas Fourikis*

FUNDAMENTALS OF MICROWAVE TRANSMISSION LINES · *Jon C. Freeman*

MICROSTRIP CIRCUITS · *Fred Gardiol*

HIGH-SPEED VLSI INTERCONNECTIONS: MODELING, ANALYSIS, AND SIMULATION · *A. K. Goel*

HIGH-FREQUENCY ANALOG INTEGRATED CIRCUIT DESIGN · *Ravender Goyal (ed.)*

FINITE ELEMENT SOFTWARE FOR MICROWAVE ENGINEERING · *Tatsuo Itoh, Giuseppe Pelosi, and Peter P. Silvester (eds.)*

OPTICAL COMPUTING; AN INTRODUCTION · *M. A. Karim and A. A. S. Awwal*

MILLIMETER WAVE OPTICAL DIELECTRIC INTEGRATED GUIDES AND CIRCUITS · *Shiban S. Koul*

MICROWAVE DEVICES, CIRCUITS AND THEIR INTERACTION · *Charles A. Lee and G. Conrad Dalman*

OPTOELECTRONIC PACKAGING · *A. R. Mickelson, N. R. Basavanhally, and Y. C. Lee*

ANTENNAS FOR RADAR AND COMMUNICATIONS: A POLARIMETRIC APPROACH · *Harold Mott*

INTEGRATED ACTIVE ANTENNAS AND SPATIAL POWER COMBINING · *Julio J. Navarro and Kai Chang*

FREQUENCY CONTROL OF SEMICONDUCTOR LASERS · *Motoichi Ohtsu (ed.)*

SOLAR CELLS AND THEIR APPLICATIONS · *Larry D. Partain (ed.)*

ANALYSIS OF MULTICONDUCTOR TRANSMISSION LINES · *Clayton R. Paul*

INTRODUCTION TO ELECTROMAGNETIC COMPATIBILITY · *Clayton R. Paul*

INTRODUCTION TO HIGH-SPEED ELECTRONICS AND OPTOELECTRONICS · *Leonard M. Riaziat*

NEW FRONTIERS IN MEDICAL DEVICE TECHNOLOGY · *Arye Rosen and Harel Rosen (eds.)*

NONLINEAR OPTICS · *E. G. Sauter*

FREQUENCY SELECTIVE SURFACE AND GRID ARRAY · *T. K. Wu (ed.)*

ACTIVE AND QUASI-OPTICAL ARRAYS FOR SOLID-STATE POWER COMBINING · *Robert A. York and Zoya B. Popovic (eds.)*

OPTICAL SIGNAL PROCESSING, COMPUTING AND NEURAL NETWORKS · *Francis T. S. Yu and Suganda Jutamulia*

Phased Array-Based Systems and Applications

NICHOLAS FOURIKIS
Defence Science & Technology Organisation
Salisbury, South Australia

A WILEY-INTERSCIENCE PUBLICATION
JOHN WILEY & SONS, INC.
NEW YORK / CHICHESTER / WEINHEIM / BRISBANE / SINGAPORE / TORONTO

This text is printed on acid-free paper.

Copyright © 1997 by John Wiley & Sons, Inc.

All rights reserved. Published simultaneously in Canada.

Reproduction or translation of any part of this work beyond that permitted by Section 107 or 108 of the 1976 United States Copyright Act without the permission of the copyright owner is unlawful. Requests for permission or further information should be addressed to the Permissions Department, John Wiley & Sons, Inc., 605 Third Avenue, New York, NY 10158-0012

Library of Congress Cataloging in Publication Data:
Fourikis, Nick.
    Phased array-based systems and applications / Nick Fourikis.
      p. cm. -- (Wiley series in microwave and optical engineering)
    Includes index.
    ISBN 0-471-01212-2 (cloth : alk. paper)
    1. Phased array antennas.   2. Microwave antennas.   I. Title.
  II. Series.
  TK6590.A6F69   1996
  621.382'4--dc20
                                                                                                   95-49567

0-471-01212-2
Printed in the United States of America

10 9 8 7 6 5 4 3 2 1

To my parents

Contents

FOREWORD	xix
PREFACE	xxiii
LIST OF ACRONYMS	xxvii

1	**Phased Array-Based Systems and Applications**		**1**
	1.1	Phased Array-Based Systems	4
		1.1.1 Radar Functions	4
		1.1.2 Radio-Astronomy Objectives	5
		1.1.3 Phased Arrays and What They Can Offer—Part 1	7
		1.1.4 Breaking the Nexus between the Aperture Real Estate and Spatial Resolution	8
		1.1.5 Breaking the Nexus between the Surveillance and Tracking Functions	9
		1.1.6 Some Archetypical Phased Array-Based Systems	9
		1.1.6.1 Directional Fixed-Beam Arrays	11
		1.1.6.2 Mechanically Steerable Fixed-Beam Arrays	12
		1.1.6.3 Electronically Steerable Arrays in One Dimension	14
		1.1.6.4 Electronically Steerable Arrays in Two Dimensions	14
		1.1.7 Historical Developments Leading to Current Phased Arrays	18
		1.1.7.1 Radar Developments	19
		1.1.7.2 Radio-Astronomy Developments	20
		1.1.7.3 Phased Arrays and What They Can Offer—Part 2	21
	1.2	Radar Systems	24
		1.2.1 The Radar Equation	25

			1.2.1.1	The RCS of Targets	26
			1.2.1.2	Atmospheric Attenuation as a Function of Frequency	26
			1.2.1.3	SNR Considerations	28
			1.2.1.4	Illustrative Examples	31
		1.2.2	The Performance of Radar Functions		33
			1.2.2.1	Volume Surveillance Radars	34
			1.2.2.2	Tracking Radars	34
			1.2.2.3	A Compromise Band for Multifunction Radars?	35
		1.2.3	Low-Probability-of-Interception (LPI) Radars		35
		1.2.4	Polarimetric Radars		36
			1.2.4.1	Theoretical Framework	36
			1.2.4.2	Implementation	38
			1.2.4.3	Rain-Induced Depolarization of EM Waves	39
			1.2.4.4	Instrumentation Requirements	39
			1.2.4.5	Benefits of Polarimetric Radars	41
		1.2.5	Bistatic and Multistatic Radars		41
			1.2.5.1	Target Position Measurements	44
			1.2.5.2	Contours of Constant Detection Range	44
			1.2.5.3	Receive Antenna Scanning Rate	45
		1.2.6	Synthetic Aperture Radars (SARs) and Inverse SARs (ISARs)		46
1.3	Basic EW Concepts				47
	1.3.1	ESM System Functions			47
	1.3.2	Concepts of Probability of Interception (POI)			49
	1.3.3	ECM Systems			49
	1.3.4	Approaches to ECCM			50
1.4	Radio-Astronomy Systems				51
	1.4.1	Fundamental Radio-Astronomy Concepts and Measurements			51
	1.4.2	Blackbody Radiation			53
	1.4.3	Derivation of Some Basic Relationships			55
	1.4.4	Image Theory Used for Radio-Astronomical Observations			57
			1.4.4.1	Picture-Point by Picture-Point Imaging Method and Applications	60
			1.4.4.2	Fourier Transform by Fourier Transform Imaging Method	65
1.5	Satellite Communication Systems				68
1.6	Future Directions and Trends				69

	1.6.1	Phased Array-Based Radar Functions		69
		1.6.1.1 Scanning or Staring Arrays?		69
		1.6.1.2 Modern LPI/FMCW Radars		72
		1.6.1.3 A Typical Modern Multifunction Radar		72
	1.6.2	Modern Approaches to the Identification Function		74
		1.6.2.1 ISARs		75
	1.6.3	Friend-or-Foe ID		78
	1.6.4	Performance of the Identification Function by Unconventional Approaches		78
	1.6.5	Future Challenges for Radars		79
		1.6.5.1 Reduction of the RCS of Airplanes		79
		1.6.5.2 Evolution of Advanced EW Systems		80
		1.6.5.3 Upgraded Weapons Requirements		80
		1.6.5.4 The Solutions Proposed by Xu		81
	1.6.6	The Case for Wideband Phased Arrays		81
		1.6.6.1 Multifunction Phased Arrays Performing Radar, EW, Radiometry, and Communications Functions		82
		1.6.6.2 Noncooperative Target Recognition		83
		1.6.6.3 Active Array-Based Radar Systems Applied to Air Traffic Control		83
		1.6.6.4 Unmanned Air Vehicle (UAV) Applications		84
	1.6.7	Advanced Phased Arrays Redefine Radar		84
	1.6.8	Developments in Radio and Optical Astronomy		86
1.7	Concluding Remarks and a Postscript			87

2 From Filled Apertures to Phased Arrays Mounted on Fully Steerable Structures — 89

2.1	General Considerations			90
	2.1.1	The Far-Field Patterns Resulting from Specific Illuminations		91
		2.1.1.1 Sum Patterns		92
		2.1.1.2 Half-Power Beamwidths of Apertures		94
		2.1.1.3 Other Illumination Functions		94
		2.1.1.4 Difference Patterns		97
		2.1.1.5 Conventional Measures of Aperture Efficiency		101
		2.1.1.6 Random Errors		103
		2.1.1.7 Systematic Errors		103
		2.1.1.8 Maximum Diameters for Conventional Apertures		104
		2.1.1.9 Unconventional Definition of Efficiency		104

xii CONTENTS

- 2.2 The Quest for More Efficient Apertures — 106
  - 2.2.1 Offset Reflectors — 106
    - 2.2.1.1 *Cancellation of the G.O. Cross-Polarization of Dual-Offset Systems of Reflectors* — 109
- 2.3 Focal Plane Imaging Systems — 114
  - 2.3.1 The Total Number of Antenna Beams — 115
  - 2.3.2 Spacing of Antenna Elements at the Focal Plane — 116
  - 2.3.3 Focal Plane Imaging Systems: Applications — 116
    - 2.3.3.1 *Radio-Astronomy Applications* — 116
    - 2.3.3.2 *Applied-Science Applications* — 117
- 2.4 Hybrid or Limited Scan Phased Array Systems — 119
  - 2.4.1 Applications — 120
- 2.5 Toward Phased Arrays — 122
  - 2.5.1 Attempts to Overcome the Fundamental Limitations of Single-Aperture Systems — 122
    - 2.5.1.1 *Phased Arrays Mounted on Fully Steerable Structures* — 123
- 2.6 Ideal Feed Horns — 127
- 2.7 Concluding Remarks — 129

3 Phased Arrays: Canonical and Wideband — 131

- 3.1 Introductory Background — 132
  - 3.1.1 Phased Array-Based Radar, EW, and Communication Systems — 133
  - 3.1.2 Astronomy, Geodesic, and Remote Sensing Systems — 134
- 3.2 Theoretical Considerations — 135
  - 3.2.1 The Far-Field Radiation Pattern of a Linear Phased Array — 135
  - 3.2.2 Array Grating Lobes — 142
  - 3.2.3 Phased Array Beamwidth and Bandwidth — 143
  - 3.2.4 Directivity of a Linear Phased Array — 145
  - 3.2.5 The Half-Power Beamwidth × Directivity Product for a Linear Phased Array — 145
- 3.3 Linear Array of Equispaced Line Sources of Nonuniform Amplitude — 146
  - 3.3.1 The Binomial Array — 146
  - 3.3.2 Array Synthesis Procedures — 147
    - 3.3.2.1 *The Synthesis of Dolph–Chebyshev Arrays* — 147

		3.3.2.2	The Schelkunoff Circle	151
		3.3.2.3	Elliott's Synthesis Procedure	153
		3.3.2.4	Taylor's One-Parameter Synthesis Procedure	156
		3.3.2.5	The Taylor $\bar{n}$ Distribution	158
		3.3.2.6	The R. C. Hansen Synthesis Procedure for Circular Distributions	160
		3.3.2.7	The Taylor $\bar{n}$ Circular Distribution	161
		3.3.2.8	The Bayliss (Difference) Distributions	161
		3.3.2.9	Other Synthesis Procedures	164
3.4	Far-Field Radiation Pattern of Planar Arrays			164
	3.4.1	Grating/Side Lobes: System Considerations		169
	3.4.2	HPBW and Directivity of Planar Phased Arrays		171
	3.4.3	The Half-Power Beamwidth $\times$ Directivity Product for Planar Arrays		172
	3.4.4	Input/Output SNR of Arrays		173
3.5	Cylindrical and Circular Phased Arrays			173
	3.5.1	The J. P. Wild Procedure to Minimize the Sidelobes of Circular Phased Arrays		174
3.6	Modern Array Synthesis Procedures			179
	3.6.1	The J. J. Lee Synthesis Procedure		179
	3.6.2	Spatial Tapors: The R. E. Willey Approach		182
		3.6.2.1	Spatial Tapers: Applications	184
		3.6.2.2	The Mailloux–Cohen Approach	187
		3.6.2.3	The Frank–Coffman Synthesis Procedure or Active Hybrid Arrays	187
		3.6.2.4	Derivative Approaches	191
3.7	Additional Quantization Errors			192
	3.7.1	Phase Quantization		192
	3.7.2	Time Delay Quantization		195
	3.7.3	Applications: The Utilization of Phase and Time Delay in Large Phased Arrays		196
3.8	Random Errors			196
	3.8.1	Realistic Examples of the Impact of Random Errors on MSSL		198
	3.8.2	The Statistical Prediction of Peak Sidelobes		199
3.9	Active and Passive Phased Arrays			201
	3.9.1	The Radar Equation Revisited		204
	3.9.2	The EIRP and G/T of Passive and Active Phased Arrays		205
	3.9.3	Design Options and Applications		207

3.10 Array Architectures	210
3.10.1 The Two Basic Array Architectures	211
3.10.1.1 Brick Architecture	211
3.10.1.2 Tile Architecture	211
3.10.2 Interconnect Approaches	212
3.11 Array Design Considerations	215
3.11.1 Array Costs	215
3.11.2 The Array Area/Volume and Thermal Problems	216
3.11.2.1 Array Cooling	217
3.11.2.2 The Humble Power Supplies	218
3.11.2.3 Typical Phased Array Switched-Mode Power Supplies	220
3.11.3 Applications: Brick or Tile Architectures?	220
3.11.3.1 Brick Architecture—Applications	220
3.11.3.2 Tile Architecture—Applications	221
3.11.4 The Integration of Modern Antenna Elements to the MMICs	223
3.11.4.1 Antenna Element Tunability	223
3.12 Wideband Phased Arrays	224
3.12.1 General Considerations	224
3.12.2 Array Considerations	226
3.12.3 Examples of Wideband Arrays	229
3.12.4 Phased Arrays for ESM and ECM Functions	229
3.12.4.1 Phased Arrays for the ESM Function	230
3.12.4.2 Phased Arrays for the ECM Function	231
3.12.4.3 Applications of ESM and ECM Arrays	233
3.12.5 The U.S. Navy's Airborne Early Warning (AEW) Radar	234
3.12.6 Multifunction, Wideband, or Shared-Aperture Systems	235
3.12.7 The Cottony and M. N. Cohen Approaches for Linear Wideband Phased Arrays	236
3.13 Special-Purpose Phased Arrays	238
3.13.1 Minimum- and Null-Redundancy Arrays	238
3.13.2 Applications of M/NRAs	240
3.14 Beamformers	241
3.14.1 General Considerations	243
3.14.2 The Formation of Stationary (Staring) Beams at RF and IF and Applications	244
3.14.2.1 Realizations Using Cables	244
3.14.2.2 Photonics-Based Beamformers	245
3.14.2.3 Realizations Using Resistive Networks	247

		3.14.2.4 Realizations Using the Blass and Butler Matrices—Applications	248
	3.14.3	The Formation of One Agile Beam—Applications	250
	3.14.4	The Formation of Several Agile Beams—Applications	251
	3.14.5	Photonics-Based Wideband Beamformers	252
	3.14.6	Digital Beamforming	253
	3.14.7	Nonlinear Beamformers	253
3.15	Array Performance Monitoring, Fault Isolation, and Correction Approaches		254
3.16	Affordable Phased Arrays: Systems Approaches		255
3.17	Concluding Remarks		257

4 Transmit / Receive Modules 259

4.1	Vacuum-Tube Amplifiers and Oscillators		261
	4.1.1	Slow-Wave, Linear-Beam Tubes (LBTs)	261
		4.1.1.1 The Klystron Family	261
		4.1.1.2 The TWT Family	264
	4.1.2	Slow-Wave, Cross-Field Tubes (CFTs)	267
	4.1.3	Fast-Wave Devices	267
		4.1.3.1 Gyrotrons	269
	4.1.4	R & D toward High-Power Vacuum-Tube Devices and the Microwave Power Module (MPM)	269
		4.1.4.1 High-Power Gyrotrons and Magnetrons	270
4.2	Solid-State Transmitters and Low-Noise Amplifiers		271
4.3	Important Comparisons between SSDs and Vacuum Tubes		274
	4.3.1	Power Output as a Function of Frequency	274
	4.3.2	Noise Figure and Thermal Power Noise Density of Devices	277
	4.3.3	The MTBF Issue	280
	4.3.4	Relative Initial and Life-Cycle Costs	281
	4.3.5	The PAE of Devices	282
	4.3.6	The MTI Stability	286
	4.3.7	Graceful Degradation Considerations	286
		4.3.7.1 A Definition of the Combining Efficiency under Realistic Conditions	288
	4.3.8	Duty-Factor Considerations	290
	4.3.9	Power Output Improvement per Decade	291
	4.3.10	Specific-Weight Factor	292
	4.3.11	Miscellaneous Considerations	293

4.4	Recent Developments toward High-Power Modules		293
	4.4.1	The MPM Module: Concepts, Applications, and Goals	294
		4.4.1.1 The Vacuum Power Booster (VPB)	*295*
		4.4.1.2 The MMIC Amplifier	*295*
		4.4.1.3 The Electronic Power Conditioner (EPC)	*296*
		4.4.1.4 System Considerations	*297*
4.5	Summary, Discussion, and Future Trends		297
	4.5.1	Advantages Offered by Solid-State-Based Systems	298
	4.5.2	Advantages Offered by Systems Utilizing Tubes	299
	4.5.3	The Future	299
4.6	The Solid-State T/R Modules		300
	4.6.1	MMIC Options	301
	4.6.2	MMIC T/R Module with or without the Antenna?	302
	4.6.3	T/R Module Realization Approaches	303
	4.6.4	The Yield of MMICs	305
		4.6.4.1 Statistical Approaches to Increase Yield and Decrease Cost	*305*
		4.6.4.2 Deterministic Approaches to Increase Yield and Reduce Cost	*306*
		4.6.4.3 Design Guidelines for Increased Yield	*306*
		4.6.4.4 The MSAG Fabrication Process for MESFETS: A High-Yield Process	*308*
		4.6.4.5 Other Approaches to Maximize Yield	*310*
		4.6.4.6 MMIC Cost Minimization Approaches	*312*
		4.6.4.7 Double- and Single-Chip T/R Modules	*312*
		4.6.4.8 RF-Wafer-Scale Integration: Another High-Yield Approach	*315*
		4.6.4.9 Recent Experience with Multichip T/R Modules	*316*
		4.6.4.10 T/R Realization Issues: Concluding Remarks	*317*
4.7	The Constituent Parts of T/R Modules		318
	4.7.1	Baseline Characteristics of Power and Low-Noise Amplifiers	318
		4.7.1.1 Recent Power Amplifiers	*318*
		4.7.1.2 Recent Low-Noise Amplifiers (LNAs)	*323*
		4.7.1.3 Power and Low-Noise Amplifiers: Concluding Remarks	*324*
	4.7.2	The Receiver Protector or Limiter	324
	4.7.3	Programmable Phase Shifters and Vector Modulators	326
		4.7.3.1 The Ferrite Phase Shifter	*327*
		4.7.3.2 Solid-State Phase Shifters	*327*

			CONTENTS	xvii

	4.7.4	Phase Shifters: At RF or IF	331
	4.7.5	Circulators	331
		4.7.5.1 Ferrite Circulators	334
		4.7.5.2 Miniature Circulators	334
		4.7.5.3 MMIC-Based Circulators	335
4.8	Concluding Remarks		335

5 Antenna Elements — 337

5.1	Outline of the Requirements	338
5.2	Candidates for Phased Array Antenna Elements	339
	5.2.1 Antenna Elements for Canonical Arrays, Scanning in One Dimension	340
	5.2.2 Antenna Elements for Canonical Arrays, Scanning in Two Dimensions	342
5.3	Patch Antennas	342
	5.3.1 Patch Dimensions	342
	5.3.2 The Electric Field of the Patch Antennas	346
	5.3.3 Bandwidth and Efficiency of Patch Antennas	348
	5.3.4 The Worst-Case Cross-Polarization Level	350
	5.3.5 Mutual Coupling Between Microstrip Antennas and Scan Blindness	353
	5.3.6 Losses Due to Surface Waves	358
	5.3.7 Microstrip-Based Phased Arrays: Design Guidelines	358
	5.3.8 Methods of Feeding Patch Antennas	359
	5.3.9 The Limitations of Conventional Patch Antennas	360
5.4	Microstrip Dipoles	360
	5.4.1 Dipoles versus Patches?	361
	5.4.2 Applications: Wideband Printed Dipoles Suitable for Brick Array Architectures	363
	5.4.3 Applications: An Experimental Array Utilizing Crossed Printed Dipoles	364
5.5	The Quest for High-Quality Dual-Polarized Antennas	365
	5.5.1 Single-Element Approaches and Applications	368
	5.5.1.1 Conventional Patches	368
	5.5.1.2 The Electromagnetically Coupled (EMC) Patch	372
	5.5.2 Subarrays of Antenna Elements	372
	5.5.2.1 Subarrays of Similar Patches Having the Same Orientation and Applications	374
	5.5.2.2 Sequential Subarrays and Applications	375
5.6	Work in Progress	375

5.7	Wideband Antenna Elements	378
	5.7.1 Log-Periodic, Patch-Based Antenna	378
	5.7.2 Spiral-Mode Microstrip (SMM) Antenna	379
	5.7.3 Planar and Antipodal Tapered Slotline Antennas	381
5.8	Concluding Remarks	385

References 387

Index 420

Foreword

These are the dawning days of a new breed of advanced radar systems that utilize the gallium arsenide solid-state monolithic microwave integrated circuits (MMIC) in the front end. These circuits are small, highly integrated, and reliable. They operate at microwave and millimeter wave frequencies—more than an order of magnitude higher than can be achieved with silicon technology. Operating bandwidths can be designed that are a full octave. Other outstanding characteristics are very high-power conversion efficiency and frequency stability. MMIC is to the radar system front end what VLSI and VHSIC silicon technology has been to radar signal processing and the digital computer world. The front end encompasses the output power amplifier for the transmit function, the low-noise amplifier for the receive function, as well as other radar microwave functions such as gain control, phase shifting and control, and transmit/receive switching.

A miniature high performance solid-state front end is the key to achieving ever-increasing new radar system applications. Conventional radar, communication, and EW systems depend on a centralized transmitter tube that is the least reliable unit in the entire system; it is measured in hours. The tube type transmitter is bulky and has limited existing radar systems to 10% prime power efficiency and 10% bandwidth. A centralized solid-state transmitter can be built with a large array of low-power MMIC amplifiers that are combined to produce very high output power. Such a transmitter is inherently fail soft because many of the amplifiers must fail before the transmitter degrades below its specified power output. Its reliability is measured in years.

MMIC is the technology that enables the long sought after active electronic scanned array (AESA). The antenna of this system is composed of an array of radiators backed by MMIC-based transmit/receive (T/R) modules. Each of 100's or 1000's of modules contains a low-power transmitter and the receiver LNA. The active array need not scan mechanically; scanning is accomplished electronically by adjusting the phase within each T/R module (i.e., radiating element) to focus the wave front in the desired direction. All the functions: transmit, LNA, phase shifting, gain control, and T/R switching are contained on one or more miniature MMIC chips. The recent generation of lower cost T/R modules from companies such as ITT, Raytheon, TI, Westinghouse, and Hughes has made the active array a reality by virtue of its cost effectiveness. Dr. Fourikis places major emphasis on the T/R module by dedicating a whole chapter (Chapter 4) to this subject.

Today's advanced airborne fighter and ground-based ballistic missile defense system developments have incorporated active arrays because they enable the radar to out perform the advanced military threat challenge. Key performance advantages are the wide (octave) bandwidth, spectral purity, and the agile beam pointing capabilities. They lead to some significant system advantages. The active array radar is capable of true multimode operation, e.g., both search and track functions can be interleaved instantaneously. The phased array exhibits rapid beam agility for accurately tracking many targets while simultaneously scanning the search volume for other targets. Advanced waveforms depend upon the agile beam to allow immediate data update after initial target detection so that a firm track is initiated within milliseconds. An alert/confirm algorithm allows 50% longer detection and track range for the same radiated power. More threat platforms can be detected and tracked at longer range making the active array a true force multiplier. Also, the same antenna can be used for radar, communication, or EW since all phased arrays share the same basic principles. The wideband multifunction benefits of the active array for military and civilian use is a recurring theme of this book.

Other applications are being realized because of other advantages such as reliability and prime power efficiency. Active arrays are being incorporated into air traffic control radars, where safety is extremely important. The active array is inherently fail-soft because of the large number of parallel devices. A total failure of 5% of the T/R modules randomly distributed within an array results in a degradation of antenna sidelobes of only 3 dB. Detection range is reduced to only 90% of full performance when as many as 10% of the T/R modules fail completely. If a T/R module fails in only its transmit or receive channel, then even less of an impact on system performance is observed. T/R module MTBF is between 50,000 and 100,000 hours, which means that the resulting radar system reliability is measured in years instead of hours.

Prime power efficiency also allows the military to develop systems that are highly mobile. The overall module RF to dc power efficiency is approximately twice the efficiency of tube or silicon solid-state transmitters for microwave systems. Prime power is further conserved by the reduced RF plumbing (waveguide) losses, adding up to between 2 and 8 dB depending on the system being compared. Consequently, the deployment of future systems should be greatly simplified. For example, for an average radiated power of 25 kW, the volume and weight are each reduced by a factor of three and the volume-weight product is reduced by a factor of nine.

The wide bandwidth and beam agility provide the flexibility and capability to design new, unique waveforms tailored to specific system requirements (e.g., long-coded pulses for low probability of intercept, high duty cycle for Doppler unambiguous detection and combinations for detection and tracking of advanced threats.) Flexible waveform signal modulation of the T/R module enables variable control of duty cycle from very short pulse duration to CW. Also, the waveform amplitude, phase, frequency and timing can be dynamically controlled as a function of the threat and environment, whether it contains land or sea clutter, rain, ECM or friendly interference.

Looking further into the future, the active array will become part of the surface structure of the host platform, such as an aircraft fuselage. This application is known as "smart skins" because the array will be able to automatically reconfigure itself when part of the array is damaged. The self-healing aspect of the active array requires sophisticated system calibration and control. The concept is that the array contains real-time test and diagnostics so that the radar processor continuously monitors the state of cach of its T/R modules. It is also important that the T/R modules are capable of being reconfigured to provide other system functions. Very wide bandwidths are required; e.g., a portion of the array could shift from C-band to X-band.

Another complex system concept is adaptive beamforming. Instead of combing the output of each T/R module and then processing the combined signals like in a conventional radar, each T/R module output would be converted by in-phase and quadrature mixing to baseband and digitized within the module. The antenna beam is then formed in the digital processor along with the typical Doppler processing and detection.

Cost reduction has driven the T/R module technology developments over the last 5 years. Over 1 billion dollars has been invested by the U.S. government for gallium arsenide MMIC and related T/R module technology development during the last ten years. Also, several manufacturing technology programs were completed which provided the seed incentive to develop low-cost module production lines. Costs in large quantities have approached the magic $1000 price range.

One of these companies, ITT, has advanced a low-cost T/R module production concept based on a high level of functional integration of the MMIC chip. During the MIMIC program, one of the most highly integrated chips ever produced was provided by ITT; it contains all the functions of a T/R module on a single chip (both high-power circuits and small signal circuits). Subsequently, modules have been produced that contained only a single MMIC chip that included all the active microwave circuitry and the digital interface logic. With this technology modules can be produced that have less than 50 total parts. Low parts count means low material cost, low assembly cost and low test cost.

With these and greater design opportunities within reach, Dr. Fourikis has undertaken a timely but challenging effort. His book is very comprehensive in addressing detailed design issues. He includes an in depth survey of important system applications. But most importantly, he provides the engineer with a detailed technical roadmap that is foundational in the development of modern active array systems. The key to the development of future radar systems will be to match the revolutionary signal processing capabilities with the evolving flexibility of active aperture architectures to solve the specific and multifaceted challenges of emerging military and civilian requirements that can be met by the transfer of the expertise gained in the military arena under the dual use technology programs.

TOM BRUKIEWA
Manager, Active Array Radar Systems, ITT Gilfillan

August 1996

Preface

Phased arrays have been used in fields as diverse as radar, communications, electronic warfare (EW), and radio astronomy; furthermore, researchers plan to use phased array-based systems to enhance airport safety and traffic efficiency, take astronomical observations of unprecedented spatial resolution at optical and radio wavelengths, remotely sense the environment, and manage the Earth's resources. Given that the first phased array was used to improve the communication links between the East Coast of the United States (Holmdel, New Jersey) and the United Kingdom in 1937, phased arrays have enjoyed an unparalleled popularity over a relatively short time.

Phase arrays continue to contribute to the defense of many countries and valuable platforms such as ships and airplanes critically depend on phased array-based radars for their survival. Similarly, it is not an exaggeration to state that phased arrays established radio astronomy as a new science that is as respectable as optical astronomy. Before the arrival of phased arrays that offered high spatial resolution, radio telescopes utilizing conventional apertures produced blurry images of celestial sources of dubious value. Radio astronomy unraveled the radio universe, which turned out to be as interesting as the universe that many generations of optical astronomers explored. Our knowledge of the cosmos has been enriched by the complementary nature of knowledge derived from radio and optical astronomy.

This book is dedicated to the radar engineer, researcher, and/or practitioner with an interest in phased array-based systems and applications and is founded on the premise that invaluable knowledge can be derived by exploring diverse systems that not only share the same fundamentals but also perform complementary functions. At a more operational level there are two main thrusts that further support our approach.

It is very hard, if not impossible, to design radar systems without a knowledge of EW that includes the passive sensing of the environment, specifically electronic support measures (ESM), electronic countermeasures

(ECM), and electronic countercountermeasures (ECCM); additionally, radar systems have generic similarities to communication systems. Finally, ESM systems and the systems dedicated to remotely sense the environment bear a remarkable resemblance to radio-astronomy systems. The other dimension of complementarity is based on the view that one does not want to reveal one's presence in the course of sensing the environment with the aid of a radar. So ESM and radar systems are used over long and short periods of time, respectively, in low-probability-of-interception (LPI) applications.

The other thrust emanates from system considerations which we shall trace after a definition of the scope of the book. The scope of the book includes the many diverse applications and phased array-based systems operating at centimeter-range (cm) and millimeter-range (mm) wavelengths. However, only the radio-frequency (RF) technology aspects of narrowband and wideband phased arrays are included in the scope of the book. Thus signal processing techniques are considered only if they have an impact on the RF aspects of the array design.

Thomas F. Brukiewa, Manager of Advanced Sensors at ITT, succinctly stated[1] the reasons for this book's orientation: "The focus for the next-generation advanced capability is now the radar front-end, i.e., the RF high frequency analog portion of the radar" and "The key to the development of future radar systems is to match the revolutionary signal processing capabilities with the evolving flexibility of active aperture architectures to solve the specific and multifaceted challenges of emerging threats."

What follows is a thematic description of the essential contents of the book, while the contents list is a chapter-by-chapter description of the topics covered in the book. The cost of phased array-based radars limited their use to military applications for a long time; similarly, the majority of radio telescopes have been realized under the aegis of national organizations because of the costs involved. Costs, however, significantly decreased mainly because of the efforts of a large and diverse group of talented researchers, engineers, and technologists who derived innovative ways to manufacture affordable monolithic microwave integrated circuits (MMICs), which are used to realize the transmit/receive T/R modules of the array.

Given that an array can have 100–100,000 modules and that these modules constitute 50% of the array cost, affordable phased arrays will soon be used in many civilian applications of considerable applied research value. The MIMIC (MIllimeter and MIcrowave Integrated Circuits) Program significantly contributed to progress in this area.

While the MMICs are undisputably the preferred candidates for low-noise amplification, medium-power amplification, high-power amplification at frequencies below 20 GHz, and various signal processing functions, tubes are the preferred candidates for the generation of high powers at millimeter

---

[1] Thomas F. Brukiewa, "Active Arrays: The Key to Future Radar Systems Development," *Journal of Electronic Defense*, p. 91, Sept. 1992.

wavelengths. Naturally there is an in-between territory in the frequency–power domain that is being contested by the two camps. Recently researchers combined the advantages afforded by solid-state and vacuum technologies to produce remarkable power modules under the aegis of the Microwave Power Module Program.

Antenna engineering, often regarded by some software engineers as a dead art, has been described by J. R. James as "a vibrant field which is bursting with activity, and is likely to remain so in the foreseeable future," in 1990.[2]

Microstrip, dual-polarization antenna elements in a variety of configurations and designs are now lightweight and electromagnetically efficient and can conform to the surface of a platform of an airplane or a vehicle; furthermore several solutions of the problem of interfacing them to MMIC-based T/R modules have been proposed.

Photonics, a relatively new field, can reduce the weight of a phased array by distributing a variety of RF and digital signals to the many array elements and T/R modules. Similarly, the bulk of beamforming networks of scanning and staring phased arrays can be significantly reduced by the use of photonic techniques. Conventional beamforming techniques and approaches, on the other hand, offer considerable versatility to match a variety of requirements.

It is now appropriate to consider the other fundamental thrust which supports the approach we have taken in this book; two examples will suffice to demonstrate the emerging trend. Future affordable phased array-based systems will be truly wideband so that each array will be capable of performing a set of interrelated and interdependent functions with unprecedented efficiency and reliability on a time-sharing basis.

Platforms having a shared aperture system will have one phased array supporting the radar, EW (ESM and ECM), communications and navigation functions, utilizing the GPS, global positioning system; additionally the platform will have ECCM capabilities. Similarly the many functions now performed by different radars in airports, such as airport surveillance, precision approach and terminal Doppler weather radar, can be performed in the future by one single wideband phased array. The latter, next generation systems will meet the future challenges imposed by an ever-increasing traffic in civilian and military airports and the need to increase system reliability which is inherently related to effective traffic control and safety on a 24 hour/day, 365-day/year basis. It is no longer reasonable to treat systems performing interrelated and interdependent functions in different books.

The social benefits resulting from current and novel phased array-based systems, currently in the planning stage, are enormous and defy a quantitative assessment. What is as exciting, however, is that phased arrays have redefined the radar functions, are used to remotely sense the Earth's environment with

---

[2] J. R. James, "What is New in Antennas," *IEEE Antennas and Propagation Magazine*, vol. 32, no. 1, p. 6, Feb. 1990.

unprecedented sensitivity and spatial resolution and radio telescopes yield images of celestial sources having spatial resolutions in the sub-milli-arc-second range. Finally, phased array-based optical telescopes are in the planning stage and phased arrays are to be used on board communication satellites positioned on the geostationary orbit and on board low-/medium-height orbits satellites that will usher in the wireless revolution. Is it any wonder that phased arrays are the subject of current leading-edge research and development?

The Defence Science and Technology Organisation, DSTO, allowed me to undertake the task of writing this book but the views expressed in it are my own.

It is a pleasure to acknowledge the help of Drs. Nick Nick Shuley and John Cashman, who reviewed parts of the manuscript and offered many helpful suggestions. Drs. David Jauncey and Rene Grogniard, provided several important references, and Dr. Roy Hughes produced some figures included in the book. Robyn E. Fourikis was instrumental in many improvements to the text and Koula Valiotis managed the production of the manuscript. Maureen Prichard was commissioned to produce the dust jacket illustration.

NICHOLAS FOURIKIS
Ascot Park, Adelaide, South Australia

February 1996

List of Acronyms

AAR	active array radar
A/D	analog-to-digital
AESA	active electronically scanned array
AEW	airborne early warning
AGF	array geometric factor (pattern)
AMRAAM	advanced medium-range air-to-air missile
AOA	angle of arrival
AQ	amplitude quantization
ARM	antiradiation missile
ASCM	antiship cruise missile
ASDE	airport surface detection equipment
AWACS	airborne warning and control system
BJT	bipolar junction transistor
BWO	backward-wave oscillator
CBIR	cross-beam interferometer
CFA	cross-field amplifier
CFT	cross-field tube
C^3I	command, control, communications intelligence
CIM	computer-integrated manufacturing
C/N	carrier-to-noise ratio
COMINT	communications intelligence
CW	continuous wave (FM: frequency-modulated FM)
ECCM	electronic counter countermeasures
ECM	electronic counter measures
EDFA	erbium-doped fiber amplifier
EFIE	electric-field integral equation

EIKA	extended-interaction klystron amplifier
EIKO	extended-interaction klystron oscillator
EIRP	effective isotropic radiated power
ELINT	electronic intelligence
EMC	electromagnetically coupled
EMI	electromagnetic interference
EMP	electromagnetic pulse
EOM	electrooptic modulator
EPC	electronic power conditioner
ERP	effective radiated power
ESM	electronic support measures
ESTAR	electronically scanned thinned-array radiometer
EW	electronic warfare (E–W—east–west)
FET	field-effect transistor (MESFET-metal–semiconductor FET)
FMA	ferrite microstrip antenna
FOV	field of view
FPA	final power amplifier
GBR	ground-based radar
g.o.	geometric optics
GPS	Global Positioning System
HBT	heterojunction bipolar transistor
HDMP	high-density microwave packaging
HEMT	high-electron-mobility transistor
HPA	high-power amplifier
HPBW	half-power beamwidth
HWSIC	hybrid wafer-scale integrated circuit
ICBM	intercontinental ballistic missile
IFF	identification friend-or-foe
IMPATT	impact ionization avalanche transit time
IRD	image rejection down converter
J/S	jammer-to-signal ratio
KGD	known good die
LBT	linear-beam tube
LCC	life-cycle cost
L/RHC	left-/right-handed circular
L/RHS	left-/right-hand side
LITA	longitudinal integration and transverse assembly

LNA	low-noise amplifier
LOS	line of sight
LPI	low probability of interception
MAFET	microwave and analog front-end technology
MAG	maximum available gain
MAM	multibeam array model
MBA	multibeam array
MCA	multichip assembly
MCM	multichip module
MHDI	microwave high-density interconnect
MIMIC	millimeter and microwave integrated circuit (program)
MMA	millimeter array
MMIC	monolithic microwave integrated circuit
M/NRA	minimum-/null-redundancy array
MPM	microwave power module
MRA	minimum-redundancy array
MSAG	multifunction self-aligned gate
MSSL	mean-square sidelobe level
MTBCF	mean time between catastrophic failures
MTBF	mean time between failures
MTI	moving-target indicator
NPD	noise power density
OTHR	over-the-horizon radar
PAE	power-added efficiency
PAR	precision approach radar
PCA	printed-circuit antenna
PCB	printed-circuit board
PILOT	Philips indetectable low-output transceiver
PIN	positive–intrinsic–negative
PM	pseudomorphic
POI	probability of interception
PRF	pulse repetition frequency
PSM	polarization scattering matrix
RAPPORT	rapid alert and programmed power (management) of radar targets
RBF	radial basis function
RCS	radar cross section
RIN	relative-intensity noise
RWR	radar warning receiver

SAG	self-aligned gate (MSAG-multifunction SAG)
SAR	synthetic aperture radar (ISAR-inverse SAR)
S/C	signal-to-scatter (ratio)
SDH	selectively doped heterostructure
SFDR	spurious-free dynamic range
SIS	superconductor-insulator-superconductor
SLL	sidelobe level
SMM	spiral-mode microstrip
SMPS	switched-mode power supply
SNR	signal-to-noise ratio
SSD	solid-state device
SSPA	solid-state power amplifier
TDRSS	Tracking and Data Relay Satellite System
TDWR	terminal Doppler weather radar
TEG	two-dimensional electron gas
TEM	transverse electromagnetic mode
TILA	transverse integration and longitudinal assembly
TOA	time of arrival
T/R	transmit/receive
TSA	tapered slotline antenna
TWT	traveling-wave tube (TWTA: TWT amplifier; CCTWT: coupled-cavity TWT)
UAV	unmanned air vehicle
URR	ultrareliable radar
VCO	voltage-controlled oscillator
VLA	very large array
VLBA	very long-baseline array
VLBI	very long-baseline interferometer (OVLBI-orbiting VLBI)
VPB	vacuum power booster
VSWR	voltage standing-wave ratio

CHAPTER ONE

Phased Array-Based Systems and Applications

> When you see something you say why? But sometimes I dream of things and think "why not"?
>
> George Bernard Shaw (1856–1950)

The collecting area of a conventional aperture is a continuous structure, whereas the collecting area of a typical radar array is made up of several antenna elements that take the form of conventional dipoles, slotted waveguides, horns, or microstrip antennas. In both cases the half-power beamwidths (HPBWs) are governed by diffraction.

In another realization the array antenna elements take the form of conventional reflectors that are either fixed on the ground or are transportable on railway lines. These are the classical radio-telescope realizations. There is, indeed, a large diversity in the realizations of phased arrays because the requirements for the many applications differ.

Array apertures and solar concentrators share the same principle of operation illustrated in Figure 1.1(*a*), which is a photograph of the Sandia Laboratories' Solar Thermal Test facility. The 222 heliostats that reflect the solar sunlight onto one point located on the tower are shown, and each mirror is made up of 25 small mirrors, illustrated in Figure 1.1(*b*). The collected solar power is used to generate electricity [1.1]. The power collected by all the constituent mirrors is added at one point in space.

There are many stories attributed to Archimedes (b. 290–280, d. 212/211 B.C.). According to one story, he burned the Roman ships that contributed to the siege of Syracuse, his native city state, sometime between 215 and 212 B.C. Although the story is apocryphal [1.2], Sakas [1.3] successfully reenacted the achievement attributed to Archimedes in 1963. An artist's impression of how the soldiers of Syracuse could have achieved their objective is shown in

2 PHASED ARRAY-BASED SYSTEMS AND APPLICATIONS

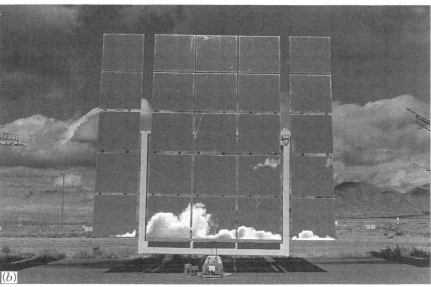

FIGURE 1.1 (*a*) The solar Thermal Test Facility, located in Albuquerque, New Mexico, is operated by the Sandia National Laboratories. The facility has 222 heliostats that are computer controlled to direct the reflected solar energy on to the receiver housed in the tower that is 200 ft high. (*Courtesy*: Sandia National Laboratories.) (*b*) One of the Facility's heliostats, which consists of 25 mirrors. (*Courtesy*: Sandia National Laboratories.)

Figure 1.2. The soldiers holding highly reflective flat shields directed the sunlight toward one Roman ship at a time.

The many principles on which phased arrays are built will be explored in this chapter after synoptic formulations of the requirements and aims attributed to modern radars and radio astronomy facilities, are outlined. We shall then explore how well phased arrays meet these requirements and aims by considering a number of current archetypical phased array-based systems.

Analytic work will follow in order to define the key parameters of the current and next-generation phased array systems. This chapter will end with the descriptions of future phased array-based systems and the underlying reasons for this movement toward phased arrays.

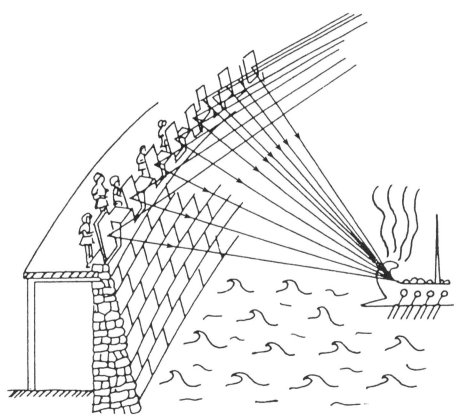

FIGURE 1.2 Artist's impression of how the Roman Ships could have been burned during the siege of Syracuse (215–212 B.C.).

1.1 PHASED ARRAY-BASED SYSTEMS

The impetus for phased array developments came from the radar and radio-astronomy communities; it is therefore useful to begin our account of current phased array-based systems by descriptions of radar and radio-astronomy requirements and systems.

1.1.1 Radar Functions

A radar emanates electromagnetic (EM) energy and detects the echo or reflection from (reflecting) objects of interest or targets. Several important target characteristics, such as its range (distance between the radar and target) and angular location, can be derived from the reflected signal. The range is derived by measuring the time interval between the transmission of the EM pulse and the reception of its reflection, while the angular location of a target can be derived by measuring the angle of arrival of the reflected signal; here we have assumed that the radar's aperture generates a highly directional beam. The simple equations that yield the radar range R and the beamwidth θ corresponding to the radar's aperture having a diameter D are $R = (\Delta t)c/2$ and $\theta = b\lambda/D$, where Δt is the time delay between the transmit/receive pulse, c the velocity of light, λ the wavelength of operation, and b a beam broadening factor that generally depends on the illumination of the aperture.

If a target is moving, the reflected radiation will exhibit a frequency shift better known as *Doppler* (after C. J. Doppler, 1803–1853) shift, which is used not only to determine the radial velocity of the target but also as a discriminant of moving targets from clutter (unwanted returns in the current context) resulting from stationary masses of land and/or sea. If Δf is the measured Doppler shift of a target $\dot{R}$, the radar–target range rate, or radial velocity, is given by the equation $\dot{R} = -(\Delta f)\lambda/2$; with this convention, positive Doppler shifts result in negative values of $\dot{R}$ (closing targets).

Conventional radars are used to "see" (detect) targets, such as ships, aircrafts, and missiles through clouds, mist, smoke, rain, and haze. The targets are first detected and subsequently tracked. Although most of these functions can be performed by passive sensors, radars can measure the range of the target from the radar.

In summary, then, the conventional radar functions are to

- Survey a certain volume of space
- Detect and track as many targets as possible
- Determine as many parameters (e.g., range, bearing, and radial velocity) related to these targets as possible

These functions are required not only during clear weather conditions and in electronically benign environments but also during inclement weather conditions and in the presence of intentional and unintentional jammers; while intentional jammers need no elaboration, unintentional jammers can be neighboring radars or broadcasting transmitters. The protective measures a radar can have through appropriate electronic counter-counter measures, ECCM design, are outlined in Section 1.3.4.

After the Gulf incident, when an Iranian airbus was mistakenly identified as an F-14 fighter and downed [1.4], radar users are no longer satisfied in seeing blips on radar screens; they would like to see and recognize images of targets (aircrafts and ships). So another important radar function has been added to the preceding list: identification. As we develop the appropriate framework, other functions will be added to the preceding list.

To perform its functions the radar's aperture should be adequate to yield the required resolution; whether the aperture is fully or partly populated by active antenna elements depends on a tradeoff between cost and the range requirements. More specifically, the maximization of the radar's range can be attained by fully populating the array aperture and connecting every antenna element to a high-power transmit/receive, T/R module.

While one radar performs the surveillance and target tracking functions, the identification of one important target is usually undertaken by another radar. The radar functions listed above are only preliminary and are put forward to initiate discussion. We shall explore the many subtleties related to the approaches taken to perform these and other additional radar functions in this chapter. In Table 1.1 we have tabulated the many frequency bands together with their designations adopted by different regulatory bodies and organizations.

1.1.2 Radio-Astronomy Objectives

One of the main objectives of radio astronomy is to deduce the prevalent emission mechanisms taking place in and around celestial sources of interest. This aim can be reached if high-spatial-resolution maps of sources taken at different wavelengths are obtained. By comparing the received energy emanated by a celestial source at different wavelengths and a knowledge of typical energy–frequency relationships attributed to different emission mechanisms, we can deduce which emission mechanisms are at work in and around the celestial source.

The resolution of the early radio telescopes was too low and the resulting radio images too blurry to be useful. More explicitly, the early radio telescopes had spatial resolutions of the order of 1 degree of arc, so many sources supporting different emission mechanisms were included in their field of view (FOV).

How high should the spatial resolution be? As high as possible—but what is a reasonable figure to aim for derived from previous knowledge? The majority of radio sources outside our galaxy (extragalactic radio sources) have an angular extent of less than 1 arc minute and conventional optical tele-

TABLE 1.1 Designations of Frequency Bands

FREQUENCY	λ	NATO	FREQUENCY BAND	RADAR BAND DESIGNATION
250 MHz	1.2 cm	A	MF / HF / VHF	
500 MHz	60 cm	B	UHF	
750 MHz	40 cm	C	UHF	
1000 MHz	30 cm	C	UHF	
2 GHz	15 cm	D		L
3 GHz	10 cm	E	3 GHz	S
4 GHz	7.5 cm	F		S
5 GHz	6 cm	G		
6 GHz	5 cm	G		C
7 GHz	4.3 cm	H	SHF	C
8 GHz	3.75 cm	H	SHF	
9 GHz	3.33 cm	I		X (12 GHz)
10 GHz	3 cm	I		X (12 GHz)
20 GHz	1.5 cm	J		K_u (18 GHz) / K (27 GHz)
40 GHz	0.75 cm	K	30 GHz	K_a
60 GHz	0.5 cm	L		
80 GHz	0.375 cm	M	EHF	mm
100 GHz	0.3 cm	M	EHF	mm

Millimeter Waves: 30, 42, 46, 54, 62, 92, 96, 137, 143 — Q, V, W, D (1 cm to 1 mm, 300 GHz) scopes have a resolution of 1–0.5 arc second, which is deemed as minimum. Spatial resolutions of the order of sub-milli-arc seconds are required to study compact radio sources such as the cores of galaxies and interstellar masers—see Section 1.4.1.

The quest for spatial resolutions of the order of 1 arc second guided the efforts of radio astronomers for several decades; they aimed to match the resolutions of radio and optical images for comparative studies, at affordable costs. If we assume that the resolution of an aperture of diameter D operating at a wavelength λ is given by the relationship λ/D, a 1-arc-second resolution can be attained by a radio telescope operating at 1 or 10 GHz, if its diameter is 61.88 km or 6.188 km, respectively. Considering that the largest monolithic radio telescope has a diameter of 305 m (hole-in-the-ground variety, located in Arecibo, Puerto Rico), these diameters are very large, indeed. Conventional apertures therefore could not meet this fundamental requirement and phased arrays could.

The last reason why high spatial resolution is required is more technical, and we shall revisit this topic in Section 1.4.3. For the time being it is sufficient to state that we need to measure the angular extent of a source and the energy it emanates at many frequencies before we can make any definitive statements related to the nature of the emission mechanism at work in or near the source. Phased arrays can yield the required spatial resolution.

1.1.3 Phased Arrays and What They Can Offer—Part 1

We have already seen that an aperture can be synthesized by summing up the contributions of many small apertures. For solar concentrator applications the phasing of the contributions of the many apertures is not necessary and the concentrator is said to operate in a "light bucket" mode.

The contributions of the many antenna elements of a phased array have to be phased, as the term implies, and summed to yield a synthesized beam. If the phases of all antenna elements are equal, the resulting beam points in the direction of the aperture's boresight axis; conversely, the beam can be pointed in other directions, $\theta_1, \theta_2, \theta_3, \ldots, \theta_n$ if the phase sets $S_1, S_2, S_3, \ldots, S_n$ are introduced between the antenna elements and the summing point. Since the phases can be changed electronically, an inertialess beam is formed that can be directed toward any direction within the array FOV, defined by the radiation pattern of the antenna elements. Although we have assumed a receive mode of operation here, similar arguments hold for the case where the array operates in the transmit mode.

For radar applications the two properties of phased arrays mentioned above—the power addition of all contributions from the many array elements and the realization of one agile and inertialess beam—are of considerable import. The third attractive characteristic of phased arrays is related to their conformity to the "skin" of an aircraft or vehicle. Thus the addition of a phased array-based system on to a platform does not affect its aerodynamic properties.

For radio-astronomical applications phased arrays offer affordable spatial resolutions; indeed, the first radio telescopes that yielded spatial resolutions of a few arc minutes, were realized by the "smell of an oil rag"—a phrase attributed to Professor W. N. Christiansen. Naturally the latest phased array-based telescopes are more expensive to construct but well within the budgets of many nations.

From the foregoing considerations it is not difficult to conclude that the essential reason for the popularity of phased arrays is their ability to meet the diverse requirements of several fields of scientific and engineering endeavors.

1.1.4 Breaking the Nexus between the Aperture Real Estate and Spatial Resolution

Phased arrays break the conventional nexus between the aperture real estate and spatial resolution. Given that aperture real estate is usually directly related with cost, tradeoffs can be made between the required resolution and cost. The designer can therefore attain the required spatial resolution by realizing a thinned array, i.e., not fully populated by antenna elements, and the extent of the thinning can be defined by the budget available for the radio telescope. Let us examine a couple of extreme cases that will add value to the comparisons between phased arrays and conventional apertures.

Even if one could realize a conventional reflecting aperture having a diameter of several kilometers, there are no known methods to mechanically move it toward the desired directions. The conventional aperture could be a hole in the ground, in which case some coverage can be attained by utilizing the Earth's rotation and/or the limited scanning properties of reflectors. Even then the telescope's FOV would be limited. If one required a resolution of the order of 1 milli-arc second or less, the conventional approach would have required an aperture having one or two Earth diameters—clearly an impossible task. By contrast, Earth-bound antennas working in conjunction with satelliteborne antennas, can yield the required spatial resolution—see Section 1.4.4.2.2. The above-mentioned resolution is not hypothetical and is required to study, inter alia, the centers of galaxies. It is therefore not an exaggeration to state that phased arrays established radio astronomy as a new science that is as respectable as optical radio astronomy. Radio astronomy unraveled the radio universe, which turned up to be as interesting as the universe that many generations of optical astronomers explored. Our knowledge of the cosmos has been enriched by the complementary nature of the knowledge derived from radio and optical astronomy.

1.1.5 Breaking the Nexus between the Surveillance and Tracking Functions

Large reflecting apertures rotating in azimuth are a common sight in civilian and military airports. The apertures mechanically rotate over 360° in azimuth to monitor the surveillance volume centered around the airport.

As the conventional radar rotates, it detects and tracks objects of interest, such as planes and rain fronts. The surveillance function is therefore inextricably coupled to the tracking function of conventional radars. By contrast a phased array-based radar yields an inertialess antenna beam that can track up to 1000 targets on a time-sharing basis. With this arrangement the user decouples the surveillance function from the tracking function of the radar. More explicitly, the proportions of time spent to perform the tracking and surveillance function are negotiable and depend on the operational and environmental requirements.

A plethora of advantages, including longer ranges and/or better system signal-to-noise ratios (SNRs), over conventional radars stem from this exceptional characteristic of phased array-based radars. Lastly systems utilizing mechanically steerable apertures waste time traveling from target A to target B to target Z, whereas systems utilizing electronically steerable phased arrays waste no time visiting targets $A, B, \ldots, Z$. The shortcomings of systems utilizing apertures that can be mechanically steered are as important. For a start the tracking of some 1000 targets scattered all over the surveillance volume at high speeds by conventional systems is not a trivial task, some will consider it impossible. As the traffic density increases with the passage of time, the same task will present insurmountable difficulties. The other aspect of the problem is reliability; it is well known that the reliability of conventional mechanically steerable apertures is low—see Section 1.6.7.

The only disadvantage of phased arrays, when compared to conventional apertures, is that the shape of their resulting beam and their HPBW generally vary as a function of the direction to where the beam is pointed—see Section 3.4.2. By contrast, the beamshape and beamwidth of a conventional aperture is at least theoretically invariant—see Section 2.1.1.7.

1.1.6 Some Archetypical Phased Array-Based Systems

In this section we shall consider some archetypical phased arrays and a variety of phased arrays in Chapter 3.

Phased arrays are conveniently divided in two broad categories: linear and planar arrays. Linear arrays yield fan beams, while pencil beams result from planar arrays. Fan beams are broad in the direction perpendicular to the length of the array and narrow in the direction along the length of the array (hence the name). Figure 1.3 illustrates a linear phased array consisting of several unit collecting areas. The resulting array pattern, resembling a diffraction grating, and the array field of view are also shown. The far-field radiation pattern, illustrated in the figure, is confined within the array FOV,

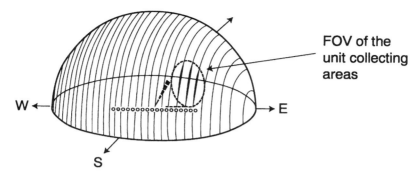

FIGURE 1.3 The grating lobes of an EW linear array are shown. Within the field of view, FOV, of the unit collecting areas used, the main beam and two greating lobes are seen.

and consists of the main beam located in the center of the FOV and two gratings, also termed grating lobes, which are located on either side of the main beam. Pencil beams resembling Gaussian beams result from arrays occupying circular areas.

Several arrays performing the electronic warfare, EW functions are linear but here we shall consider a linear array used for AEW (airborne early warning). The requirement for an airborne platform to perform the early warning function stems from the limitation of ground-based radars. While the latter radars perform well at high elevation angles, their performance deteriorates at low elevation angles; intruder planes and helicopters therefore can enter a territory unobserved. However if the radar is airborne, or on board an aerostat or a satellite, low flying aircrafts entering a territory under surveillance are easily detected.

The Ericsson's AEW system, illustrated in Figure 1.4, consists of two collinear arrays, 8 m long, that are mounted on top of a Fairchild Metro III airplane. The radar operates in the E/F bands (2–4 GHz) and performs its functions over either side of the airplane.

The principal requirements of the Ericsson AEW system are: high-performance capability; multirole air, sea, and ground surveillance; and long-range radar function with three-dimensional capabilities in the air surveillance mode [1.5].

The array utilizes some 200 T/R modules and the resulting beam can be electronically scanned over either side of the aircraft over $\pm 60°$ in azimuth. Given that the radar's scan is electronic, the scan rate is almost infinitely and instantaneously variable. It therefore possesses all the advantages attributed to phased arrays, that is, the surveillance and tracking functions are decoupled. The running cost for this AEW system is low because the aircraft is small (it weighs one quarter as much as a Grumman E-2-C) and the data gathered is downlinked to the ground-based C^3I (command, control, and communications intelligence) network.

FIGURE 1.4 Ericsson's AEW (airborne early warning) phase array. (*Courtesy*: Ericsson.)

Planar arrays are further divided in the following categories: directional fixed-beam arrays, mechanically steerable fixed-beam arrays, and electronically steerable arrays in one and two dimensions.

1.1.6.1 Directional Fixed-Beam Arrays If all the antenna elements of a phased array are connected in phase to a receiver or transmitter, the resulting antenna beam will be along the boresight axis of the array. An easy way to realize the phased array described above is to connect all the antenna elements to one receiver/transmitter via equal lengths of cable. In one realization, the 64 patch antennas shown in Figure 1.5 are connected to the receiver/transmitter at point A, via minimal loss connections. Patch antennas can be realized on printed-boards and what is shown in Figure 1.5 is the artwork used for the etching process. The microstrip lines connecting each patch antenna to point A have equal lengths.

These realizations are phased array substitutes of conventional monolithic apertures, such as parabolic reflectors and lenses. The advantages over their conventional counterparts are that they are volumetrically attractive, less expensive to realize, and have lower inertia; the last characteristic is important when the antenna is mechanically pointed to several targets at high speeds. If the resulting beam has to be steered, these arrays can be consid-

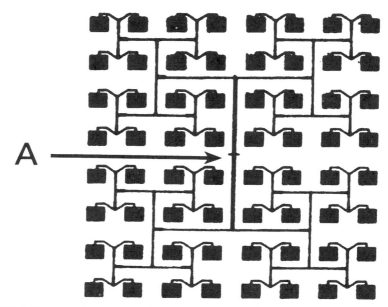

FIGURE 1.5 Layout of an array having 64 antenna elements, patch antennas, which are interconnected by equal lengths lines to a receiver/transmitter at point A.

ered as intermediate steps between continuous apertures and electronically steerable phased arrays.

Examples of directionally fixed-beam arrays are illustrated in Figures 1.6a and 1.6b. The arrays shown are the SEASAT and SIR-A, 128-element L-band planar array antennas, which were parts of a spaceborne imaging radar system used for global ocean and land surveillance [1.6]. The spatial resolutions for SEASAT and SIR-A were 25 and 40 m, respectively. The images obtained by the SEASAT system contributed to studies of the large-scale features of oceans.

The dimensions of the arrays are 10.7×2.2 and 9.4×2.2 m, respectively, and their gains at 1275 and 1278 MHz were 34.9 and 33.8 dB, respectively; overall losses were 2.3 dB for SEASAT and 2.9 dB for SIR-A [1.6, 1.7].

1.1.6.2 Mechanically Steerable Fixed-Beam Arrays Mechanically steerable fixed beam arrays result if any of the arrays described above is gimbaled in two dimensions. These arrays have all the attractive features of the arrays in the previous category plus a wider FOV, which is attained by the mechanical movement of the array. Two important disadvantages of these arrays are:

1. Mechanical systems require many parts, which historically have high failure rates. As an example, the antenna of the F-15's AN/APG-63 is the radar's highest failing component [1.8].

FIGURE 1.6 Examples of directional fixed-beam microstrip arrays: (*a*) the SEASAT array; (*b*) the SIR-A array. (*Source*: Mailloux et al. [1.6], © 1981 IEEE.)

2. The inertia of mechanically scanned arrays is a fundamental constraint, especially when tracking multiple high-velocity targets and/or when the radar has to perform several functions. Electronically steerable arrays do not share this important limitation.

Most U.S. fighters, such as the F-15, -16, and -18 aircraft, utilize radar arrays of this kind mounted on the nose of the aircraft; the radar performs a multitude of functions that ensure the survival of the fighter. The F-16 radar could have the following generic characteristics [1.8]:

1. The antenna size is about 750 × 500 mm
2. The mean time between failures (MTBF) is 150–200 h.
3. The weight is approximately 140 kg.
4. The flyaway price is around $1.8 million (M) (with limited spares), although the lifetime cost can typically be 2–3 times the initial cost over a 15–20-year period [1.9].

Figure 1.7 illustrates the nose array radar of the F-18. The nose radar on board the B1-B overcomes many of the above limitations—see Section 1.6.5.3.

1.1.6.3 Electronically Steerable Arrays in One Dimension Electronically steerable arrays in one dimension are relatively easier to realize and less costly than arrays that are electronically steerable in two dimensions. Usually these arrays are electronically steerable in the elevation and mechanically steerable in azimuth, so that a 360° surveillance volume is continually monitored.

A typical example of radar arrays in this category is the S-band AN/APY-1 fitted to the E-3A AWACS (airborne warning and control system), which consists of a 707 civilian aircraft, on top of which the antenna system, enclosed in a rotodome, is mounted. The oblate ellipsoid-shaped radome rotates with the antenna in azimuth and is electronically scanned in elevation. The E-3A is designed for airborne warning and control of airborne assets in conflict. The radar array has the following characteristics [1.8]:

1. Approximate azimuth and elevation beamwidths are about 1° and 4.75°, respectively.
2. Ultralow sidelobes in azimuth (e.g., ≤ −40 dB, probably −50 dB).
3. The cost of an E-3A aircraft system is about U.S. $268 M, 80% of which is attributed to the radar system.
4. 360° coverage in azimuth.
5. The antenna size is 7.5 × 1.5 m.

Figure 1.8a illustrates the antenna of the E-3A, which consists of a planar waveguide slot array of 4000 slot radiators, illustrated in Figure 1.8b. The low sidelobes of the array essentially decrease the false alarm rate of the system.

1.1.6.4 Electronically Steerable Arrays in Two Dimensions Electronically steerable arrays in two dimensions have N_{mn} antenna elements, and each element has a programmable phase shifter Φ_{mn} (or a delay line D_{mn}). Under computer control the phases ϕ_{mn} (or delay lines d_{mn}) are adjusted so that the

FIGURE 1.7 The F-18 radar array, which is mechanically steerable. (*Courtesy:* DSTO.)

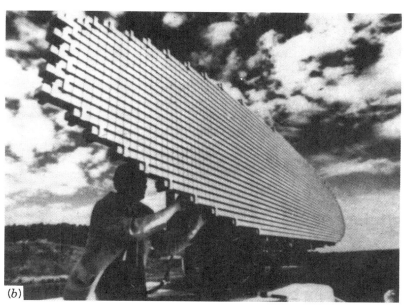

FIGURE 1.8 The E-3A AWACS radar phased array: (*a*) the E-3A AWACS in flight; (*b*) the E-3A low-sidelobe antenna. (*Courtesy*: Westinghouse.)

resulting antenna beam is electronically steered to the required position. For narrowband operation phases are introduced while delay lines are used for wideband operation—see Section 3.2.3 for a quantitative discussion on this topic.

The bulk of phased array-based systems are in this category, but here we shall consider only two. The very-large-array (VLA) radio-astronomy facility and the PAVE PAWS system, one of the largest phased array-based radars built in the United States.

1.1.6.4.1 The VLA Radio-Astronomy Facility The VLA, commissioned in the early 1980s, illustrated in Figure 1.9, is situated in the Plains of San Augustin (west-central New Mexico) 80 km west of the town of Socorro [1.10]. It consists of twenty-seven, 25-m-diameter antennas arranged in a Y-shaped array. Each arm of the Y is approximately 21 km long, and the antennas can be moved to various predetermined positions on the arms by a rail-mounted transporter. The interconnections between antenna stations is implemented with the aid of low-loss large-diameter waveguides operating in the TE_{01} mode.

At the time of its commissioning all antennas were equipped with dual-polarization receivers operating at the wavelengths of 1.3, 2, 6, 17 and 22 cm. It is the largest radio-astronomy compact array, and its cost in 1977 was U.S. $78 M.

The array geometry allows the observer to attain variable spatial resolution by adjusting the overall dimensions of the array and its interelement spacings. A plethora of high-sensitivity and high-resolution observations have been taken by the VLA.

1.1.6.4.2 The PAVE PAWS System The PAVE PAWS phased array system consists of two phased arrays each having a 120° FOV so that the total coverage of the system is 240° in azimuth [1.11]. Only one face of the array is shown in Figure 1.10(a). The system is designed to provide early warning of attacks by submarine-launched ballistic missiles and to aid in tracking satellites at the UHF (ultra-high-frequency) band.

Each face has a diameter of 31.09 m and 1792 antenna elements followed by solid-state T/R modules, each of which can generate 400 W. As the total number of antenna element positions is 5354, the array is not fully populated; a future increase of array power is therefore possible.

Assuming that a target has a radar cross section of 10 m^2, the radar range is about 5556 km; PAVE PAWS systems have been built at the following Airforce bases: Otis, in Massachusetts; Beale, in California; Robins, in Georgia; and Goodfellow, in Texas. The array is an example of good engineering practices; we have already mentioned that the array power can be increased. Similarly the system can theoretically accommodate the two polarizations, without any modifications to the antenna elements of the array, whenever the need arises and funds become available.

18 PHASED ARRAY-BASED SYSTEMS AND APPLICATIONS

FIGURE 1.9 The very large array (VLA), located in Socorro, New Mexico (USA). (*a*) Aerial view. (*b*) Close-up view of one arm of the Y. (*Courtesy*: NRAO.)

1.1.7 Historical Developments Leading to Current Phased Arrays

Was the importance of phased arrays appreciated overnight universally, or was the route to acceptance long and torturous? We are inclined to think that phased arrays always had their critics and alternatives to phased arrays have been considered and defended with some vigor; on the other hand, however, any novel approaches or discoveries are initially resisted. In what follows we

FIGURE 1.10 The PAVE PAWS UHF solid state radar phased array. (*a*) One face of the array. (*b*) Close-up of the array elements. (*Courtesy*: Dr. Eli Brookner.)

shall briefly trace the historical developments leading to current phased arrays in the fields of radar and radio astronomy.

1.1.7.1 Radar Developments There were several papers and patents related to phased arrays [1.12–1.14] between 1899 and 1926, and P. S. Carter first recognized the role of mutual coupling in arrayed antenna elements as early as 1932 [1.15].

The first use of phased arrays was reported in 1937, in a reference that carefully described previous work related to phased arrays [1.16]. Six rhombic antennas were used along a straight line about 1.2 km long. The observed SNR improvement attained by combing the six antennas with appropriate phases, was 7–8 dB in the communication channels between Holmdel, New Jersey (USA) and England, in accord with the theoretical predictions for the improvement of $10 \log N$ where N is the number of antennas used.

In 1948 H. A. Wheeler calculated the radiation resistance of a planar array of cophased dipoles that formed the basis of many later papers [1.17].

Most of the major advances in the theory of radar phased arrays and in the implementation occurred in the 1960s. One of the early phased array-based radars, the AN/SPY-1, also known as the RCA's AEGIS system, was reported in 1974 [1.18]. It has been fitted to several U.S. Ticonderoga-class cruisers [1.19].

In 1974 L. Stark wrote [1.20]:

Phased-array antennas are here, operating and proving their usefulness in carrying out functions that conventional antennas could not. Phased arrays extend the capability of radar by improving vastly the number of targets which can be seen and tracked. The improvement comes about because the radar beam can be pointed to a new direction in microsecond speeds and it may be widened or narrowed at microsecond speeds, thus providing a great amount of beam agility.

The development of phased arrays had to wait for the availability of fast-acting phase shifters, and computers to control the phased array [1.20]. Computer power was also needed to process and display the data phased arrays produced.

Costs associated with large, fully steerable, phased arrays limited their use to defense applications.

1.1.7.2 Radio-Astronomy Developments In the radio-astronomy context several competing systems were proposed and used with limited success before phased arrays were finally accepted. The major requirement has always been spatial resolution at an affordable cost. What follows is a brief description of some of the rival systems.

The Moon's occultation with the Sun [1.21] was used to increase the resolution of monolithic telescopes. Such systems yielded many important astronomical results, including the correlation of intense solar burst activity with the sunspots on the solar disk.

Another approach toward the quest for high spatial resolution was to use one antenna on top of a cliff situated near the sea to form a two-element interferometer (sea-cliff interferometer or Lloyd's mirror in optical engineering terminology) [1.22–1.24]. This system was used to determine the accurate position of sunspots and of several strong radio sources. The optical identification of the newly discovered radio sources was made possible by the relatively high resolution afforded by these techniques [1.25, 1.26].

These ingenious techniques devised and used by radio astronomers to make important astronomical discoveries could not, however, gain a wider acceptance because the range of sources seen by these systems was not great.

A natural extension of the latter systems was the realization of many interferometers arranged in one [1.27, 1.28] or two [1.29, 1.30] dimensions to form phased arrays (or diffraction gratings in the optical engineering terminology). It is important to note that these developments took place in the early 1950s.

Despite the successes of these phased arrays, acceptance was not widespread. Phased arrays were thought to have many conceptual limitations such as narrow bandwidth and an inability to track sources over a long period of time. Such notions were derived from the embryonic state of the early phased arrays.

The Kraus- [1.31, 1.32] and Ratan- [1.33] type telescopes were seen to be free of these limitations and economical solutions that offered considerable collecting area, reasonable resolution, and adequate steerability.

Technological and scientific matters aside, the radioastronomers of that generation had an either (phased arrays) or (single dish telescopes) mindset. The supporters of phased arrays referred to large single dishes as "windjammers" or "windbreakers" while the supporters of single dishes preferred other methods to attain spatial resolution. There was intense polarization between the two camps.

Both camps had impressive lists of achievements. The synthesis camp could show radio images of unprecedented spatial resolution of galaxies and other sources, while the single dish camp could list numerous discoveries of new interstellar molecules, quasars (quasi-stellar objects) and masers.

With the passage of time, phased arrays have been widely accepted; more importantly however the complementarity of the two systems gained ground in the minds of many. Scientifically, single dish radiotelescopes explore what is there and phased arrays augment the acquired knowledge; technologically phased arrays use not one but several "windjammers."

1.1.7.3 Phased Arrays and What They Can Offer—Part 2 Figures 1.11 and 1.12 are photographs of the Mills and Christiansen crosses, respectively. Instead of building a monolithic aperture that had the required spatial resolution (a mission impossible financially and technologically), one could build a skeleton (or thinned) aperture that yielded the same resolution. Such an approach did not violate any physical laws and allowed the designer to break the nexus between resolution and aperture gain.

22 PHASED ARRAY-BASED SYSTEMS AND APPLICATIONS

FIGURE 1.11 The latest realization of The Mills Cross; each of the arms is 1.6 km long. (*Courtesy*: University of Sydney.)

There is nothing extraordinary about the shape of the thinned array. The VLA has a Y shape, and a large circular array has been built [1.34]. In theory, the first two arrays can be readily extended to further improve their resolution.

While the VLA is the largest compact phased array in operation, several arrays and/or conventional apertures (Earth-bound or on-board orbiting satellites) can be connected to form a very-long-baseline-interferometer (VLBI) or an orbiting VLBI (OVLBI) system. The latter system truly has phenomenal dimensions and resolutions of the order of sub-milli-arc seconds [1.35]—see Section 1.4.4.2.2.

In the radar context, one can attain a range of about 5000 km by using an aperture of about 30 m in diameter in the UHF band. These parameters, associated with the PAVE PAWS radar system, represent the upper boundaries of conventional radars. As the array is not fully populated, only 1792 of the total 5354 elements are active, one can further increase its radar range by fully populating the array.

Apart from the traditional radar operations, modern phased arrays can perform the following functions on a time-sharing basis.

- Multiple target track; up to a thousand targets tracked by an inertialess antenna beam

FIGURE 1.12 The two arms, 378 m long, of the Chris (Christiansen) cross. The larger antennas have been used in conjunction with the main cross to form a compound interferometer. (*Courtesy*: Prof. T. W. Cole.)

- A variety of searches, such as high-speed horizon, medium-range, and long-range
- High angle search/track
- Terrain following/avoidance
- Fire control and midcourse guidance

The last two functions need some explanation. Multipath propagation and ground clutter afford some protection to low-flying aircraft and helicopters from ground-based radars. In that scenario, the low-flying aircraft requires a ground-looking radar having a reasonably wide beam.

When an operator must direct and control antiaircraft gun systems with high precision against low flying targets, the requirement is for a fire control radar having the narrowest beamwidth to circumvent the same multipath and ground clutter problems. Phased array-based radars can meet both requirements.

24 PHASED ARRAY-BASED SYSTEMS AND APPLICATIONS

While the preceding list of diverse functions and applications of phased arrays reflects their versatility, future applications envisaged, and explored in Section 1.6.5, augment that list.

1.2 RADAR SYSTEMS

In this section we shall consider radar systems in some detail before we embark on the descriptions of other systems. While the fundamentals related to radar systems are synoptically outlined, we shall focus on current and future applications where phased arrays are either used or there are some definitive plans to use them.

Let us consider a monostatic radar—specifically, a system that uses the same aperture to transmit and receive EM radiation to and from a target. The radar's aperture has a directive gain and is defined as the ratio of the power per unit solid angle radiated in a given direction from the antenna to the power per unit solid angle radiated from an isotropic antenna supplied with the same power. The isotropic antenna is a hypothetical antenna that radiates radio waves of a constant strength in every direction, or in 4π steradians. If we assume that the radiated field in the direction (θ, ϕ) is $F(\theta, \phi)$ then the gain $G(\theta, \phi)$ is given by the equation

$$G(\theta, \phi) = \frac{F(\theta, \phi)}{P_0/4\pi} \tag{1.1}$$

where P_0 is the power supplied. If the total power radiated is P_t, the directivity of the aperture $D(\theta, \phi)$ is given by the equation

$$D(\theta, \phi) = \frac{F(\theta, \phi)}{P_t/4\pi} \tag{1.2a}$$

$$\frac{P_t}{P_0} = \eta' \tag{1.2b}$$

where η' is the ohmic efficiency, which accounts for any resistive losses; often η' is equal to unity. If A is the geometric area of the aperture and $A\eta$ is the effective area of the aperture A_e, the gain of the aperture is given by the well-known equation

$$G = \frac{4\pi}{\lambda^2} A\eta \tag{1.3}$$

For conventional reflectors the aperture efficiency is a measure of the losses caused by aperture illumination, spillover, and blocking; in the context of phased arrays, η can be a measure of the losses caused due to illumination, and those associated with the aperture not being fully populated.

1.2.1 The Radar Equation

The output SNR of a radar system is simply the ratio of the power P_r received at the input terminals of the receiver to the noise power of the system, and it is related to the range R by the following equation:

$$\text{SNR} = \frac{P_r}{kTB} = \underbrace{\frac{P_T G}{4\pi R^2}}_{\substack{\text{On-target} \\ \text{power}}} \underbrace{\frac{A_e \sigma}{4\pi R^2}}_{\substack{\text{Power at} \\ \text{the } Rx \\ \text{input}}} \underbrace{\frac{1}{kTB}}_{\substack{\text{Reciprocal} \\ \text{of noise} \\ \text{power}}} \underbrace{10^{-0.2\alpha R}}_{\substack{\text{Atmospheric} \\ \text{attenuation} \\ \text{factor}}} \quad (1.4)$$

where kTB = total receiver noise power
$\quad k$ = Boltzmann's (L. Boltzmann, 1844–1906) constant; 1.38×10^{-23} joules (J. P. Joule, 1818–89) per degree kelvin (William Thomson, 1824–1907) (J/K)
$\quad T$ = equivalent noise temperature of the system
$\quad B$ = system's bandwidth
$\quad P_T$ = total transmitted power
$\quad G$ = the aperture gain
$\quad \lambda$ = wavelength of operation
$\quad \sigma$ = average target radar cross section (RCS)
$\quad \alpha$ = the atmospheric loss attenuation coefficient and is expressed in dB/km

The four components of the radar equation—the power on target, the received power, the reciprocal of the total receiver power, and the atmospheric attenuation factor—have been separated for clarity. To the extent that other losses, including transmission losses and losses due to polarization mismatches, are negligible, we have ignored them. If these losses cannot be ignored, they would appear on the numerator of Equation (1.4).

Inserting Equation (1.3) into (1.4), we obtain

$$\text{SNR} = \frac{P_T G^2 \sigma \lambda^2 10^{-0.2 R \alpha}}{(4\pi)^3 R^4} \frac{1}{kTB} \quad (1.5)$$

where P_T is the average power transmitted over a pulse period; it is related to the peak transmitted power P_{pk} by the equation $P_T = P_{pk} d_t$, where d_t is the duty factor equated to the ratio of the pulse duration τ and the pulse period T. The duty factor is an important parameter, which we shall consider further in Chapter 3, where tubes and solid-state transmitters are compared.

Although the radar equation looks simple, there are many hidden subtleties, some of which we shall explore here.

1.2.1.1 The RCS of Targets As can be expected the RCS of a complex target such as an aircraft depends on the look angle or aspect angle [1.36]. The aircraft presents a large physical area and subsequently a large radar cross section (RCS) area when viewed from the side rather than from the nose.

Target scintillation is a well-known phenomenon; it is due primarily to the motion of scattering points that intercept the transmitted EM energy. Target scintillation takes place not only when the target moves but also when the radar frequency changes because the relative position of the target scatterers changes. Therefore, σ is both time- and frequency-dependent. Here we are considering targets having complex shapes not standard targets, such as a sphere or a corner reflector when viewed from the front.

Finally, σ is polarization-dependent (see Section 1.2.4); certain targets will have different RCSs at different polarizations. Given that we have no a priori knowledge of the shape and polarization properties of the target, polarization agility (on transmit and receive) is mandatory for a radar if all the information available related to the target is to be obtained.

For the purposes of calculating the range of a target, some designers assume that σ is the time-average RCS of the target when one polarization is received. Measurements of σ of targets at different polarizations and frequencies and approaches to minimize it, is an established field of research.

Designers have accepted the figures for RCS tabulated in Table 1.2 [1.8, 1.37] of 1, 10, and 100 m^2 as the RCS corresponding to cruise-type missiles, fighter planes, and bombers, respectively [1.8].

1.2.1.2 Atmospheric Attenuation as a Function of Frequency Figure 1.13 illustrates the one-way attenuation α of EM waves as a function of frequency at sea level and at an altitude of 4 km. The attenuation is a function of

TABLE 1.2 Commonly Accepted RCSS for Different Targets (in m^2)

Jumbo airliner	100
Large bomber or airliner	100–40
Medium bomber or airliner	20
Large fighter	10–6
Small fighter	2
Small single-engine aircraft	1
Cruise-type missile	1
Human	1
Conventional unmanned winged missile	0.5
Bird	0.01

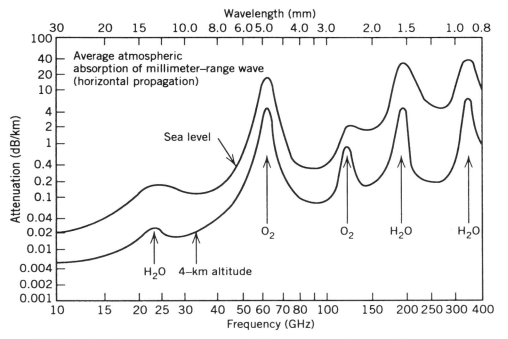

FIGURE 1.13 Atmospheric attenuation of EM waves as a function of frequency for sites at sea level and at 4-km altitude.

frequency and the peaks are due to resonances of the H_2O and O_2 molecules. It can be implied that the attenuation is also a function of relative humidity and temperature.

The transmission windows and the approximate atmospheric attenuation [1.38] experienced by EM waves in clear weather and during heavy-rain conditions (25 mm/h) are tabulated in Table 1.3 [1.38]. Figure 1.14 illustrates the attenuation experienced by EM waves by atmospheric gases, rain, and fog as a function of frequency [1.39]. The figure is useful because comparisons of attenuation experienced by systems operating at cm, mm, infrared, and optical wavelengths can be readily made.

Designers have to accept the free-space attenuation of the EM waves as well as the attenuation caused by scattering during periods of rainfalls. We shall consider the effects of rain on polarimetric systems (radar and satellite communication) further in Section 1.2.4.3.

While communication and radar systems usually operate within frequency bands where the atmospheric attenuation is minimal, secure communication and radar systems operate (over short ranges) within frequency bands where

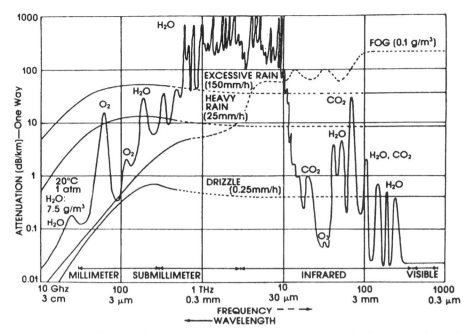

FIGURE 1.14 Attenuation experienced by EM waves by atmospheric gases, rain, and fog. (*Source*: Preissuer [1.39].)

the atmospheric attenuation is high. A knowledge of the magnitude of attenuation due to rain and the atmosphere often serves the purpose of setting safe operational margins for radar and communication systems. While atmospheric attenuation is always taken into account in this book, rain attenuation is considered only when special cases are examined.

More detailed studies of the many influences of the rain on the propagation of the EM waves have been undertaken by the mm-wave radar and satellite communications communities [1.38, 1.40], respectively. Estimates and measurements of the same attenuation under conditions of snowfalls and fog have been collated in Currie and Brown [1.38].

1.2.1.3 SNR Considerations What is an adequate SNR for a reliable detection to take place? The problem is not trivial, and the answer depends on many aspects, including the following:

The nature of the target: nonfluctuating (Marcum [1.41] target) or fluctuating

Fluctuating targets: slow/fast fluctuations (Swerling 1–4 target models)

Integration of one or N pulses
The required false alarm rate
The required probability of detection

The case where one pulse (single look) illuminates a nonfluctuating target, such as a sphere or a corner reflector viewed from the front has been treated in Hovanessian [1.36] and Lawson and Uhlenbeck [1.42] with the aid of results obtained in Rice [1.43], where the probability of detection P_s has been derived as a function of SNR, with the probability of false alarm P_n as a parameter. Gaussian noise has been assumed.

For example, a probability of detection of 85% results when the SNR is equal to 14 dB and the probability of false alarm is 10^{-8}, a typical case. Another useful way to present the relationship between the SNR and P_s is shown in Figure 1.15 (for this example P_n is set equal to 10^{-8} and the radar range is constant). In the linear range of the S-curve the SNR is proportional to the probability of detection P_s, a result which is intuitively supported.

Very seldom is a radar detection decision based on a single-pulse detection. A train of pulses is usually received from the target before one can declare a target detection. Thus acceptable detections can be declared even when the single-pulse SNR is near or below unity. The process by which the pulses are combined is known as *pulse integration*.

In an ideal receiver the energy from all the N radar returns is added in an integrator prior to envelope detection. Ideally the result is to achieve an

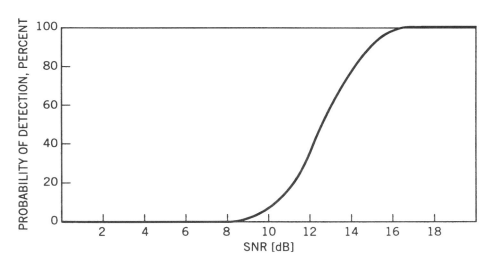

FIGURE 1.15 The system signal-to-noise ratio, (SNR), dependence on the probability of detection, when the probability of false alarm and radar range have been set.

effective value of SNR equal to the SNR attributed to a single pulse multiplied by N. This process is referred to as *coherent integration*. These benefits, however, are attained only if the radar transmitter is highly stable in phase and frequency.

Many radars utilize noncoherent integration because it is easier to implement. Compared to coherent integration, noncoherent integration is not as efficient.

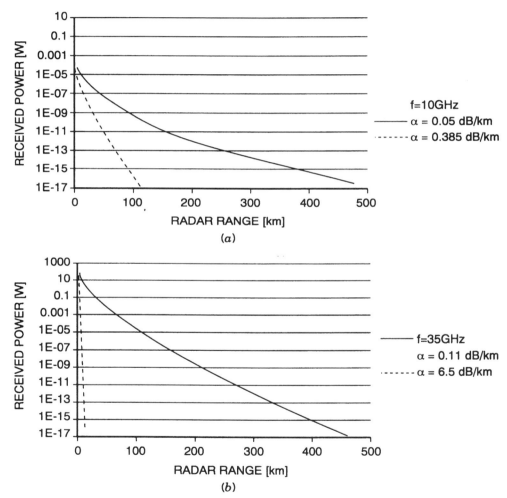

FIGURE 1.16 The relationship of the radar range to the power received at the input of a radar receiver, when the signal-to-noise ratio, is 15 dB. Solid lines correspond to clear weather conditions, while dashed lines correspond to heavy rain conditions, 25 mm/h—see Table 1.3. Ideal conditions, outlined in the same table, apply. The frequency of operation is as follows: (a) 10 GHz; (b) 35 GHz.

Swerling [1.44] and Meyer and Mayer [1.45] have explored the cases where the targets are fluctuating and when the integration of several pulses takes place. The latter reference contains a plethora of graphs to aid the designer of radar systems.

False-alarm probabilities of 10^{-6}–10^{-8} are commonly used, and in the absence of a detailed knowledge of a radar system, a SNR of 10–20 dB is commonly accepted.

1.2.1.4 Illustrative Examples An illustrative example will serve the purposes of demonstrating how the maximum range is significantly dependent on the frequency of operation, which, in turn, defines the atmospheric attenuation experienced by the EM waves.

The power available at the input of the radar's receiver, after reflection from the target, P_r, is given by the equation

$$P_r = \frac{P_T A_e^2 \sigma}{\lambda^2 R^4} \frac{1}{4\pi} 10^{-0.2\alpha R} \qquad (1.6)$$

As the range increases P_r and the resulting SNR will decrease monotonically; for the purposes of this example, we shall plot P_r as a function of R until the SNR is 15 dB. The remaining parameters we have used are tabulated in Table 1.3.

Figure 1.16a, 1.16b, and 1.16c illustrate the relationship of P_r (at the input terminals of the radar receiver) as a function of range when the

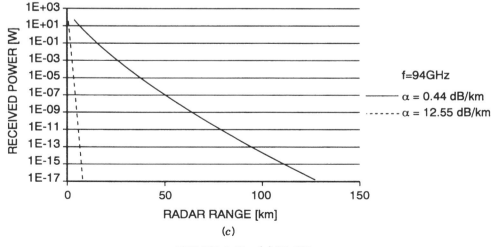

FIGURE 1.16 (c) 94 GHz.

TABLE 1.3 Parameters Used for the Illustrative Example

Parameters	Frequency (GHz)				
	10	35	94	140	240
α (dB/km)—clear weather	0.05	0.11	0.44	1.8	5.2
α (dB/km)—heavy rain, 25 mm/h	0.385	6.50	12.55		
σ (m^2)	10	10	10	10	10
B (MHz)	10	10	10	10	10
A_e (m^2)	10	10	10	10	10
T (K)					
Ideal	300	300	300	300	300
Realistic	300	400	500	600	800
P_T (kW)					
Ideal	800	800	800	800	800
Realistic	800	20	1	0.1	0.01

frequency of operation is 10, 35, and 94, respectively, for the following ideal conditions:

1. The same transmitted power, of 800 kW, can be generated at the frequencies we have considered.
2. Receivers have the same total system temperatures at the frequencies we have considered.
3. Under clear-weather conditions (solid lines) and under heavy rain conditions (dashed lines).

The assumptions we made for the bandwidth (10 MHz), the target RCS (10 m^2) and the effective area of the antenna (10 m^2) are not only realistic but also allow the designer to scale the results presented here to other values readily. For Figure 1.16a–1.16c the dashed lines illustrate the effects of heavy rain (25 mm/h) on the radar range attained at 10, 35, and 94 GHz. It is clear that rain considerably diminishes the radar range attainable in clear-weather conditions.

Figure 1.17 illustrates the received power versus radar range attainable under the realistic conditions related to T and P_T listed in Table 1.3. The radar range, at millimeter-range wavelengths, decreases considerably when these realistic conditions apply even in clear weather. By comparing Figures 1.16 and 1.17(a), one can deduce the improvement margin expected if the transmitted power and system noise temperatures of mm-wave radars are the same as those readily attainable at cm wavelengths. Figures 1.16 and 1.17 clearly illustrate the regimes of applicability in terms of range for radars operating at cm and mm wavelengths.

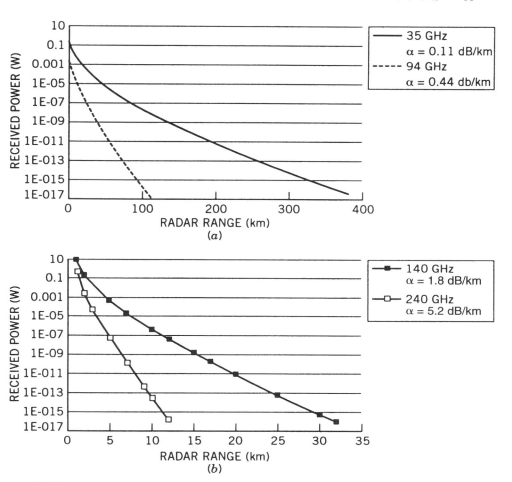

FIGURE 1.17 Received power is a function of range when the realistic conditions outlined in Table 1.3 apply. Clear weather conditions are assumed and the frequency of operation is as follows: (*a*) 35 GHz, solid line and 94 GHz dashed line; and (*b*) 140 GHz, solid line with diamonds, and 240 GHz, solid line with squares.

1.2.2 The Performance of Radar Functions

In a typical application, a radar is assigned a large surveillance volume that it scans regularly. The detection and tracking of targets of interest is performed while the radar antenna scans its surveillance volume. The same volume of space is visited every few seconds, and the resulting information forms a time history of the targets as they move from or toward the radar.

Let us consider the different functions, volume surveillance and target tracking, separately.

1.2.2.1 Volume Surveillance Radars

Let us assume that (1) the radar searches an angular volume of Ω steradians in time t_s and (2) the antenna beam Ω_b dwells a time t_0 in each direction subtended by the beam. The total scan time is $t_s = t_0 \Omega / \Omega_b$ and $P_T = t_0 P_{av}$, where P_{av} is the average power during t_0. By inserting these equations into Equation (1.5), we obtain

$$R^4 = \frac{P_{av} A_e \sigma}{\text{SNR}} \frac{1}{4\pi \Omega} \frac{t_s}{kTB} 10^{-0.2 \alpha R} \qquad (1.7)$$

if we recall that $\Omega_b = \lambda^2 / A_e$. For a volume search radar the two important parameters for maximizing range are the average transmitter power and the antenna aperture. The frequency dependence of the range is implicit in the term associated with the atmospheric attenuation [the last term on the right-hand side (RHS) of Equation (1.7)]. Additionally, the lower frequencies are preferred for volume search because it is easier to (1) generate large amounts of power at lower frequencies and (2) realize systems that have lower system equivalent-noise temperatures. Surveillance radars usually operate at L band and below, where the range can be maximized.

The other issue is related to the RCS of aircraft. The designer can in general decrease their RCS by suitable shaping of their outer contours. Further decreases in their RCS are usually valid over narrow bands. It is generally very hard to render aircraft invisible from, say, X band to L band and at the lower frequencies—see Section 1.6.5.1.

1.2.2.2 Tracking Radars

In performing this function the radar tracks the target continuously for an interval of t_0; Equation (1.5) can therefore be written as

$$R^4 = \frac{P_{av} t_0 A_e^2 \sigma}{\lambda^2 \, \text{SNR}} \frac{1}{4\pi kTB} 10^{-0.2 \alpha R} \qquad (1.8)$$

If one ignores the atmospheric attenuation factor, there is an advantage in operating at shorter wavelengths. When one considers short ranges (the case where the atmospheric attenuation factor can be ignored), this is valid. For longer ranges, however, this factor cannot be ignored for it becomes the dominant factor. Operation at X band is considered optimum for the reasons given above; one could contemplate operating tracking radars at higher frequencies (e.g., up to 35 GHz, where the atmospheric attenuation is not that high), but the power output of transmitters decreases as the frequency of operation increases.

Apart from detectability, the tracking radar should also yield good angular accuracy, which, in turn, can be attained by radars operating at the higher frequencies. Looking at Figures 1.16 and 1.17, it is not too difficult to deduce that the K-band (27–40 GHz), a band between cm and mm wavelengths, offers reasonable ranges and spatial resolutions. There are clearly many tradeoffs to be considered when a radar is designed to perform a given set of functions. Although all cases cannot be considered here, the relevant equations have been defined here for the designer to use.

The cross-range resolution—the resolution along a direction perpendicular to the line of sight (LOS) between the radar and a target—improves as the aperture of the radar increases. As was noted earlier, the accuracy by which the angle of arrival is measured improves as the radar aperture increases. Under normal conditions—when there is no relative motion between the radar and the target—resolution along the LOS improves as the pulse duration decreases (or the system bandwidth increases).

1.2.2.3 A Compromise Band for Multifunction Radars? Given that it is too expensive to have one radar operating at L-band to perform the surveillance function and another operating at X-band to perform the tracking function, a radar operating at a compromise band, the C-band is usually chosen to perform both functions. From the foregoing considerations it is clear that neither functions are performed efficiently at C-band.

In Section 1.6.6 we shall consider radars operating over 2–3 octaves. These radars will be capable of performing the surveillance and tracking functions at frequencies that are optimum for these functions. The same radars will also be capable of performing the EW and communications functions. Additionally, the radars will have ECCM capabilities. In a similar vein, wideband radar will be capable of performing a set of civilian functions related to safe air traffic control in airports—see Section 1.6.6.3. Truly wideband radars will be capable of performing a multitude of interdependent and interrelated tasks (at optimum frequencies) with unprecedented efficiency on a time-sharing basis.

1.2.3 Low-Probability-of-Interception (LPI) Radars

For some time the majority of radars were pulsed radars that in effect met the system requirements in a simple and inexpensive manner. It was, for instance, easy to transmit large powers in the shortest possible time. Similarly, the target's range was deduced from measurements of the time lag between the transmitted and received radiations. As the transmitted powers increased, with the passage of time the radar range also increased, which was desirable. The radar's presence, however, by electronic support measures ESM systems became easier to detect, which was undesirable. Recently the radar's presence has been accompanied by the threat from antiradiation missiles (ARMs).

The Gulf War certainly proved the usefulness of not only the ESM systems and the ARMs but also the vulnerability of conventional radars.

Ideally, designers aimed for radar systems that transmitted as much power as possible, in order to maintain the long-range requirement and satisfy the LPI requirement. One of the most straightforward LPI radars is the continuous-wave (CW) radar, which has a 100% duty cycle. The radar has an LPI capability because its instantaneous power is low, while the power within one cycle can be high. Thus the radar's functions (surveillance and detection) are efficiently performed.

The nearest realizable CW radar is the linear, frequency-modulated CW (FMCW) radar [1.36], where the transmitter signal is frequency modulated by a linear waveform. The received signal has the same modulation but is delayed relative to the transmitted signal. In Section 1.6.1.2, it will be shown that the target's characteristics can be easily deduced by these radars. The acceptance of FMCW radar has been delayed until recently because of the technological problems associated with its realization. Naturally, threats experienced in modern warfare accelerated the introduction of FMCW radars into many a valuable platform.

1.2.4 Polarimetric Radars

We have already seen in Section 1.2.1.1 that the target RCS, σ, is polarization-dependent. This being the case the measured σ is dependent on the polarization of the illuminator transmitter and that of the receiver. The following example will clarify this important point.

Let us assume that we are illuminating a flat plate with a radar that is set up to transmit and receive the left-handed circular (LHC) polarization. The reflected power will have the opposite polarization: right-handed circular (RHC) polarization. Under ideal conditions of high-cross-polarization isolation, the radar will not be capable of receiving the RHC polarization; thus the flat plate will be invisible to the radar. Polarimetric radars are capable of measuring the RCS of targets (or clutter) without the above-mentioned limitations. Further, the same radars have improved sensitivities because the polarimetric properties of targets (or human-made objects) are fundamentally different from the polarimetric properties of clutter, such as sea, foliage, and rough terrain.

1.2.4.1 Theoretical Framework   Chandrasekhar [1.46] set out the theoretical framework for polarimetric measurements in 1955, and Cohen [1.47] related the four Stokes parameters I, Q, U, and V to parameters that can be measured by a polarimeter in the radio-astronomy context. Parallel work on polarimetric radars began at the Georgia Institute of Technology and the Georgia Tech Research Institute in the 1960s [1.48], and a plethora of research papers originated from research carried out at the above-mentioned research establishments [1.38].

Without any loss of generality, let us consider the four principal polarizations: vertical (V), horizontal (H), LHC, and RHC. The two linear polarizations are orthogonal, while the two circular polarizations are opposite.

The reflected power from any target or clutter can be completely described by the radar scattering matrix [often called the *polarization scattering matrix* (PSM)]. The matrix describes the radar reflectivity characteristics in terms of two orthogonal transmitted polarizations and two orthogonal received polarizations. Thus, for vertical and horizontal polarizations, the received signals from a target can be expressed by the equations

$$E_{rH} = a_{HH} E_{tH} + a_{VH} E_{tV} \tag{1.9}$$

$$E_{rV} = a_{HV} E_{tH} + a_{VV} E_{tV} \tag{1.10}$$

if we ignore factors of proportionally. Here a_{HH} is a coefficient corresponding to the case where the horizontal polarization is transmitted and received. Similarly, a_{HV} is a coefficient that corresponds to the case where a horizontal polarization is transmitted and a vertical polarization is received. Subscripts r and t, respectively, stand for receive and transmit modes.

In matrix form, Eqs. (1.9) and (1.10) are written as

$$\begin{vmatrix} E_{rH} \\ E_{rV} \end{vmatrix} = \begin{vmatrix} a_{HH} & a_{VH} \\ a_{HV} & a_{VV} \end{vmatrix} \begin{vmatrix} E_{tH} \\ E_{tV} \end{vmatrix} \tag{1.11}$$

Each a_{ij} terms of the matrix is complex; that is, they have an amplitude and a phase term. Furthermore, these terms can be related to the RCS of the target by the equation

$$a_{ij} = \sqrt{\sigma_{ij}} \exp j\phi_{ij} \tag{1.12}$$

For a monostatic radar

$$a_{VH} = a_{HV} \tag{1.13}$$

so that there are only there only three complex, independent coefficients in the scattering matrix. Often this redundancy is used to calibrate polarimetric radars.

The scattering components for circular polarization can be related to those for linear polarization by the following relationships [1.38]:

$$|C_{RR}| = \left| \frac{a_{HH} - a_{VV}}{2} + ja_{HV} \right| \tag{1.14}$$

$$|C_{LL}| = \left| \frac{a_{HH} - a_{VV}}{2} - ja_{HV} \right| \tag{1.15}$$

$$|C_{RL}| = \left| \frac{a_{HH} + a_{VV}}{2} \right| \tag{1.16}$$

For the case where the circular polarizations are transmitted and the linear polarizations are received, the following equations are useful [1.38]:

$$|a_{LH}| = \tfrac{1}{2}|a_{HH} + ja_{VH}| \tag{1.17}$$

$$|a_{LV}| = \tfrac{1}{2}|a_{HV} + ja_{VV}| \tag{1.18}$$

$$|a_{RH}| = \tfrac{1}{2}|a_{HH} - ja_{VH}| \tag{1.19}$$

$$|a_{RV}| = \tfrac{1}{2}|a_{HV} - ja_{VV}| \tag{1.20}$$

The phase between the horizontally and vertically polarized components of the received electric field is called the *polarimetric phase* ϕ, which can be computed by performing a dot-product vector operation between the horizontal and vertical received components of the electric field:

$$\overline{E_{rH}} \cdot \overline{E_{rV}} = |\overline{E_{rH}}||\overline{E_{rV}}| \cos \phi \tag{1.21}$$

from which we can deduce [1.38]

$$\cos \phi = \frac{a_{HH} a_{HV} + a_{VH} a_{VV}}{\left(a_{HH}^2 + a_{VH}^2\right)^{1/2} \left(a_{HV}^2 + a_{VV}^2\right)^{1/2}} \tag{1.22}$$

1.2.4.2 Implementation The measurements of the quantities a_{ij}, which are related to σ_{ij}, are performed in two stages. For the case where the two linear orthogonal polarizations are used, the transmitter transmits one pulse of vertically polarized power, and the scattered powers that are vertically and horizontally polarized are received in two separate channels. From these measurements the coefficients a_{VH} and a_{VV} are deduced. The polarization of the next pulse changes to horizontal and the scattered powers that are vertically and horizontally polarized are measured again. From these measurements coefficients a_{HH} and a_{HV} are deduced. Similar measurements are performed when the polarizations used are opposite circular.

In more general terms, the PSM can yield information related to the following properties of stationary, simple metallic objects:

1. The maximum return from the target: the amplitude of the optimum polarization that maximizes the RCS of the target
2. A measure of the target's symmetry and orientation
3. The number of bounces of the reflected signal
4. The target's ability to polarize the incident radiation

While measurements of the full PSM yields the maximum benefits for a radar system, measurements of two elements of the PSM yield some benefits, which are usually application-dependent.

It is possible to derive an optimum transmitter/receiver polarization combination for target detection in the presence of background clutter [1.49]. Alternatively, the contrast of radar images can be maximized by the appropriate selection of polarization combinations [1.50]. In the latter approach an optimal polarimetric matched filter is sought; furthermore, this filtering task can be performed adaptively [1.51–1.55].

In another development Brown and Wang have shown that an improved radar detection capability results by using adaptive multiband polarization processing [1.56, 1.57].

Given the trend toward adaptive systems, the quest for procedures and techniques that allow accurate measurements of the parameters under consideration almost instantaneously is of considerable importance.

1.2.4.3 Rain-Induced Depolarization of EM Waves Rainfalls introduce depolarization of the EM waves, which, in turn, cause additional attenuation and a decrease of the isolation between the channels accommodating the two polarizations. The depolarization of EM waves is caused by the introduction of a differential phase and amplitude attenuation to the two channels accommodating the two polarizations of a dual-polarization system [1.58]. Methods of reducing the effects caused by the depolarization of EM waves adaptively in real time have been proposed for dual-polarization satellite communication systems [1.59, 1.60]. The application of these techniques to polarimetric radars has been proposed [1.61].

1.2.4.4 Instrumentation Requirements The most important requirements for polarimetric radars are polarization agility, adaptability, and frequency diversity. Instrumentally, polarization isolation is important for two reasons:

1. The minimization of errors in the measurements of the polarization coefficients.
2. Protection of the radar from jammers. More explicitly, jammers can render a radar that has poor cross-polarization isolation inoperative by transmitting considerable powers having a polarization state orthogonal or opposite to the radar's polarization.

Figure 1.18 relates the channel isolation due to crosstalk and was derived by assuming that the co- and cross-polarization scattering parameters are equal [1.38]. The error due to crosstalk between the two channels χ, expressed in decibels (dB), is given by the equation

$$\chi = 20 \log\left[1 + \frac{1}{10^{\rho/20}}\right] \quad (1.23)$$

where ρ is the channel isolation in dB.

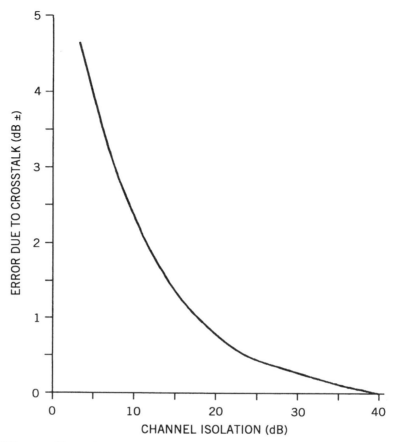

FIGURE 1.18 Error due to crosstalk as a function of the polarization isolation when both the co- and cross-polarization components are equal.

It can be seen that a polarization isolation of 25 dB will induce an error of only 0.5 dB, which is acceptable. However, if the difference between the co- and cross-polarization scattering parameters measured is, say, 20 dB, the same graph can be used to derive the induced errors with a certain adjustment. The nominal isolation in this case will be 25−20 = 5 dB, which, in turn, will induce an unacceptable error of, say, 4 dB.

More detailed calculations [1.62] related to the position and shape of the cross-polarization lobes of conventional reflectors resulted in the following conclusions. To obtain an accuracy of 0.5 dB for the cross polarized scattering coefficient lying x dB beneath the like-polarized scattering coefficient, a one-way isolation of $x + 16$ dB is required. In Ulaby et al. [1.63] it is shown that the difference between the co- and cross-polarization scattering coefficients can be up to 25 dB. This difference can be taken as a boundary condition.

Designers therefore tend to opt for cross-polarization isolation levels of 25–40 dB. Polarization isolation figures in excess of 40 dB are routinely obtained in satellite communication systems, even during rainy periods [1.59].

1.2.4.5 Benefits of Polarimetric Radars An improvement of at least 10 dB in clutter reduction was claimed by a radar using circular polarization as early as 1954 [1.64]. The radar transmitted and received power had the same circular polarization, so rain clutter returns that had the circular polarization opposite to one originally transmitted were considerably attenuated. An improvement of a few decibels was reported in Swartz et al. [1.50] in the contrast of the scenes when the polarization of the illuminating radiation and that of the receiving system were optimized. Speckle reduction in synthetic aperture radar (SAR) images was reported by the use of polarization information [1.65].

Description of the benefits dual polarization radars offer to the user is not a trivial task, for these benefits are application-dependent and other techniques are usually used in conjunction with polarimetric information. Here we shall consider the case of the airborne detection of stationary objects [1.66]. This a typical application where other conventional target discrimination techniques are not applicable.

Polarimetric radars base their discrimination performance on intracell target features and are insensitive to the internal signal-to-clutter (S/C) ratios. In Figure 1.19 two cases are illustrated when (*a*) the cross-range resolution is low, for instance, 35 m cross-range by 5 m downrange when the target is 5 m square and (*b*) the cross-resolution matches the target size. For both cases the probability of false alarm P_{FA} or P_n has been set constant.

With reference to Figure 1.19*a*, the polarimetric radar will not outperform the conventional radar-utilizing amplitude threshold when the intracell S/C ratios are either too high or too low. When the intracell S/C ratio is between these two cases, an improvement of 2–4 dB can be attained by the polarimetric radar. In Figure 1.19*b* it is clearly seen that the probability of detection is independent of the intercell S/C for the polarimetric radar. By contrast, the conventional radar-utilizing amplitude thresholding is still dependent on the intercell S/C ratio.

From the foregoing considerations it is clearly seen that the maximum benefits of polarimetric radars reside with high-resolution systems SARs, ISARs, and phased array based-radars.

1.2.5 Bistatic and Multistatic Radars

Most television viewers are cognizant of the effects of bistatic radars. The television transmitter and the receiver are not colocated, and the television

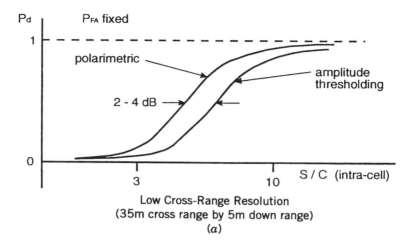

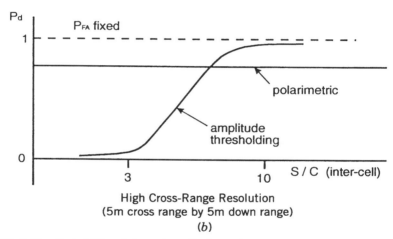

FIGURE 1.19 Probability of detection, P_d dependence on the signal-to-clutter S/C ratio when the false-alarm rate is set. (*a*) the case where the cross-resolution is low for a polarimetric and conventional radar; the abscissa (horizontal axis) is the intracell S/C ratio; (*b*) the case where the resolution matches the target for a polarimetric and conventional radar; the abscissa is the intercell S/C ratio. (*Source*: Holm [1.66], © 1990 SPIE.)

image is modulated in intensity whenever an aircraft intercepts the line of sight (LOS) between the receiver and the transmitter. A history of bistatic radars is outlined in Dunsmore [1.67].

Given that radar operation does not depend on specular reflections, one can perform the transmit function at one site and the receive function at

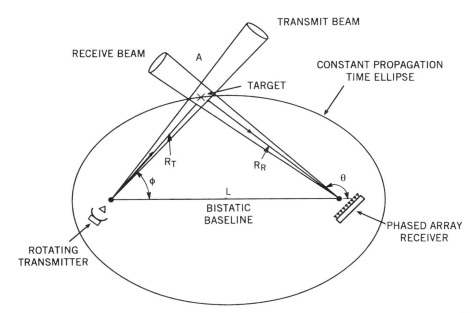

FIGURE 1.20 The principle of bistatic radar operation. The ellipse is the locus of target positions when a constant propagation time is assumed. The transmitting and receiving stations form the foci of the ellipse.

another. The separation between the two sites can be a few meters [1.68] or several hundreds of kilometers [1.67]. Such an arrangement, the bistatic radar illustrated in Figure 1.20, has the following attractions:

1. When the distance between the receive and transmit sites is considerable, third parties do not know the position of the receiving site(s). This being the case, the transmitter can be located at a safe distance from the targets, while the receivers can be located at convenient and safe locations. With this arrangement, the receivers do not suffer from ECM such as barrage and deception jamming and are protected from ARMs.
2. A target RCS advantage over the monostatic radar. Most aircraft are designed to exhibit low RCS toward the direction of monostatic radars and it is hard to design aircraft that have low RCSs in all aspect angles.
3. Illuminators of opportunity, such as television transmitters, can be used as transmitters to ensure covert operations. Naturally, the designer will have to implement several modifications to the existing hardware before such a system can be realized.

Bistatic radars are more complex than their monostatic counterparts, and a communication link between the transmitter and the receiver is usually needed. The link serves the purpose of relaying timing signals and the transmitter's waveforms and frequencies so that the complete transmitter waveform is available at the receiver. Additionally, the orientation of the transmitter is required at the receiving site. The communication link can be implemented with the aid of LOS, landline, satellite, or troposcatter systems. Finally, the receiving site requires a phased array (see Section 1.2.5.3), a requirement that adds to the complexity of the system but affords significant benefits.

Multistatic radars have one or more transmitters and/or one or more receivers. Transcontinental airlines, for instance, can be tracked by several illuminators and receivers, encountered along their flight paths.

1.2.5.1 Target Position Measurements For the bistatic radar, the radar range and bearing of a target are deduced from measurements of the time taken for the transmitted pulse to travel from the transmitter to the target and on to the receiver. In a two-dimensional case, where the transmitter, the receiver, and the target are in one plane, this measurement defines an ellipse illustrated in Figure 1.20, with the transmitter and receiver at the foci. The equal time-delay contours form ellipses with the same foci. For the three-dimensional case, the ellipses become prolate spheroids. The elevation and azimuth angles of the transmitter then define the position of the target on the surface of the spheroid.

1.2.5.2 Contours of Constant Detection Range For monostatic radars the contours of constant detection range or equivalently of constant received power are spheres. For bistatic radars we can deduce the shape of these contours by considering the radar equation [Eq. (1.5)] modified for the bistatic case, which becomes

$$P_r = \frac{P_T G_T G_R \lambda^2 \sigma_B}{(4\pi)^3 R_T^2 R_R^2} \tag{1.24}$$

where R_T and R_R are the ranges of the target from the transmitter and the receiver, respectively, and σ_B is the bistatic target RCS. Here we have ignored the atmospheric and other losses for convenience. Equation (1.24) can be recast as

$$R_T^2 R_R^2 = \frac{k \sigma_B}{P_r} \tag{1.25}$$

where

$$k = \frac{P_T G_T G_R \lambda^2}{(4\pi)^3} \tag{1.26}$$

is a constant term. To the extent that σ_B and P_r for a given target are constant, equation (1.25) can be recast as

$$(R_T R_R)^2 = \text{constant} \qquad (1.27)$$

Equation (1.27) corresponds to the ovals of Cassini, which are derived by a point P such that the product of its distances from two fixed points is a constant, i.e., $b^2 = \text{constant}$ and the two fixed points corresponding to the positions of the transmitter and receiver are separated by a distance $L = 2a$.

With reference to Figure 1.21, the ovals, circles, and the lemniscate are shown.

1.2.5.3 Receive Antenna Scanning Rate If a bistatic radar is to have a data rate comparable to its monostatic counterpart, the receiving antenna has to be able to follow the directions from which energy might be scattered toward it by any targets encountered by the transmitter's pulses. If a pulse emitted by the transmitter is to be received at a time t after it left the transmitter antenna, the equation

$$ct = R_T + R_R \qquad (1.28)$$

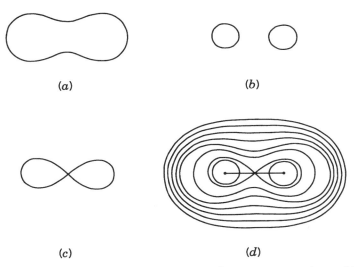

FIGURE 1.21 For a bistatic radar the contours of constant target detectability form ovals of Cassini: (a) ovals; (b) circles; (c) lemniscate; and (d) all of the above cases.

should hold. With reference to Figure 1.20, we can deduce that

$$\frac{R_T}{\sin \theta} = \frac{R_R}{\sin \phi} = \frac{L}{\sin(\theta - \phi)} \quad (1.29)$$

From equations 1.28 and 1.29 equation

$$\tan\left[\frac{\theta}{2}\right] = \frac{ct + L}{ct - L} \tan\left[\frac{\phi}{2}\right] \quad (1.30)$$

has been deduced after some manipulations [1.67]; furthermore, the maximum scan rate is given by

$$\left[\frac{d\theta}{dt}\right]_{max} = -\frac{c}{L} \frac{1 + \cos \phi}{\sin \phi} \quad (1.31)$$

Given that the maximum scan rate can exceed a million degrees per second, when $\phi \to 0°$, the use of a phased array, where the resulting beam can be electronically scanned quickly or where the array generates multiple staring beams, is mandatory.

1.2.6 Synthetic Aperture Radars (SARs) and Inverse SARs (ISARs)

The spatial resolution of SARs [1.69] and ISARs [1.70] is independent of the radar range. While SARs are normally airborne (or spaceborne) and yield very high-spatial-resolution maps of large areas of the ground, ISARs can yield high-resolution images of selected targets, such as aircraft, for the purposes of identification.

The operation of a SAR depends on the motion of the airborne radar with respect to the targets, while the operation of the ISAR depends on the motion of the target with respect to the radar.

The SAR technique yields high-spatial-resolution maps (e.g., of the order of ≤ 1 m) in the azimuth direction (perpendicular to the LOS) and is obtained by the synthesis of radar data obtained over a considerably longer aircraft flight than the azimuth resolution cell. More explicitly, the along the track resolution of a SAR is equal to half the real length of the phased array-based radar regardless of the range, aircraft height, and frequency of operation. The assumption is here made that the Earth is flat, a reasonable assumption for airborne SARs.

The present generation of European spaceborne SARs (ERS-1 and -2) uses a fixed-beam antenna with slotted waveguides as radiators and a centralized receiver and transmitter. The next-generation SAR systems require dual-polarized antennas with steerable beams [1.71]. This requirement can be met by steerable dual-polarized phased arrays [1.71].

In Section 1.6.2.1, the operation of ISARs is outlined and their domain of applicability is defined.

1.3 BASIC EW CONCEPTS

Electronic warfare (EW) is defined as actions taken through the use of the EM energy or its properties to determine, exploit, reduce, or prevent the hostile use of the EM radiation [1.36]. Apart from ECM (electronic countermeasures), the other components of EW are ESM (electronic support measures) and ECCM (electronic countercounter measures).

ECM define actions taken to prevent or reduce the effectiveness of hostile EM radiation; ECM include jamming, disrupting, and deceiving any sought-after information. Standoff and escort jamming is used to protect aircraft from ground-based radars while blinking jammers deny ARMs their guidance signal during their off-times.

Passive ECM methods include chaff and decoys; chaff involves dispensing a large number of reflectors, such as metallic foilstrips in bundles from aircraft for the purposes of deceiving the radar operator. Decoys that can be launched by the aircraft or simulated on the ground constitute other forms of passive ECM devices.

By contrast ECCM define actions taken to ensure the friendly use of the EM radiation in the presence of ECM. Radar systems are seldom considered in isolation from EW systems, mainly because it is difficult, if not impossible, to correct a radar that has been designed without the flexibility required to counter the ECM threats.

1.3.1 ESM System Functions

Electronic support measures define actions taken to intercept, identify, analyze, and locate the sources of hostile EM radiation. This information is then used by the ECM systems. Alternatively, ESM systems are used for spectrum surveillance. The systems operate on a receive-only mode, but unlike the receiving part of radar, ESM systems operate over very wide bands of frequency—specifically, over a few octaves.

Any transmissions emanating from radars, communicators, broadcasters or jammers are received and analyzed by ESM systems. During peacetime the analyses are usually centered around issues related to whether the many users adhere to their bandwidth allocations. The spatial coverage the many users enjoy might also be of interest to spectrum managers. Given that spectrum is a valuable commodity, spectrum surveillance will continue to be a valuable management tool.

During war (cold or conventional) the different ESM systems are assigned different tasks. ELINT (electronic intelligence) and COMINT (communications intelligence) systems are complex and costly installations where electronic and communications intelligence is gathered. Reconnaissance/surveillance receiver systems intercept, collate, analyze and locate radar signals in near real time. For communications ESM systems the deciphering of the intercepted transmissions follows the collation of the information received and other complementary information.

Radar warning receivers (RWRs) are used on board aircraft, ships, and submarines and by ground-forces personnel. RWRs perform the following functions on a real-time basis to counter threats from surface to air, air-to-air missiles, and antiaircraft gun systems.

- Intercept as many radar emissions possible—a requirement that can usually be met by a system having a high probability of interception (POI).
- Derive as many radar parameters as possible, for example, TOA (time of arrival), AOA (angle of arrival), frequency of operation, pulse width and amplitude, and PRF (pulse repetition frequency).
- Sort or deinterleave the radar pulses. Even a 10° FOV in azimuth can encompass millions of pulses per second [1.72].
- Assign the derived emanations to a number of radars.
- Match the radar characteristics thus derived to the characteristics of specific radars stored in the library.
- Assign platforms to the radars.
- Define and prioritize threats.
- Display as many essential parameters of the threat as possible, such as range and bearing.

Usually ESM systems operate in conjunction with ECM systems.

Apart from the stated differences, ESM systems share many similarities with the receiving sections of radar. More explicitly, ESM systems require highly directional beams and sensitive receivers, which allows them to intercept, detect, and accurately derive the many characteristics of weak and/or distant transmissions.

The basic equation on which RWRs are designed is

$$\text{SNR}_{\text{RWR}} = \underbrace{\frac{P}{4\pi R^2}}_{\substack{\text{Radar} \\ \text{power} \\ \text{isotropic}}} \underbrace{G_R}_{\substack{\text{Radar} \\ \text{antenna} \\ \text{gain}}} \underbrace{A_{\text{eff}}}_{\substack{\text{RWR} \\ \text{aperture}}} \underbrace{\frac{1}{kTB}}_{\substack{Rx \text{ noise} \\ \text{power}}} \underbrace{\frac{1}{L}}_{\text{Losses}} \qquad (1.32)$$

where SNR_{RWR} = is the SNR at the RWR
P = radiated power in watts
R = the range between the radar and the RWR
G_R = the gain of the radar aperture
A_{eff} = the effective area of the RWR
kTB = the noise power of the RWR
L = the system losses at the RWR

It is worth noting that the SNR_{RWR} is inversely proportional to the square of the range; for a radar the SNR is inversely proportional to R^4.

1.3.2 Concepts of Probability of Interception (POI)

In this section we shall conceptualize the different probabilities of interception for the cases involving a mechanically rotating radar and an ESM system. The same concepts apply to phased array-based radars.

The wide-frequency-coverage requirement for the ESM system dictates either a frequency scanning receiver or a channelized receiver; the channelized receiver operates efficiently in the presence of intentional or unintentional jammers. The following cases can be discerned when (1) both the ESM system and the radar rotate, and a beam-to-beam POI can be derived; (2) the radar rotates and the ESM system scans in frequency or vice versa, and a beam-frequency POI can be derived; and (3) the radar frequency hops and the ESM system scans.

The requirement to maximize the POI of an ESM system in the frequency–spatial domains can be met by wideband channelized receivers and staring phased arrays yielding enough independent and contiguous antenna beams to fill the required surveillance volume.

1.3.3 ECM Systems

Unprotected radar and communication systems can be rendered useless by ECM systems capable of transmitting significant jamming powers (CW, noise, or any other combination) within the bandwidths of the systems described above.

Given that radar and communication systems can operate at center frequencies widely separated, ECM and ESM systems have to operate over wide bandwidths, even though their instantaneous bandwidth often matches the bandwidths of the radar and communication systems. An overview of surface navy ESM and ECM development has been reported [1.73] and phased array-based ECM systems that provide high effective radiated power (ERP) in the vicinity of 1 MW, for airborne platforms [1.74] have been realized.

Let us formulate the fundamental reason why ECM systems can successfully disable an unprotected radar. The power received by a radar from a target can be derived from Equation (1.5). Similarly, the power received from a jammer that transmits power, P_j through an antenna having a gain G_j is given by the equation

$$P_r|_j = \frac{P_j G_j G_r \lambda^2 10^{-0.1\alpha R}}{(4\pi)^2 R^2} \quad (1.33)$$

The ratio of powers received from the jammer and that reflected from the target, also known as the *jam-to-signal* (J/S) ratio, is therefore given by the equation

$$J/S = 4\pi \frac{P_j}{P_r} \frac{G_j}{G_r} \frac{R^2}{\sigma} 10^{0.1\alpha R} \quad (1.34)$$

Equations (1.33) and (1.34) illustrate the significant advantage a jammer has over the radar. More explicitly, the jamming signal travels one way only (jammer–radar path) in contradistinction to the radar signal, which travels from the radar to the target and back.

The implicit assumption made here is that the boresight axes of both the radar and the jammer are coincident, which is at best valid for only a small fraction of the time as the radar antenna moves in azimuth to cover the surveillance volume. For the rest of the time the jamming signal enters the radar receiver through the sidelobes or the cross-polarization lobes of the radar's antenna. For these more realistic cases, some attenuation of the jamming signal, dependent on the characteristics of the radar's antenna, is incurred.

1.3.4 Approaches to ECCM

The moment a radar is switched on, a third party can measure its center frequency, its PRF, and all its pertinent characteristics; similar comments can be made about communication systems. Furthermore, the jammer has the advantage we have outlined in the previous section. It is abundantly clear that these systems require some protection.

In what follows we shall outline some protective actions from ECM and will emphasize the often forgotten dictum that it is more economic to incorporate ECCM in the initial radar/communication design rather that modify a radar/communication system to protect them from ECM at a latter stage.

The requirement for radar/communication antennas to have low sidelobes and polarization purity can be easily drawn. A radar should have agility and flexibility of as many of its pertinent parameters as possible, including frequency, PRF, polarization-state, pulse width, and intrapulse modulation on a pulse-to-pulse basis. Additionally, the radar can use pulse compression, spread-spectrum techniques, and pulse integration techniques against ECM actions. Pulse compression can typically add a 30-dB advantage to the radar, while pulse integration can add a 10–17-dB advantage (the integration of 10 or 50 pulses is assumed) [1.75].

Spread-spectrum communication systems [1.76] came into existence in an effort to preserve the integrity of communication systems under severe jamming conditions. By spreading the band of the communication signals, the jammer is forced to spread the power available over a wider frequency band; the intended damage is therefore minimized. The underlying principle here is that for a given transmitter the power–bandwidth product is constant.

In relation to spread-spectrum techniques it is useful to introduce here the concept of processing gain (or time–bandwidth product), which is the ratio of the overall system bandwidth to the instantaneous bandwidth. Let us consider an X-band radar operating at 10 GHz. The radar's overall bandwidth will be 1 GHz, while its instantaneous bandwidth is, say, 10 MHz. The radar's processing gain will therefore be 100, which is too low. By comparison, modern communication systems operating in the frequency-hopping (FH) mode can have processing gains of 7000 [1.77]. Communication systems having processing gains of 19,000 and 58,000 have also been considered [1.78].

1.4 RADIO-ASTRONOMY SYSTEMS

We urge the reader who has no interest in celestial bodies to read this section for the fundamental concepts and techniques developed by radio-astronomers are now used in several systems devoted to applied science applications. How a celestial body looks to an observer depends on the predominant physical processes taking place within it or in its vicinity and the instrumentation used. In the early years of radio astronomy, when radio telescopes had very broad antenna beams, the cosmos appeared blurry at radio wavelengths. Discrete radio sources were observed with rudimentary radio telescopes that consisted of mechanically movable conventional paraboloids. The quest for high spatial resolution by radio astronomers has been outlined. Let us derive some fundamental quantities often used in radio astronomy and applied science applications.

1.4.1 Fundamental Radio-Astronomy Concepts and Measurements

The nature of the emission mechanisms taking place in and around celestial sources of interest can be deduced from measurable astronomical quantities that, in turn, are derived when a radio telescope is pointed in the direction of the sources under consideration. The flux density of a source S is defined as the power received by a radio telescope per square meter of its collecting area per hertz or jansky unit (after K. G. Jansky, founder of radio astronomy, 1905–1950), is one of the fundamental quantities. From a knowledge of the flux density of the source, as a function of frequency, and the source's angular extent, one can usually deduce the physical processes at work near or at the source.

Typically the radio telescopes used to take radio-astronomical observations consist of one or more conventional, fully steerable, reflectors, each equipped with a highly sensitive radiometer and signal processing subsystems. The radiometer is a specialized receiver, the characteristics of which we shall outline in Section 1.4.3.

Radio-astronomical observations can be broadly divided into three categories:

1. Spectral observations of sources also known as *continuum observations*. Here the flux density of the source is measured at spot frequencies extending over 2 or more octaves. At each spot frequency, the instantaneous band of the radiometer is as wide as possible. In the past the fractional bandwidth of the system was limited to 10% by the cryogenically cooled parametric front-end amplifiers. (It is recalled here that the fractional bandwidth is the ratio of the system's bandwidth and the system's center frequency.) When cryogenically cooled, low-noise transistor front ends are used, the system bandwidth is usually limited by the signal processing subsystems for phased array-based radio telescopes. For conventional radiometers connected to one aperture, the bandwidth can be as wide as 1–2 GHz—see Section 1.4.3.

2. Spectral line observations (or simply line observations) where the line radiation emanates from clouds of atoms, radicals, and/or molecules located in the interstellar medium. Line radiation is confined into a very small band of frequencies centered around the transition frequencies of the atoms, radicals, and/or molecules The radical OH, and the H_2O and SiO molecules have transitions, at the cm and mm wavebands, which mase. Powerful emanations are therefore received from very compact sources of these masers; other molecules discovered in the interstellar medium include CO, CS, H_2CO, H_2CS, CH_3CHO, and CH_3NH_2 [1.79].

3. Variable continuum or line observations.

The observed or apparent brightness of a source B_0 is another fundamental quantity and is defined as the received power per unit effective collecting area per unit solid angle per unit bandwidth.

There is one more parameter of importance—the equivalent antenna temperature T_A, which corresponds to the "signal" in communication theory terms. We shall first derive expressions for T_A (and S) before we derive what corresponds to the "noise" in communication theory terms.

1.4.2 Blackbody Radiation

All bodies that are at a temperature above zero degrees Kelvin (0 K) emanate thermal or blackbody radiation. Some of the notable sources supporting blackbody radiation are the quiet Sun, the Moon, and Mars. The term *quiet Sun* refers to periods when the radio Sun is quiet, that is, when the Sun does not support burst activity. The entire Universe appears to be within a blackbody "enclosure" where the temperature is about 2.7 K [1.80]. Let us examine how the brightness of celestial bodies varies as a function of frequency for thermal sources.

The brightness B of the radiation emanated by the blackbody at a temperature T is given by Planck's (Max Planck, 1858–1947) law

$$B = \frac{2h\nu^2}{c^2} \frac{1}{[\exp(h\nu/kT) - 1]} \qquad (1.35)$$

where ν and h are the frequency and Planck's constant, 6.62×10^{-34}, respectively, and k is Boltzmann's constant (1.381×10^{-23} J/K). If operation at radio wavelengths is considered, the photon energy $h\nu$ is much smaller than kT; hence the Rayleigh–Jeans (Lord Rayleigh, J. W. Strutt, 1842–1919; J. H. Jeans, 1877–1946) approximation to Planck's law can be applied to transform Equation (1.43) into

$$B = \frac{2kT}{\lambda^2} \qquad (1.36)$$

Figure 1.22 illustrates the celestial sphere that has a brightness distribution $B(\theta, \phi)$ as a function of θ and ϕ, the usual spherical coordinates being observed by a radio telescope that has an antenna pattern, $P_0(\theta, \phi)$. The power per unit bandwidth w received from a solid angle Ω of the sky if one polarization is received is given by the equation

$$w = \frac{A_e}{2} \iint_\Omega B(\theta, \phi) P_0(\theta, \phi) \, d\Omega \qquad (1.37)$$

It is assumed here that the radiation received is incoherent and unpolarized. If the antenna is placed inside a blackbody enclosure at a temperature T, and $B(\theta, \phi)$ is constant, uniform, and equal to

$$B_c = \frac{2kT}{\lambda^2} \qquad (1.38)$$

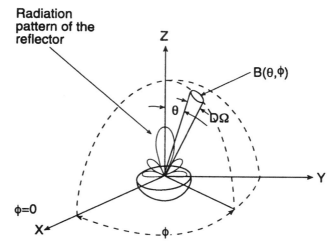

FIGURE 1.22 An antenna used to measure $B(\theta, \phi)$.

then Equation (1.37) can be written as

$$w = \frac{kTA_e\Omega_A}{\lambda^2} \tag{1.39}$$

where Ω_A is the antenna solid angle. If we introduce the well-known equation $\Omega_A A_e = \lambda^2$ into Equation (1.39), we deduce that $w = kT$, which is exactly equal to the noise power per unit bandwidth available at the terminals of a resistor R at temperature T.

In this regime of operation at cm and mm wavelengths, the power received by the antenna in the frequency interval $d\nu$ is frequency-independent and is equal to $kTd\nu$. Temperature T is then the equivalent antenna noise temperature, T_A, and the power received by the antenna is equal to $kT_A d\nu$.

Recalling that the average power received by the radiometer when it is switched to a resistor at a physical temperature T is also $kTd\nu$, a straightforward calibration procedure for T_A is available, since there is a one-to-one correspondence between T_A and T. Furthermore, T_A can be produced by any emission mechanism at work in or around the celestial source under consideration. Here we assumed that there is a perfect match between the input of the radiometer and the resistor at temperature T. An antenna connected to a radiometer therefore measures the equivalent antenna temperatures of radio sources that are within the antenna's FOV.

1.4.3 Derivation of Some Basic Relationships

For cases where the source is much smaller in extent than the beam area, the following relationships have been deduced [1.81]:

$$P = kT_A \, d\nu = \tfrac{1}{2} S A_e \, d\nu \qquad (1.40a)$$

or

$$S = \frac{2kT_A}{A_e} \qquad (1.40b)$$

T_A is therefore proportional to S and A_e. Given that T_A is measurable, S can be deduced.

If the size of the source Ω_s is known, then its flux density can be defined as the integral of the radiation brightness b over the source or

$$S = \int_S b \, d\Omega = \frac{2k}{\eta_b \lambda^2} T_A \Omega_s \qquad (1.41)$$

For a single aperture η_b, the main beam efficiency is the ratio of the power received through the main beam to the power received through the main beam and through the sidelobes of the aperture [1.82]. For phased arrays the latter power consists of the same component as for the single aperture, plus the power received through the grating lobes. In deriving Equation (1.41), it is assumed that T_A is uniform over the source, which we have assumed to emanate thermal radiation. As can be seen, the size of the source and the induced T_A are required before we can assign a value to the source's flux density.

While this book is devoted to phased arrays capable of yielding very high spatial resolutions, it is only fair to add that the resolution is needed only if the sources are compact. For extended sources like molecular clouds one is not disadvantaged in using single dish radiotelescopes. Indeed many important discoveries of interstellar molecules were made by these systems. The added advantage of course is that one can change the frequency of operation of a radio telescope simply by changing its receiving system. This allows the observer to explore for instance the existence of certain transitions of important molecules. For a phased array one has to change N receiving systems, a more expensive task. The two types of systems are however complementary in the sense that single dish telescopes explore what is there and elaborate synthesis systems are used to augment the acquired knowledge.

In general, $S \propto \lambda^n$, where n is an index corresponding the emission mechanism related to the source. If the source is a thermal source (blackbody), then $n = -2$ as per Equation (1.38). For sources where the emission mechanism is predominantly synchrotron (one of the important nonthermal emission mechanisms), S varies approximately as $\lambda^{0.7}$. This emission mechanism takes place under the following conditions. When an electron, traveling at relativistic speeds, is injected at right angles to a steady and uniform magnetic field, it executes circular motion about the magnetic field lines. While its orbital speed around the circle may be constant, it experiences a centripetal acceleration and as a result emits EM waves. The first sources from which synchrotron radiation have been observed were the Crab Nebula and from a large galaxy near the constellation of Virgo [1.83–1.86]. Other emission mechanisms and their spectra are described in Bekefi and Barrett [1.86].

The antenna temperature is the "signal" (in communication theory terminology) received at the antenna terminals when the antenna is pointed toward a radio source. The "noise" (in communication theory terminology) is determined by the root mean square (rms) of the minimum detectable signal of the radiometer ΔT used. With reference to Figure 1.23, we distinguish two kinds of radiometers: the total power radiometer (a), where the radiometer is permanently connected to the antenna; and the Dicke-switched radiometer [1.87] (b), where the antenna is alternatively switched between the radiometer and a constant temperature load, acting as a calibrator and where synchronous detection is implemented. The latter radiometer configuration

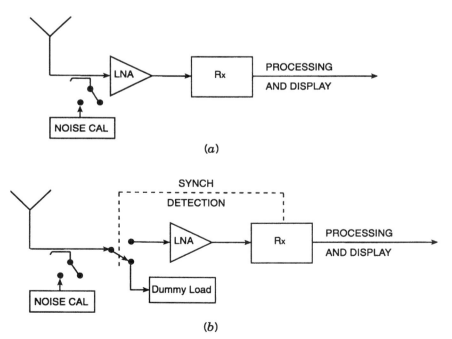

FIGURE 1.23 Block diagrams of radiometers: (a) total power; (b) Dicke-switched.

greatly minimizes the gain variations inherent in any radiometer. Their sensitivities in terms of their rms minimum detectable signal Δt_{min} are given by the equations

$$\Delta t_{min}|_{\text{total power}} = [T_A + T_s] \left[\frac{1}{B\tau} + \left[\frac{\Delta G}{G} \right]^2 \right]^{1/2} \quad (1.42)$$

$$\Delta t_{min}|_{\text{Dicke-switched}} = 2[T_A + T_s] \left[\frac{1}{B\tau} \right]^{1/2} \quad (1.43)$$

respectively, and where B is the system's bandwidth, T_s is its equivalent noise temperature, and τ is the integration time. The ratio $\Delta G/G$ is a measure of the gain variations of the radiometer and is typically equal to 0.01% for modern radiometers [1.88].

By equating the ΔT_{min} given by Equations (1.42) and (1.43), the following equation was derived [1.88] for the integration time

$$\tau = \frac{3}{B[(\Delta G/G)]^2} \quad (1.44)$$

required for the case where the ΔT_{min} obtained by these two radiometers is equal.

When $B = 2$ GHz and $\Delta G/G = 0.01\%$, τ is equal to 0.15 s. It is clear that under these conditions, the Dicke system is superior to the total power system for integration times longer than 0.15 s. The reverse is true if the integration time is shorter than 0.15 s.

The bandwidth of 2 GHz is typical for radiometers connected to one aperture. Although larger bandwidths are usually available from wideband amplifiers, it is well known that T_s for wider bandwidth amplifiers is significantly higher than that corresponding to narrowband amplifiers.

1.4.4 Image Theory Used for Radio-Astronomical Observations

We have seen that a thinned array can have the same resolution as a monolithic aperture having the same dimensions. Let us explore this concept in same detail. With reference to Figure 1.24, suppose that two identical antennas, having a radiation pattern $A(\theta, \phi)$, are placed in positions $a(0, 0)$ and $b(m\lambda, n\lambda)$. Both antennas are pointed to the same source, which has a brightness distribution $B(\theta, \phi)$. Let the spacing between them be ab and assume that their outputs are combined in such a way that the sine S_{ab} and

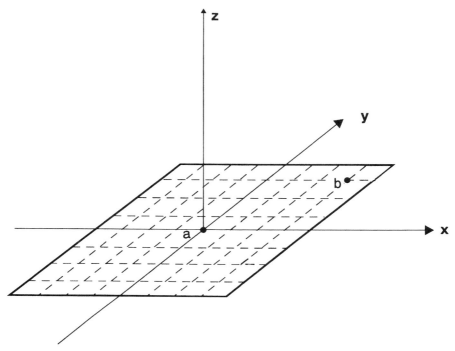

FIGURE 1.24 Illustration of the principle of aperture synthesis; initially two antennas are placed in positions a and b.

cosine C_{ab} terms of the resulting waveform are recorded (amplitude and phase). The output of the receivers may be written in a general way as

$$C_{m,n} + jS_{m,n} \propto \iint A(\theta, \phi) B(\theta, \phi) \exp[-2\pi j(m\theta + n\phi)] \, d\theta \, d\phi \quad (1.44)$$

Equation (1.44) has the form of a Fourier transform. It is therefore possible to determine the smoothed Fourier transform of $B(\theta, \phi)$ by moving one antenna from one location to the other and record the resultant terms before deriving a map of $B(\theta, \phi)$ by Fourier inversion. When this operation is repeated over many spacings, the brightness distribution of the source may be obtained from [1.89–1.90]

$$B(\theta, \phi) \propto \frac{1}{A(\theta, \phi)} \sum_{-M}^{M} \sum_{-N}^{N} (C_{m,n} + jS_{m,n}) \exp[2\pi j(m\theta + n\phi)] \quad (1.45)$$

where $\pm M$ and $\pm N$ are the extreme positions of the antennas in two perpendicular dimensions. Given that $A(\theta, \phi)$, is known, the brightness distribution, $B(\theta, \phi)$, can be derived.

It is very seldom that $B(\theta, \phi)$ is derived in the manner just described; usually the outputs of several pairs of antennas (interferometers) forming a cross, a Y, or a circle are taken simultaneously and the derivation of $B(\theta, \phi)$ is attained faster. In this mode of operation the brightness distribution of a source is obtained on a Fourier transform-by-Fourier transform basis, a method that is referred to as aperture synthesis. Alternately one can obtain one set of Fourier components of an image when a set of interferometers occupies a set of positions and then repeat the procedure when the same antennas occupy another set of positions.

The other method of obtaining the brightness distribution of a source is more familiar to radar and applied science researchers. It calls for the formation of independent and simultaneous antenna beams separated by an angular distant that ensures no loss of information is incurred. The output of each beam then yields the $B(\theta, \phi)$ of a pixel of the source. This picture point-by-picture point method of forming an image, is referred to as image synthesis.

Although the hardware employed in acquiring $B(\theta, \phi)$ by the two methods is different, both methods share the same theory. Similarly, Earth rotation can be used by the first method to obtain more Fourier transforms or scan a source by the second method.

The relentless quest for higher spatial resolution led to the developments of the very long-baseline interferometry (VLBI) method, where several antennas widely separated are interconnected in phase coherence to form an array. When the antennas observe the same source, maps of sources having spatial resolutions of a few milli-arc seconds are derived. In many experiments the antennas used are located in different continents. The natural progression was to electrically connect some antennas on board orbiting satellites to Earth bound antennas in order to further increase the spatial resolution. The spatial resolutions attained by this method, the Orbiting VLBI, OVLBI, are less than 1 milli-arc second.

Given that the collecting area of phased arrays is thinned, the minimum detectable signal, Δt, has to be re-defined. If $T_{sys} = T_s + T_A$, Δt for phased arrays is given by the equation [1.91]

$$\Delta t = \frac{T_{sys}}{\sqrt{Bt}} \frac{A_{syn}}{nA_e} \qquad (1.46)$$

where A_{sys} and A_e are the equivalent areas of the synthesized antenna and the actual antennas used for the measurement, respectively, and n is the square root of the number of independent antenna baselines in the measurement. Under very reasonable assumptions n is equal to the number of antennas used in the measurement (or the total number of independent antenna positions in a system where the antennas might be moved to obtain the desired baselines). When n is equal to the number of antennas used in the array, nA_e is equal to the actual collecting area employed in the measurement. In the limit when the array is completely filled, $nA_e = A_{syn}$. The minimum detectable signal for phased arrays is therefore equal to the product of the minimum detectable signal derived from a total power radiometer and the ratio of the equivalent area of the synthesized antenna over the array effective area [1.91].

1.4.4.1 Picture-Point by Picture-Point Imaging Method and Applications
In this section we shall outline two methods of attaining picture-point by picture-point images as well as citing astronomical and applied science applications utilizing these methods.

1.4.4.1.1 Mills Cross and Radioheliograph Approaches—Astronomical Let us consider how one or several pencil beams are formed in the Mills cross [1.92]. If all unit collecting areas of the east–west arm of the cross are connected in phase, the resulting fan beam $\mathbf{F}_{EW}(\theta, \phi)$ will extend from north to south. Similarly, the north–south arm would yield a fan beam $\mathbf{F}_{NS}(\theta, \phi)$ along the east–west direction. Here $\mathbf{F}_{EW}$ and $\mathbf{F}_{NS}$ are complex response patterns given by the equations

$$\mathbf{F}_{EW} = f_1 + jg_1$$

$$\mathbf{F}_{NS} = f_2 + jg_2 \tag{1.47}$$

If the power output of the system, due to a unit source with coordinates θ, ϕ is $P_{0°}$ when the antennas (E-W and N-S) are connected in phase and $P_{180°}$ when the phase is changed to 180°, we shall have

$$P_{0°} = k(\mathbf{F}_{NS} + \mathbf{F}_{EW})(\mathbf{F}_{NS} + \mathbf{F}_{EW})^*$$

$$P_{180°} = k(\mathbf{F}_{NS} - \mathbf{F}_{EW})(\mathbf{F}_{NS} - \mathbf{F}_{EW})^* \tag{1.48}$$

where the asterisk denotes the complex conjugate and k is a constant of proportionality. The normalized output power after the square-law detector has been shown [1.92] to be

$$P = f_1 f_2 + g_1 g_2 \tag{1.49}$$

Furthermore, for a symmetrical array $g_1 g_2 = 0$. With this technique one pencil antenna beam results, pointing toward the zenith. Another way of looking at the result is that there is output only in the crossover area of the two fan beams. Multiple pencil beams are formed when the unit collecting areas are phased so that staring beams result in the desired directions—see Section 3.14.2. It can be shown that the resolution of each beam resulting from a T.

There is a perception that image formation of celestial sources is restricted to radio-astronomy applications because in non-radio-astronomy applications one is interested in obtaining a radio image after a short integration time—say, seconds or less, and not days. While long integration times are required to obtain images of weak sources, the same does not hold for radio images of the Sun.

The Culgoora Radioheliograph, commissioned in 1967, [1.34 and 1.93] was a radio telescope that consisted of 96 steerable paraboloids 13 m in diameter and operated at 80 MHz. The antennas formed a circle 3 km in diameter and the radiotelescope had a spatial resolution of 3.5 arc minutes, at the zenith. The radioheliograph yielded radio pictures of the Sun in the two senses of circular polarization at intervals of 1 s and the observations taken significantly contributed to the understanding of short- and long-duration solar burst radiations.

The pictures were formed when 48 staring antenna beams generated along a north–south line [1.94] were swept along an east–west direction by changing the phases of the local oscillators used in the system. Unfortunately this unique system is no longer in operation.

In Figure 1.25$a-i$ one the most spectacular solar events, recorded on November 22, 1968 is illustrated [1.95]. Each image is formed in 1 s and the two opposite circular polarizations are recorded; the white circle indicates the optical disk.

The event depicts a magnetic arch, clearly seen in Figure 1.25 d, that radiates by virtue of the energetic electrons trapped within it. With the passage of time the arch expands and develops into three discrete sources before it fades and gives way to a compact long-enduring storm source (shown in the last picture, Figure 1.25i); the polarization of the sources varies as the storm develops.

1.4.4.1.2 Mills Cross Approach—Applied Science Applications Passive microwave remote sensing of the Earth's surface has progressed considerably in recent years. Surface parameters of fundamental geophysical interest have been recorded and monitored by spaceborne and airborne microwave radiometers. These parameters include sea-surface temperature, salinity, windspeed, soil moisture content, and arctic sea-ice concentration (see reference [1.91] and references cited therein).

A decade ago airborne radiometry was performed by using one radiometer to scan the scene as the aircraft was moving [1.96, 1.97]. Neither the resolution nor the sensitivity of these early radiometers was adequate at millimeter range wavelengths. The resolution was inadequate (1° HPBW) because a small rotating flat surface was used to scan the scene; similarly the sensitivity was low because the integration time was short. What is required is an airborne phased array capable of generating several staring beams forming a line that is perpendicular to the flight direction. As the plane flies, the line of staring beams is swept and an image is formed. We shall cite a couple of approaches to meet this requirement.

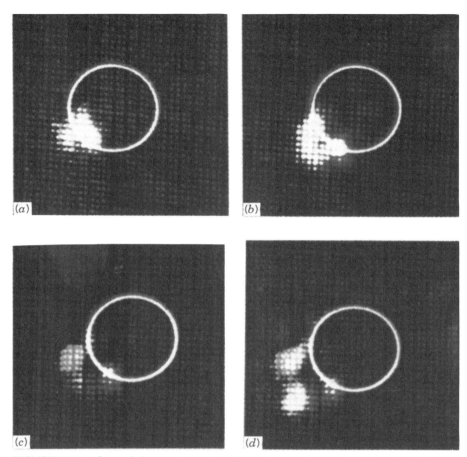

FIGURE 1.25 One of the most spectacular solar events recorded by the Culgoora Radioheliograph. The nine radio pictures (a to i) of the radio Sun were taken at 80 MHz, on the November 22, 1968, at the Universal Times shown. Each picture is formed in 1 s and the two opposite circular polarizations are recorded; the white circle indicates the optical disk. The event depicts the development of a magnetic arch, which radiates by virtue of the energetic electrons trapped within it. With the passage of time the arch expands and develops into three discrete sources before it fades and gives way to a compact long-lasting source shown in the last picture (i). (a) $01^h01^m.5$ UT. (b) 01^h13^m. (c) 01^h19^m. (d) $01^h23^m.5$. (*Source*: Wild [1.95] and by permission of the Australian Academy of Science.)

One approach to image formation by the process of obtaining as many picture points as possible instantaneously includes the proposal for radiometric sensing of the sea-surface temperature at 5 GHz with the aid of a satelliteborne cross-beam interferometer radiometer (CBIR), shown in Figure 1.26 [1.98].

RADIO-ASTRONOMY SYSTEMS 63

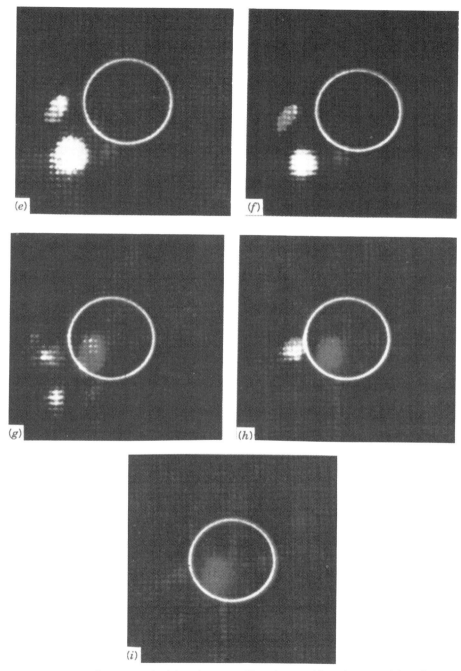

FIGURE 1.25 (Continued) (e) $01^h33^m.5$. (f) 01^h35^m. (g) $01^h44^m.5$. (h) $01^h51^m.5$. (i) $01^h52^m.5$.

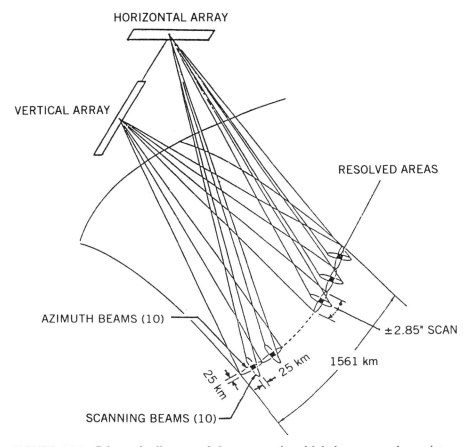

FIGURE 1.26 Schematic diagram of the proposed multiple-beam cross-beam interferometer radiometer (CBIR) and its footprint on Earth. (*Source*: Malliot [1.98], © 1993 IEEE.)

According to the proposal the CBIR, in the form of a T, is on board an orbiting satellite at a height of 833-km and operates at 5 GHz. Ten collinear quasi-pencil beams are formed, and their footprints, having a resolution of 25 km in a 1561-km swath, sense the surface temperature to a sensitivity of 0.5 K as they move along a track in a push-broom mode. The staring beams are formed with the Butler matrix, and the Mills periodic 0–180° phase-switching technique is used to form the instantaneous pencil beams. The information obtained is of interest to the meteorological and oceanographic research communities.

A CBIR operating at 1.43 GHz may be used for sea-surface salinity and moisture sensing; furthermore, the CBIR concept may be extended to high-resolution radiometry at EHFs from platforms located at the geostationary orbit.

1.4.4.2 Fourier Transform by Fourier Transform Imaging Method This method is of particular importance for radioastronomical observations. If we exclude the Sun, other celestial sources are weak and become weaker as their distance from our Galaxy increases. Long on-source integration times are therefore the norm. For a given array where the positions of all its unit collecting areas are fixed, the distances between them will change as the antennas follow the source. With this arrangement therefore the array acquires additional Fourier components as it follows the celestial source. For applied science applications such the remote sensing of the ground or sea by airborne phased arrays, all Fourier components must be measured instantaneously because the scene changes as the airborne platform moves along; hence there is a strong incentive to measure as many Fourier components as possible by utilizing a given set of antenna elements.

1.4.4.2.1 Images Resulting from Minimum Redundancy Arrays—Applied Science Applications A phased array of a special kind has been used on board an aircraft to monitor the surface soil moisture [1.91, 1.99–1.101]. In Ruf et al. [1.100] a theoretical framework is developed to support radiometric observations taken from airborne sensors pointed toward the Earth.

Soil moisture is optimally measured at a long wavelength (e.g., 21 cm) and a spatial resolution of about 10 km is required in order to achieve meaningful understanding of the global hydrologic cycle and its coupling to the atmosphere and energy cycle (see Ref. [1.91] and references cited therein).

The FOV of the array [electronically scanned thinned-array radiometer (ESTAR)] is a fan beam perpendicular to the direction of flight and a number of staring beams are generated within the array FOV. Finally the beams, generated by cross-correlating the outputs of the array elements, scan the scene in a broom-sweep fashion. The ESTAR is shown in Figure 1.27.

The phase array is a minimum-redundancy array (MRA) (see Section 3.13.1), which is capable of generating the maximum number of nonredundant spacings by using the minimum number of antenna elements. MRAs were first proposed and realized by radio astronomers in the 1950s. In Chapter 3 the theoretical framework of MRAs and the ESTAR system are further considered.

From the foregoing considerations we can conclude that images formed by the picture-point by picture-point method can be used not only for radio astronomical observations, but also for applied science research, namely, the typical applications related to remotely sensing the Earth's surface and atmosphere from space- or airborne platforms. In that context phased arrays offer the required high spatial resolution and sensitivity, an option that was not attainable with previous techniques and approaches.

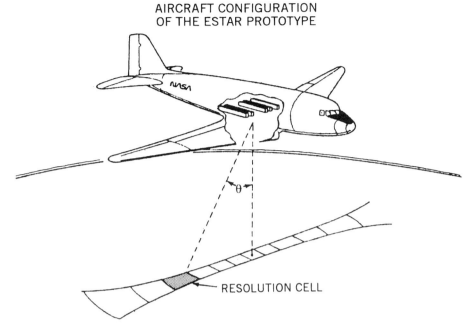

FIGURE 1.27 The airborne electronically scanned thinned-array radiometer (ESTAR) and its footprint on Earth. (*Source*: Swift et al. [1.101], © 1991 IEEE.)

1.4.4.2.2 From the VLA to VLBI and OVLBI While the resolution of conventional radio astronomy phased arrays, such as the VLA (also known as *compact arrays*), is in the vicinity of 1 arc second (depending on the frequency of operation), higher spatial resolutions can be attained only if the many collecting areas of an array are widely separated. With this arrangement, the VLBI, the data are recorded at each site where phased synchronized clocks maintain the phase coherency of the recorded data. In this context the baseline between two antennas is the vector between them.

Postobservation processing of data received at the different sites follows at one site where special processors are available. This technique, first demonstrated in 1967 [1.102], yields spatial resolutions of the order of 1 milli-arc second routinely when the antennas are separated by a distance nearing one Earth diameter. A review of the early VLBI work is outlined in Klemperer [1.103], and a recent book [1.104] provides thorough coverage of the many aspects of the VLBI techniques and experiments. For a long time after the first demonstration of the technique, radio telescopes located in national radio-astronomy centers in different continents were connected together for short-time VLBI observations.

Recently a dedicated VLBI network, the very long baseline array (VLBA), has come into operation [1.105–1.107]. It consists of 10 antennas 25 m in diameter, located at sites ranging from Hawaii to the Virgin Islands. The VLBA is built to image celestial sources at frequencies ranging from 300 MHz to 45 GHz with a capability for extension to 86 GHz. The spatial resolution ranges from 20 to 0.2 milli-arc seconds. Furthermore, the VLBA can be linked with the VLA and other VLBI networks in Canada and Europe to (1) increase its spatial resolution, (2) attain longer on-source integration times; and (3) yield images of celestial sources having a diverse range of baselines between antenna elements.

Inter alia, the VLBA is used to observe compact celestial masers (e.g., OH, H_2O, SiO), pulsating sources, and the centers of galaxies and for geodesic research. For the latter research, simultaneous observations at two frequencies are required for the determination of ionospheric delays.

All VLBA antennas are remotely controlled, and a wideband high-density recording system allows unattended operation at a sustained data rate of 128 Mb/s (megabits per second) for a day, and peak rates of up to 512 Mb/s. A single operator in the Array Operations Center controls the operations of the VLBA, at Socorro, New Mexico, where the correlator (performing the correlations between different antenna pairs) supports processing of all the observations with the array.

To further increase the spatial resolution of radio telescopes, some of the antennas have to be on-board satellites orbiting the Earth. Observations of many radio sources have been taken with antennas on board the geostationary Tracking and Data Relay Satellite System (TDRSS) satellite [1.108] and antennas located in two continents (Australia and Japan) at 2.3 GHz [1.109, 1.110] and 15 GHz [1.111]. The TDRSS was designed to relay data between ground stations and satellites in low Earth orbit via satellites in geosynchronous orbit. For communications with satellites TDRSE (E for east, deployed at 41°W) has two 4.9-m-diameter antennas, both operating at 2.3 and 15 GHz. There is a smaller antenna for the uplink (15 GHz) and the downlink (14 GHz) with the ground control station located at White Sands, New Mexico. Radio-frequency signals from celestial sources at 2.3 or 15 GHz are amplified, coherently translated to 14 GHz, and relayed to Earth. A tone from a frequency standard on the ground is transmitted to the satellite, where it is used to phase-lock all the on-board oscillators.

The OVLBI technique yields resolutions of the order of sub-milli-arc second, depending on the frequency of operation and the orbit of the orbiting satellite used. The attained resolution allows observers to better probe energy generation and motions at the heart of exotic objects like quasars (quasi-stellar) objects. More accurate measurements of the relative positions and motions of water-vapor maser emission features within star-forming regions can also be made. These measurements aid in the establishment of the fundamental distance scale of the universe [l.109].

While the method of image formation described above is particularly suitable for radio astronomy and geodesy, it could find applications in other fields as well.

1.5 SATELLITE COMMUNICATION SYSTEMS

Propagation at centimeter- and millimeter-range wavelengths takes place usually along a line of sight; this constraint restricts communications between sites separated by a distance of approximately 50–100 km. In order to establish communications between sites separated by longer distances, one needs a transponder (a module that receives the signals at one frequency and transmits them at another after amplification) placed on top of a hill or on a pole.

Satellites placed at a distance of 35,786 km above Earth (in a geostationary orbit) are always seen by sites separated by about 150° in latitude, which corresponds to a distance of about 12,752 km. The angle subtended by Earth at this height is 17.34°. Thus communications between sites widely separated can take place on a 24-h basis during the lifetime of a satellite, which normally is about 7 years. Earth-orbiting satellites that can perform this function are sometimes regarded as "sky hooks" [1.112].

Let us examine the transmission of signals between a satellite and an Earth station through free space when the gains of the transmit and receive antennas are G_t and G_r, respectively. The resulting carrier-to-noise (C/N) ratio at the receiver is given by the equation

$$\text{C/N}|_r = P_t G_t G_r \frac{1}{kT_s B} \left[\frac{\lambda}{4\pi R} \right]^2 \tag{1.50}$$

where P_t = transmit power
R = the distance between the transmitter and the receiver
λ = wavelength of operation
T_s = equivalent system noise temperature
B = system bandwidth
k = Boltzmann's constant

The inverse of the term in brackets is usually referred to as the *free space propagation loss*.

A detailed knowledge of the modulation system is required to convert the derived C/N into a SNR. In the early days of satellite communications (1965–1967, INTELSAT-I) G_t was very low (no directivity in the orbit plane), so G_r had to be high to derive an adequate C/N ratio (usually much higher than 10 dB to include margins to offset the attenuation of the signals by rain). As more directive beams were used, the diameter of the receiving antennas decreased accordingly; from Earth stations having antennas about 30 m in diameter (INTELSAT Standard A), or 11 m (INTELSAT Standard B) to a few meters [1.113]. As can be expected, the availability of devices yielding ever-increasing output powers and ever-decreasing equivalent system noise temperatures contributed to an acceleration of this trend.

Graphically this trend was depicted in the following terms. Initially most of the limited power available on board the satellite was dissipated in warming the cosmos and the fish living in oceans and seas separating the communications nodes. Thus the satellites of that era were aptly described as "fish warmers" [1.112]. As an example, considerable power was directed toward the Atlantic Ocean for communications between the United Kingdom and the United States. Currently we are witnessing a situation where almost all power is directed toward the communication nodes; this being the case, the receiving antennas look like hub caps [1.112].

Inspection of Equation (1.50) will lend some insights into satellite communication systems; R (for a satellite in the geostationary orbit) and k are constants; the wavelength of operation is determined by spectrum availability and the signal attenuation due to tropospheric effects; and P_t and G_t are determined by weight and volumetric considerations, which, in turn, are affected by the technology of the day. Similarly, the bandwidth is defined by channel requirements that are affected by the modulation methods. The quantity G_r/T_s or simply the G-over-T ratio is a significant parameter that characterizes different systems.

For some time offset paraboloids have been used as the apertures for satellites; in Section 2.4.1 we shall see that phased arrays are being considered as suitable substitutes. We have already noted that the angle subtended by a geostationary satellite is 17.34°, thus the scanning angle of the phased array would typically be less than $\pm 8.5°$, an easy specification to meet.

1.6 FUTURE DIRECTIONS AND TRENDS

In this section we shall summarize current system capabilities and explore future applications where phased array-based systems will play a significant role. Given that phased arrays are redefining the radar functions, it is worth reconsidering some of these functions.

1.6.1 Phased Array-Based Radar Functions

If we are considering phased arrays for a given application, we have the option of designing either scanning or staring phased arrays. Let us consider the applications where one of these options is more suitable than the other.

1.6.1.1 Scanning or Staring Arrays? Staring arrays are required for ESM systems because they provide a high POI. Similarly, staring phased arrays are required for radio-astronomical observations because that is the fastest way to attain a radio map of a celestial source. Speed is an important factor because national observatories usually receive applications for telescope time far exceeding the time available for observations. If the observations are taken at mm wavelengths, the weather frequently further limits the time available for scheduled operations.

The radar application is an interesting case. The rationale on which phased array-based radars are scanning arrays is based on the premise that all available power should be directed where it is needed most, that is, at the pixel (picture element) containing an incoming target, of unknown origin. The question of sensitivity therefore arises. It can be shown that the resulting output SNRs for scanning and staring arrays are identical when nonfluctuating targets are considered. The assumptions made here are that all available power is directed toward one pixel (scanning case) and that the same power floods the surveillance volume (staring case). While the illumination power per pixel is initially low for the latter case, by the time the whole surveillance volume is scanned by the scanning array, the power per pixel is exactly the same for both systems. Integration of pulses is linear when the target is nonfluctuating.

If the target is fluctuating, the improvement due to integration is not proportional to the integration time and some loss in the SNR of the staring array is incurred. Furthermore, this loss will depend on the nature of the fluctuating target and the integration time required, usually the timeframe required to scan the entire surveillance volume. It is therefore easy to conclude that the scanning system will yield SNRs that are equal (theoretical/ideal) or better than those obtained by a staring system.

While these considerations are valid, scanning systems are usually preferred because they also offer flexibility of operations in an ever-changing environment. Modern multifunction radars, as we shall see in Section 1.6.1.3, do not scan the surveillance volume in a television raster manner. In one scenario long-range surveillance is allocated a significant portion of the timeframe required to scan the entire surveillance volume. We have already mentioned that phased arrays track targets $A, B, C, \ldots, Z$ without spending any time in the angular distances between A and B, B and C, and so on. These are only two examples of the flexibility afforded by modern scanning multifunction phased arrays.

Can we easily dismiss staring arrays for radars? Definitely not, because they possess LPI capability. If this requirement is paramount, one can flood-illuminate the entire surveillance volume and receive the returns with the aid of a staring receiving array [1.114, 1.115].

Scanning and staring arrays can be seen as extreme or boundary conditions and arrays having a number of staring beams that are scanned [1.115, 1.116] mechanically or electronically occupy the middle ground. The motivation for generating some staring beams is closely related to saving valuable time within the timeframe of the radar. If the inertialess beam of a modern multifunction radar dwells for too long toward high clutter directions, the remaining time is not sufficient for the radar to perform its other assigned functions. A concept of "load" has therefore emerged [1.115] for phased array-based radars. If a cluster of 5–10 staring beams is generated, the time taken to complete the surveillance of a given volume is shortened; furthermore, this cluster can be moved anywhere within the array's FOV.

With reference to Figure 1.28, the Martello radar array generates a transmit fan beam of length equal to 6–8 beamwidths along the elevation. In the receive mode it generates 6 or 8 independent and simultaneous antenna beams, again along the elevation. The whole array is mechanically rotated to cover the remaining surveillance volume [1.117]. Again, costs and complexity were the main reasons the designers of this array opted for a system where mechanical scanning is used.

In a remarkable paper Keizer [1.118] proposed, inter alia, a bistatic radar (where the transmit and receive apertures are side by side) that can generate, for example, four independent and simultaneous transmit and four independent and simultaneous receive beams. Naturally the transmit and receive

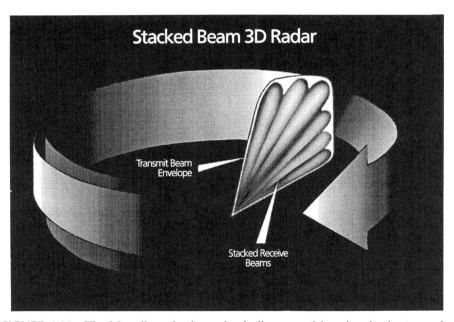

FIGURE 1.28 The Martello radar is mechanically scanned in azimuth; the transmit beam envelope is shown together with the six elevation receive antenna beams. (*Source*: [1.117].)

beams move in unison, and each pair can be scanned independently. Such arrays can provide more flexibility to future multifunction phased arrays. In the same paper Keizer proposed a scheme where a common transmitter generates frequencies f_1, f_2, f_3, and f_4, which are utilized by the four faces of the phased array capable of scanning 360° in azimuth on a time-sharing basis. Thus considerable savings result in the transmitter cost. We shall consider these proposals that will contribute toward the realization of affordable and efficient phased arrays, in Section 3.16.

1.6.1.2 Modern LPI/FMCW Radars We have already considered FMCW radars that have a 100% duty cycle and LPI properties; the comparison is here made with pulsed radars [1.119–1.120]. The block diagram of a linear FMCW radar utilizing different antennas for the receiver and transmitter is shown in Figure 1.29a while Figure 1.29b illustrates a similar radar utilizing one antenna for the receiver and transmitter.

An excellent account of the developments related to FMCW radars is given in Reference [1.121]. The difficulties in the realization of FMCW radars utilizing the same antenna for the transmit and receive function have been centered around the isolation required between the transmit and receive EM waves. Typically the transmit signal is of the order of watts, and the returned signal is of the order of picowatts. It is only recently that a satisfactory solution to this problem has been proposed and implemented in the Philips Indetectable Low Output Transceiver, PILOT Mk2 [1.122].

Figure 1.29c shows the transmitted signal in solid lines, while the returned signals from targets 1 and 2 are shown in dashed lines. Δf is the total frequency deviation of the waveform that determines the range resolution δR given by the equation $\delta R = c/2\,\Delta f$, where c is the velocity of light. The difference signal, δf_1, between the transmitted and received signal is proportional to the time delay d_1 and the slope of the linear FM, SL, or $\delta f = (\mathrm{SL})d_1 = (\mathrm{SL})2R/c = [\Delta f/T]\,2R/c$. Normally, many signals from targets at different ranges are received, the resulting waveform is hence, complex. With reference to Figure 1.29b, the received waveform passes through an A/D converter before it is resolved into its frequency components with the aid of an analyzer utilizing a fast Fourier transform (FFT) process.

In his proposals Keizer [1.118] utilizes pulsed CWs as the transmitting signals for the four-faced phased array. Other approaches of attaining LPI capabilities include spread-spectrum techniques.

1.6.1.3 A Typical Modern Multifunction Radar Operational circumstances and scenarios usually define the scanning schedule of the inertialess antenna beam. Typically the radar beam searches for targets at long ranges for a considerable fraction of the radar's timeframe for it is important to detect

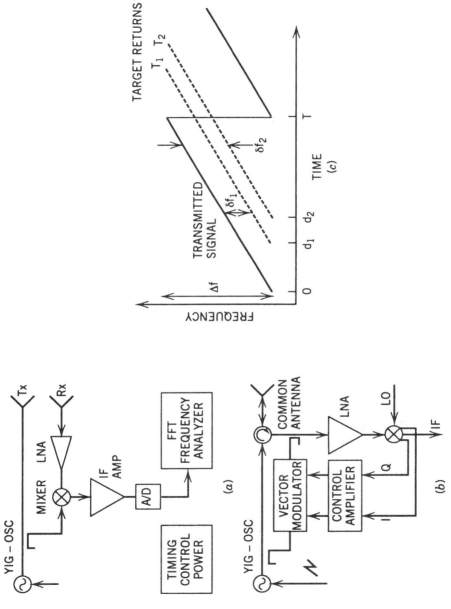

FIGURE 1.29 Typical FMCW radars utilizing two antennas (*a*) and one antenna (*b*). (*c*) The time versus frequency plots of a FMCW radar. The transmitted signal is depicted in a solid line while the target returns for targets 1 and 2 are shown in dashed lines.

and track these targets as early as possible. Some time has to be allocated to tracking targets at short ranges and for the performance of the surveillance function. Ultimately the entire surveillance volume has to be scanned at intervals that are negotiable by operational requirements.

The skills of operators have been infused into complex computer programs, so that detections, track initiations, tracking and other tedious (but necessary) tasks are performed with speed and precision automatically by computers. Modern day operators assume the roles of directing operations, in consultation with their commanders and supervisors, and select the operational menus appropriate to the different mission requirements.

Keizer [1.118] described a typical modern multifunction radar for maritime operations having the following pertinent characteristics.

Band of operation	C-band because this band provides a good combination of low-angle coverage and all-weather performance, in clutter
Radar faces	Four
The number of targets it tracks simultaneously	Hundreds
Supports the engagement of	10 targets
Detection range	50, 100, and 180 km when the target has an RCS of 0.1, 2, and 10 m^2, respectively
Coverage	Full hemisphere
Number of radiating elements	4900 per array
RF power	8 kW (solid state)
Antenna gain for a broadside beam	38 dB
HPBW	2°
Sidelobe level	-35 dB

The time allocation for the four-face active phased array is shown in Figure 1.30; as can be expected, the long-range search takes up the highest proportion of the timeframe. Naturally these are only typical estimates that are negotiable and are scenario-dependent.

1.6.2 Modern Approaches to the Identification Function

The importance of the identification function cannot be overemphasized, for the most sophisticated weapon systems cannot be used if a reliable identification of the incoming target is not attained.

We have already made a case to add the identification function to the list of functions for modern arrays—see Section 1.1. We have also seen that high-resolution images of targets can be obtained by SARs and ISARs—see Section 1.2.6. Let us explore the differences that exist in images taken at different wavelengths.

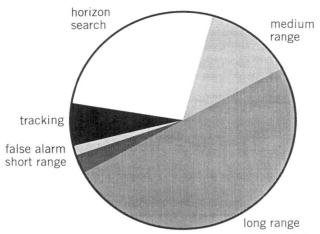

FIGURE 1.30 Typical time allocation for the four-face-active multifunction radar, after Keizer [1.118].

The application of Shannon's theory of information [1.123] to imaging is outlined in Perrin [1.124]. An object immersed in a radiation field can be sensed by that field with no finer resolution than the order of wavelength of the radiation. Images formed at shorter wavelengths, such as optical wavelengths, therefore have a higher information content, which, in turn, increases our ability to recognize objects and targets of interest.

While information content is of paramount importance, optical systems have several drawbacks. More explicitly these systems have short ranges and can operate efficiently only during clear weather conditions; additionally these sensors are passive and cannot yield the distance of the targets from the sensor. One therefore can propose sensor complementarity in the sense that a radar acquires targets (the surveillance function) while the imaging of some targets of interest is performed with sensors operating at infrared or optical wavelengths (the reconnaissance function).

Sensor complementarity can be seen in another way; if the surveillance function is performed at centimeter- and millimeter-range wavelengths by a SAR unaided by optical/IR sensors, the false identification rate will be high. The use of the additional sensor/s on some targets will decrease the false identification rate during clear weather conditions. The other dimension to sensor complementarity is related to FOV considerations. While the FOV of SARs is large, for example several tens of kilometers, the FOV of optical/IR sensors is adequate to identify the ordinary range of targets.

1.6.2.1 ISARs An ISAR provides an inexpensive means of identifying a selected target. The ISAR has a single aperture and has an instantaneous bandwidth Δf, the center frequency of which f_c changes continually. The image is then formed by utilizing stored information taken at frequencies f_c. While Δf is defined by the equation $\Delta f = c/2\ell$ the total bandwidth B is limited by the instrumentation available. In this equation, c is the velocity of light and ℓ is the range depth of the target's ensemble of scattering elements.

The slant- and cross-range resolutions Δr_s and Δr_c attainable are given by the equations [1.125]

$$\Delta r_s = \frac{c}{2B} \quad \text{and} \quad \Delta r_c = \frac{\lambda}{2\omega T} \tag{1.51}$$

where ω and T are the target's rotation rate and integration time, respectively. It is noted that the slant-range dimension is along the LOS of the radar and target while the cross-range dimension is along a direction perpendicular to the LOS. As can be seen, the slant- and cross-range resolutions decrease as the total system bandwidth is increased and the wavelength of operation decreases, respectively. Operation at millimeter-range wavelengths, therefore, holds the promise of improved slant and cross resolutions because of the inherent increased absolute bandwidth potential of these systems.

1.6.2.1.1 ISARs Operating at mm Wavelengths Assuming that ω and T are set, one can improve the slant-range resolution by increasing the system bandwidth readily; while operation at mm wavelengths can improve the cross-range resolution, there are problems related to the radar range defined by the equation [1.126]

$$R^4 = \frac{PG^2\lambda^2\sigma T_1}{(4\pi)^3 k T_s (\text{SNR})} \frac{10^{-0.2\alpha R}}{L} \tag{1.52}$$

where T_1 is the transmitted pulse width and the other terms have been defined in Section 1.2.1.

Let us consider the range R of an ISAR imaging system that produces a synthetic range profile of a target by utilizing a series of bursts of pulses; each burst consists of n pulses stepped in frequency from pulse to pulse by a fixed frequency step size df ($ndf = B$). The resulting range is given by the equation [1.126]

$$R^4 = \frac{PG^2\lambda^2\sigma T_1}{(4\pi)^3 k T_s (\text{SNR})_v} \frac{10^{-0.2\alpha R}}{L} \left(\frac{n}{m}\right) \tag{1.53}$$

where m is the number of resolution cells resolved by the radar and $(\text{SNR})_v$ is the processed pixel SNR required for image pixel visibility. Typically, if an imaging radar can first detect a target, the same target can be imaged after

FUTURE DIRECTIONS AND TRENDS 77

detection when the common radar parameters in Equations (1.52) and (1.53) remain the same for both detection and imaging.

Operation at mm wavelengths will decrease the radar range because (1) λ is in the nominator of Equation (1.52), (2) α increases as the wavelength decreases, and (3) P decreases as the wavelength of operation decreases—see Chapter 4.

Taking into account the power available from modern solid-state sources, practical antenna sizes and the clear-weather attenuation of signals at mm-wavelengths, the maximum range is limited to about 20 km for sizable targets such as aircraft. This is not an adequate range for many applications.

A proposal to increase the on-target power generated at mm wavelengths by utilizing a circular array of antennas, described in Section 2.5.1.1.2, can substantially increase the range ISARs operating at mm-wavelengths [1.127].

Table 1.4 illustrates the improvement in the range and cross-range resolution for ISARs operating at 34 and 94 GHz when a circular array of eight unit collecting areas are used. The cross range resolution improvements, expressed as ratios. $\rho|_{5 \text{ GHz}} = f/5$ and $\rho|_{10 \text{ GHz}} = f/10$ are with respect to systems operating at 5 and 10 GHz, respectively.

If the requirement is for maximum radar range, operation at 34 GHz is appropriate; if cross-range improvement is the major requirement operation at 94 GHz is preferable.

1.6.2.1.2 Phased Array-Based ISARs Currently multifunction phased array-based radars are used to perform many radar functions, and an ISAR to image targets of interest. A proposal has been advanced to use one staring phased array to perform all the radar functions including the identification function [1.128]. The phased array generates about 1000 staring beams that fill the surveillance volume ($\pm 45°$ in azimuth and $\pm 22.5°$ in elevation), and the identification function is performed by using the inverse synthetic aperture technique already described. The scene can be flood illuminated and the staring beams are formed on the receive mode; with this approach the system has an LPI capability.

TABLE 1.4 Improvements in the Radar-Range and Cross-Range Resolution for ISARs Operating at Millimeter-Range Wavelengths by the Method Proposed in [1.126]

Frequency (GHz)	R_0^a (km)	R_1^b (km)	$\rho_{5 \text{ GHz}}$	$\rho_{10 \text{ GHz}}$
34	20	53.4	6.8	3.4
94	20	39.4	18.8	9.4

[a] R_0 = range when one unit collecting area is used.
[b] R_1 = range when eight unit collecting areas are used.

While the realization of 1000 staring beams seems difficult at present, the use of photonics techniques, described in Chapter 3, Section 3.14.2.2, will render the task less daunting.

1.6.3 Friend-or-Foe ID

Another conventional method of performing the identification function is the use of an active identification friend-or-foe (IFF) interrogator mounted on the radar to trigger coded responses from a transponder aboard each aircraft. This approach to the identification function is inexpensive and effective in times of peace. The concern, however, is that existing IFF techniques can be spoofed, jammed, or otherwise rendered ineffective in times of hostile engagements.

1.6.4 Performance of the Identification Function by Unconventional Approaches

ESM systems can be used for aircraft identification because

1. Airplanes are literally "glowing" in the electromagnetic spectrum.
2. The airplane's emanations are highly characteristic of the particular on-board emitters.

There are, however, two problems with this approach:

- Range information cannot be usually derived by ESM systems that provide ID and azimuth information.
- The airplane might switch off its jammers in order to deny the ESM user one of its signatures.

Given that a radar can provide range and azimuth information, an ESM system coupled to a radar form a powerful complementary sensor combination capable of performing the identification function [1.72]. Furthermore, the radar is switched on for relatively short periods of time, so that other ESM systems do not readily detect its presence. The generic requirements for the ESM system and possible realization approaches are outlined in Reference [1.72]; the WJ-1780 AOA/frequency matrix surveillance and identification system has been described in the same reference.

Considerable effort has been expended in attaining target location by using two or more passive sensors, such as ESM systems. This approach is especially suitable for ELINT and COMINT stations. It calls for two sensors (ESM systems) widely separated to locate a noise jammer by using the principle of triangulation. In more realistic scenarios ghost noise jammers can reduce the system effectiveness. In Briemle [1.129], the *passive jammer locator* developed at (what was) the Telefunken System Technic in Ulm, Germany is described as well as some experimental results supporting the feasibility of the system. As can be expected, the accuracy of the positional information of jammers has to increase as the number of jammers increases.

1.6.5 Future Challenges for Radars

It is not difficult to imagine a future environment where the air traffic increases and requirements for safety and equipment reliability become more stringent with the passage of time. Technological breakthroughs based on multidisciplinary research and development (R & D), on the other hand precipitate new challenges and threats. Here we shall outline some of the essential developments and the challenges they impose for modern radars. We shall begin with the military challenges before we outline the civilian challenges.

1.6.5.1 Reduction of the RCS of Airplanes Here we shall follow the approach taken by Xu [1.130]. As a result of multidisciplinary R & D encompassing the fields of electromagnetics, air dynamics, and strength of structures and materials, the RCS of airplanes has decreased dramatically. Table 1.5 illustrates the estimated RCS of some well-known aircraft [1.131]. More explicitly, the reduction of the RCS of stealth aircraft resulted by controlling its shape to eliminate "bright points" formed by joints, sharp edges, corner reflectors, and large areas of low radius of curvature. Additionally, use is made of composite materials coated with absorbing coatings [1.132].

TABLE 1.5 Estimated RCS of Some Well-Known Aircraft [1.131]

Aircraft Type	Estimated RCS (m^2)
B-52	100
Blackjack (Tu-160)	15
FB-111	7
F-4	6
Mig-21	4
Su-27	3
Rafale-D	2
B-1B	0.75
B-2	0.1
F-117A	0.025

This phenomenal decrease in the RCS of aircraft has the following consequences:

1. The defense time against stealth aircraft is shortened; similarly, it is more difficult to identify the aircraft because the radar returns are weaker.
2. If the transmit power of the defense radar is increased, in order to detect the stealth aircraft, then the detection of the radar by ESM systems is rendered easier.
3. The self-screen power for the stealth aircraft is lower; similarly, the effectiveness of decoys is enhanced.

In case the situation looks too depressing for the designer of defense radar systems, we ought to mention that it is very difficult to design aircrafts that are invisible at all frequency bands, such as from UHF right to IR (through L, S, C, X, K, etc. bands) and at all look angles. A strong case can therefore be made for multisensor defense systems or for wideband radars—radars that operate over 2–3 octaves. Finally, bistatic and multistatic radars illuminate stealth aircraft at angles where their RCS is not minimum.

1.6.5.2 Evolution of Advanced EW Systems Currently EW systems are being integrated in order to improve their overall system effectiveness. Their integration takes two forms: (1) the integration of several subsystems and (2) the integration of EW functions to the structure of the aircraft and to its other functions.

The following subsystems are integrated: radar/threat warning, missile attack, and laser illumination warning, as well as the transmission of various types of jamming. The frequency coverage is usually in the following bands: 2–18 GHz, millimeter-range waves, IR, and at optical wavelengths.

In order to minimize the number of antennas on board an aircraft, a concept of aperture sharing has emerged. The same aperture supports as many functions as possible (see Chapter 3, Section 3.12.6); for example, the radar, EW and communications functions are accommodated in one aperture and the overall aircraft RCS is decreased because multiple apertures are eliminated. A high-speed data bus accommodates the avionics data with the data generated by the many sensors.

1.6.5.3 Upgraded Weapons Requirements There is need to replace air-to-air missiles such as Sidewinder, Sparrow, Skyflash, and Aspide with missiles that are more accurate, effective, and maneuverable. The capability of attacking multiple targets simultaneously is a significant characteristic of advanced medium-range air-to air missiles (AMRAAMs). The new weapons therefore require next-generation radars to detect low RCS targets farther, faster, and lower as well as to track multiple targets simultaneously. A range of at least 100 km detection range will be required for the airborne fire control radars, AFCRs operating in a multitarget track mode. This is not a trivial task, considering that future aircraft will have very low RCSs.

Currently the MTBFs for the AFCRs is around 100 h: MTBFs of several hundred to several thousand hours are required. A large-scale logistic infrastructure is at present required to maintain current AFCRs. The AFCRs typically account for 50% or more of an avionic system's cost; the ever-shrinking defense budgets can no longer support expensive AFCRs.

The nose radar of the B-1B consists of a steerable phased array operating at X band and its antenna is flat, oval shaped and can be rolled over to three different positions: forward (its normal position) and either side of the aircraft. This arrangement permits the aircraft to look off the side without changing course. The radar is used for navigation, penetration, weapon delivery, and air-refueling [1.133].

1.6.5.4 The Solutions Proposed by Xu The threats and developments mentioned above pose a formidable set of challenges for the design engineer; we have already mentioned some ways of meeting these challenges. Data fusion from several sensors operating over widely different wavelengths, such as decimeter-range (dm), cm, mm IR, and optical, will buy some insurance toward the detection of "low observables." Similarly, bistatic radar will achieve the same objective and the transmitter can be ground-based or airborne. In the latter case the combat aircraft would carry a passive (receive only) phased array-based receiver. This arrangement renders the combat aircraft radio-silent. In a more sophisticated scenario, multistatic radars are used and the "blinking" technique protects the transmitters from antiradiation missiles.

SARs have been considered not only for the surveillance function but also as an add-on mode for fire control radars on board some upgraded fighters and bombers. ISARs will continue to be used as an aid to identification.

Future active phased arrays should have LPI capabilities and enhanced multiple track modes. Conventionally the antenna of AFCRs is installed in the nose of the aircraft; because of space restrictions, few of the antennas have diameters larger than a meter. By using conformal phased arrays, one can better utilize the space available on the nose of the aircraft, the array can be larger and the effects of the radome are eliminated. With this arrangement the radar's range and angular resolution will increase to meet future challenges.

1.6.6 The Case for Wideband Phased Arrays

While narrowband phased array-based radars perform a limited set of functions, including surveillance, detection, and tracking, truly wideband phased arrays perform a larger set of interdependent functions more efficiently because:

1. They operate at the optimum frequency for each particular function—see Sections 1.2.2.3 and References [1.134] and [1.135].
2. The economies and efficiencies resulting from the processing of interdependent data resulting from the same phased array [1.136].
3. The radar arrays possess ECCM capabilities [1.134, 1.135].

We shall discern here the two concepts of wideband operation. In the first instance, the array instantaneous bandwidth is equal to the total bandwidth and extends over 2–3 octaves; we shall refer to these arrays as *type A*. In the second instance the instantaneous bandwidth is narrow, that is, the array has a 10% fractional bandwidth or less while the total bandwidth extends over 2–3 octaves; we shall refer to these arrays as *type B*.

Zhang [1.137] supports the view that type B arrays are likely to be more popular than type A arrays. The efficiency of the type B arrays is bound to be higher than that corresponding to type A arrays, because resonant circuits offer greater efficiencies; additionally, type A arrays require protection from intentional or unintentional jammers because their instantaneous bandwidth is very wide. Jammer excision in the frequency or spatial domain can protect this array. Low-loss frequency mutiplexing can offer the same protection.

Type A arrays have LPI capabilities because their transmitted power can be spread over 2–3 octaves.

1.6.6.1 Multifunction Phased Arrays Performing Radar, EW, Radiometry, and Communications Functions

We have already seen in Section 1.2.2.3 that the optimum frequency bands for the surveillance and tracking functions are L- and X-bands, respectively. Proposals for a common aperture supporting the radar functions as well as the ESM, ECM, passive radiometric sensing function, and communications on a time-sharing basis have been advanced [1.134–1.136]. For airborne platforms where real estate is at a premium, the sharing of a common aperture is attractive. Additionally, efficiencies result because the integration of the radar and EW functions is possible. If all the required functions are performed by the same array, then power management and handover functions become more efficient and timely; the sharing of displays is also possible.

We have already considered the use of radiometers for radio-astronomical observations and for a range of applied science applications. Radiometers have also been used for measuring the temperature of ground targets [1.138–1.139]. The great attraction of radiometers is that they are passive sensors, which is an advantage for covert operations.

The array will naturally have ECCM capabilities because the operator will have the choice to perform the above-mentioned functions at frequency bands that are not jammed; to a certain extent suitable software can automatically protect the multifunction phased array from ECM threats.

Graceful degradation of the array performance, due to the malfunction of a number of modules, will ensure the survivability of the platform.

1.6.6.2 Noncooperative Target Recognition We have already considered many approaches to perform the identification function, which is often referred to as the *recognition function*. Until recently there were no books that systematically outlined progress related to methods and techniques of recognition and identification, even though extensive research and experimental design work is in progress in a number of countries, resulting in large numbers of published patents and papers.

The book by Nebabin [1.140] fulfilled this need recently. It covers active, sensors, passive sensors, and a combination of both. Most techniques are based on wideband sensors for the natural resonances of a target can extend over a wide frequency range.

The defense against a missile that employs decoys represents an extreme case. If the missile has a nuclear warhead, 500 decoys are expected. In order to distinguish the missile from the decoys, highly sensitive radars operating over the 100–1000 GHz frequency range are expected to be designed (Ref. [1.140] and references cited therein).

1.6.6.3 Active Array-Based Radar Systems Applied to Air Traffic Control
The systems used in many airports (civilian and military) have been designed in the 1960s. Given that air traffic control is a 24-h/day, 365-day/year undertaking, downtime for maintenance should be minimum. Even a 53-h maintenance downtime and other glitch fixes, for the ASR-9 radar, is unacceptable [1.141].

Increased traffic flow, mobility, and safety in airports on one hand and the decreasing costs of T/R modules on the other hand will lead the way for future air traffic control functions to be performed by wideband multifunction phased array-based systems [1.141]. The modernization program currently in the proposal stage includes civilian and military airports and airfields as well as the provision for mobility for the military systems; typically installation of the latter systems should not take more than an hour, for rapid deployment.

What follows is a listing of the airport radars and the frequency bands of operation:

1. The airport surveillance radar, ASR, currently operating at S-band
2. The precision approach radar (PAR) operating at X-band
3. The terminal Doppler weather radar (TDWR) operating at S, C, or X-band
4. The airport surface detection equipment (ASDE) operating Ku-band

The TDWR provides automatic detection of microbursts and low-level wind shear [1.142]. Microburst-induced wind shear can result in a sudden, change in air speed, which has had disastrous effects on aircraft performance during landing and takeoff. For instance, 149 aircraft accidents that resulted in more than 450 fatalities between 1975 and 1985 were attributed to wind shear.

The proposed wideband multifunction system is to operate in the X- to Ku-band to perform most of the functions cited; if the frequency band of the ASR is shifted to C-band and the frequency coverage of the MMIC-based T/R module can be extended to C-band, all the above-mentioned systems can be performed by the same system.

Apart from the obvious advantages of running one system instead of four and the increased reliability of the active phased array, the following advantages are attainable from the proposed array. Beam agility, available in the proposed system, allows increases in the radar's track range performance by 50%; the comparison is made with respect to conventional, mechanically scanned systems. Additionally, weather conditions and aircraft can be tracked simultaneously with parallel processing channels, saving the cost of another radar system. The performance advantage is locating the aircraft and weather conditions on the same displays.

The Department of Defense National Airspace System Modernization Program is part of the dual-use technology thrust [1.143].

1.6.6.4 Unmanned Air Vehicle (UAV) Applications UAVs extend the horizon of a platform (e.g., of a ship or helicopter) and hence the battle space. If UAVs are equipped with SARs having ultrawide bandwidths, the radars can penetrate foliage or ground [1.144]. Thus a commander can perform the surveillance function and detect targets under the foliage or ground.

1.6.7 Advanced Phased Arrays Redefine Radar

In this section we shall examine in some depth the advantages of phased array-based radars when compared to mechanically scanned radars and conclude that the former systems are forcing designers to redefine the radar functions.

Airborne applications of active phased arrays operating at cm wavelengths are envisaged pro tem as the basis of our comparisons. Active phased arrays it is noted here are arrays where minimal losses are incurred between the arrays antenna elements and the low-noise amplifiers (LNAs) and the power amplifiers. More explicitly, a T/R module follows every antenna element of the array. By contrast, some losses are incurred between the antenna elements and its active circuits in passive phased array-based systems. The additional limitation of passive arrays is that one high-power transmitter, which is usually a traveling-wave tube (TWT), can feed the entire array. In one realization (see Section 1.1.6.2) such an array is mechanically steered.

Passive phased arrays have been deployed in the U.S. Air Force's B-1B bomber, the Russian Mig-31 and the French "Rafale." At lower frequencies active and passive phased arrays have been used in variety of weapons systems such as the U.S. Navy's Aegis cruisers and destroyers [1.145].

Active phased arrays, also referred to as *active electronically scanned arrays* (AESAs), are the latest development in the ongoing evolution to replace conventional, mechanically scanned radar systems. It is estimated that about 2000 T/R solid-state modules are required for an airborne phased array [1.145] in a typical fighter application. AESAs have the following advantages over the conventional mechanically scanned radar systems [1.145]:

1. The generation of one inertialess beam is the most important advantage of AESAs because it increases the system's MTBF from 1000 h, (corresponding to the conventional system) to 100,000 hours, variable dwell time is spent on targets located in positions $A, B, C, D, \ldots, Z$ and no time is wasted in traveling from A to B, from B to C, and so on. Typically the conventional mechanically scanned array might take up to 8 s to scan the entire surveillance volume while the AESA takes 8 ns or less to move its beam from one position to the next. Distributed logic is used to speed up the beamsteering function of the array. The ability to perform high-speed update rates and burst modes approaching 100,000 beam positions per second offers new possibilities in multimode radars.

2. An AESA offers graceful degradation in the sense that the array can endure random failures of up to 10% of its T/R modules before performance degradation becomes noticeable. By contrast, a passive array that is mechanically scanned and uses one TWT, has no graceful degradation. Additionally it is claimed that the elimination of the high power TWT radar transmitter and its associated high-voltage power supply will provide an order of magnitude increase in reliability at the total system level.

3. As the mechanically scanned antenna operates, its specular reflection moves with it and scans the same large volume of airspace as the radar beam. Furthermore, the specular reflection can present a very large RCS, which can be detected and exploited by third parties. By comparison, the specular reflection of an AESA is stationary and can be directed toward a point in space where it is less likely to be intercepted.

4. The AESA can have radiation patterns to suit the required operation; for instance, a wide beamwidth is required for terrain avoidance and minimum sidelobe levels are required toward certain directions where ESM systems are known to be located.

Although this is a very impressive list of advantages, the following developments will lead to unprecedented efficiencies of operation at reasonable costs: the realization of several independent and simultaneous transmit/receive beams and the evolution of truly wideband, multifunction active phased arrays.

86 PHASED ARRAY-BASED SYSTEMS AND APPLICATIONS

The two outstanding hurdles for the wider adoption of AESAs are the cost of T/R modules, which is slowly decreasing, and the poor efficiency of solid-state transmitters at millimeter-range wavelengths.

1.6.8 Developments in Radio and Optical Astronomy

Goldsmith [1.146] described some single-aperture antennas operating at mm wavelengths and Kellerman [1.147] outlined in some detail OVLBI observations and a U.S. proposal for a millimeter array (MMA); an artist's impression of the proposed array is shown in Figure 1.31. In the same paper plans by Japan and the former Soviet Union to launch satellites dedicated for OVLBI observations are noted. The maximum spacing achieved by using OVLBI was two Earth diameters in 1992.

Plans by different groups of optical astronomers to build modern Michelson-type inteferometers are described by Davis [1.148].

FIGURE 1.31 Artist's impression of the U.S. proposed millimeter array (MMA). It consists of 40 antennas, 8 m in diameter, which can be positioned in circles 250, 900, and 3000 m in diameter, depending on the spatial resolution required [1.147].

1.7 CONCLUDING REMARKS AND A POSTSCRIPT

Radar functions and radio-astronomy objectives are outlined before some archetypical systems drawn from both disciples are considered.

In the radar context phased array-based systems generate reasonable amounts of spatially distributed powers that are added on target; the same systems break the nexus between the surveillance and the tracking functions.

In the radio-astronomy context, phased array-based radio telescopes break the nexus between the aperture real estate and spatial resolution. The popularity of phased arrays is therefore based on their ability to meet diverse requirements set by different disciplines. The theoretical frameworks outlined in this Chapter enabled us to:

1. Define the regimes of application for radars operating at cm and mm wavelengths.
2. Derive the optimum bands of frequency for radars to perform the surveillance and tracking functions.
3. Appreciate the importance of deriving the maximum information content from a target by the use of polarimetric radars having polarization agility and diversity.
4. Gain knowledge of EW concepts which in turn can be used to optimize their effectiveness so that valuable platforms such as ships and airplanes are protected or to aid the radar designer in the realization of systems having ECCM capabilities.

Apart from monostatic radars, other systems such as ESM, ECM, SARs, and bistatic radars use phased arrays; furthermore proposals for communication satellites and ISARs to utilize phased arrays have been put forward. Lately it has become obvious that phased arrays will play a significant role in ushering in the "wireless revolution," which is based on constellations of Earth-orbiting satellites such as Iridium, Globalstar, Odyssey, Leonet 1, and Leonet 2 systems set in orbit at heights ranging from 785 to 10,355 km [1.149]. The Iridium constellation of satellites for instance utilizes a high-efficiency, linear, lightweight, space qualified phased array panel that can simultaneously transmit or receive 16 beams [1.150].

Lastly a case is made for wide band phased arrays that can operate over 2–3 octaves. These affordable phased arrays perform a set of interelated functions effectively; for example one set of functions are radar and radar-complementary functions while the other set is related to airport safety management functions. Truly multifunctional wideband phased arrays are also known as shared aperture arrays. The assumption, to be qualified in Chapter 4, is here made that affordable T/R modules are either available now or will be available in the not too distant future.

The driving force for future requirements is the ever-decreasing RCS of platforms such as ships and airplanes. We have shown that phased arrays can not only meet these future challenges but also redefined the radar functions in the process.

Turning to radio astronomy, the radiometric imaging function, a core radio-astronomy function, is now used for the remote sensing of the environment. The techniques and methods developed by radio astronomers are therefore used for a variety of applied science applications. The same systems can be used to complement the radar functions. Interestingly radio-astronomy systems have requirements that are not too different from the requirements of future truly multifunctional phased arrays that can perform a set of interelated functions efficiently.

An important lesson can be drawn from Levy et al. [1.109], the group that utilized an antenna on board an Earth-orbiting satellite, designated to perform communications functions, for the first OVLBI radioastronomy experiments. One can propose that radio-astronomy arrays are used to perform communication or radar functions; alternatively radar arrays can be used to perform communication tasks.

When the barriers that separated radioastronomers and radar/communications, scientists/engineers for too long are lowered or removed, several possibilities become only too obvious.

CHAPTER TWO

From Filled Apertures to Phased Arrays Mounted on Fully Steerable Structures

I would rather understand one cause than be King of Persia

Democritus of Abdera

You can be excused for thinking that there is a slight of hand here; just as you thought that we shall never consider filled apertures in a book dedicated to phased arrays, filled apertures come in again through the back door. There are several valid reasons for this approach; we shall outline the important ones here.

In the PAVE PAWS phased array-based radar, dipoles are used as the antenna elements or unit collecting areas. By contrast, some phased arrays, such as the VLA, utilize fully steerable reflectors as unit collecting areas. Therefore some of their defining characteristics, for example, their radiation, patterns and cross polarization properties, are important to phased array designers.

Although there are no lingering doubts that systems utilizing single apertures cannot have the spatial resolutions offered by phased array-based systems, one would like to delineate the regime where phased arrays are used.

Whilst filled apertures have been used to generate only one antenna beam for a long time, the apertures are capable of generating several antenna beams. One would therefore like to define the regime of applications for filled apertures and for phased arrays that are also capable of generating several independent antenna beams simultaneously. Astronomical and applied science applications where these systems are used are also reviewed.

For certain applications the designer can combine the advantages of phased arrays with those often attributed to filled apertures to derive inexpensive systems. For these applications it is convenient to mount a small-sized phased array near or at the focal plane of a reflector; the resulting ensemble is then used to electronically scan a scene at fast rates, while the monolithic aperture is stationary; the same scanning capability is also available to the user in any direction in which the reflector is pointed. Here the assumption is made that the spatial resolution attainable by a single aperture is adequate. These systems are also known as hybrid phased arrays or limited scan systems.

The road to phased arrays naturally leads us to phased arrays mounted on fully steerable structures. We shall explore their characteristics and compare them to those corresponding to phased arrays.

Our coverage of filled apertures will be synoptic, however, because several references and books have been devoted to the study of filled apertures under several titles, including reflector antennas and/or lenses or continuous apertures. Modern topics related to filled reflectors, such as the cross-polarization of offset reflectors, will, however, be thoroughly treated here. Furthermore, characteristics that have an impact on a system's performance will be singled out and treated with the attention they deserve.

We have adopted a unified mathematical approach that allows us to explore the properties of circular, rectangular, and square reflectors with the aid of the lambda functions.

2.1 GENERAL CONSIDERATIONS

In this section we shall define the far-field patterns resulting when circular and rectangular apertures are illuminated by specific illumination functions; the salient characteristics are the resulting beamwidths specified by their half-power-beamwidths (HPBWs) and the sidelobe levels. Given that the sidelobe levels as a function of an angular distance away from the boresight axis of the aperture monotonically decrease for the illuminations we shall consider, the first sidelobe level will be closely monitored.

Theoretically this is an important step before we examine the impact of real-world considerations on the gain and sidelobe levels of conventional apertures. These considerations include the accuracy with which the apertures are manufactured and the impact of deformations on the same apertures due to gravity (and other causes) on the received and transmitted radiation.

Other issues we shall consider are the difficulties encountered when one contemplates the generation of multiple antenna beams by using symmetrical apertures. These considerations will naturally lead us to asymmetrical or offset structures and what they can offer to the designer.

2.1.1 The Far-Field Patterns Resulting from Specific Illuminations

In this section we shall consider illumination functions that have the same phase throughout the antenna's aperture and whose amplitude changes as a function of angular distance from the boresight axis of the antenna in a prescribed manner. These illuminations yield far-field radiation patterns, known as the *sum patterns*, where there is only one maximum centered along the antenna boresight axis. We shall also consider illumination functions that yield far-field patterns, known as the *difference patterns*, where a null results in the direction of the boresight axis of the aperture. These patterns result when one half of the aperture is fed in relative antiphase with the other half. The terms *sum* and *difference patterns* originated in the direction-finding literature where the output of two adjacent feed horns illuminating the same reflector where added or subtracted to provide the sum and difference patterns.

With reference to Figure 2.1, the far-field pattern $G(u, v)$ of a rectangular aperture of sides a, b illuminated in phase with amplitude distributions $F(x)$ and $F(y)$ along the x and y axes, respectively, is given by the equation

$$G(u,v) = \int_{-a/2}^{a/2} F(x) \exp(-2j\pi ux) \int_{-b/2}^{b/2} F(y) \exp(-2j\pi vy) \, dy \quad (2.1)$$

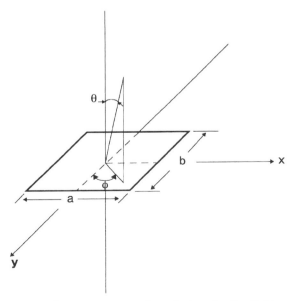

FIGURE 2.1 The coordinate system used to derive the far fields of a rectangular reflector.

where

$$u = -\sin\theta \frac{\cos\phi}{\lambda} \quad \text{and} \quad v = -\sin\theta \frac{\sin\phi}{\lambda} \qquad (2.2)$$

The coordinate system is centered at the middle of the rectangular aperture and θ and ϕ are the angles usually related to the spherical coordinates. λ is the wavelength of operation, and the rectangle is large enough, which implies that the resulting pattern angles θ are small enough so that the obliquity correction factor $1 + \cos\theta$ (which is the pattern of the elementary source $dx\,dy$) can be neglected. Given that the two integrals can be separated, it is sufficient to consider only

$$G(u) = \int_{-a/2}^{a/2} F(x) \exp(-2j\pi ux)\, dx \qquad (2.3)$$

where

$$u = -\frac{\sin\theta}{\lambda} \qquad (2.4)$$

2.1.1.1 Sum Patterns If $F(x)$ is a symmetrical, even illumination, a typical situation for a line aperture, the far-field radiation pattern, would be given by the equation

$$G_L(u) = 2\int_0^{a/2} F(x)\cos(2\pi ux)\, dx \qquad (2.5)$$

which is the Fourier cosine transform of the symmetrical illumination.

Similarly, the far-field radiation pattern of a circular aperture $G_c(u)$ of diameter D, illuminated with a circularly symmetrical amplitude distribution $f(\rho)$ in phase, is given by the equation

$$G_c(u) = 2\pi \int_0^{D/2} f(\rho) J_0(2\pi u\rho)\rho\, d\rho \qquad (2.6)$$

where ρ is a radial coordinate and $J_0(z)$ is the Bessel (after F. W. Bessel, 1784–1846) function of the first kind of order zero. If the line source and the circular aperture are uniformly illuminated, that is, if $F(x) = f(\rho) = 1$, then

$$G_L(u) = a\left[\frac{\sin(\pi au)}{\pi au}\right] \quad \text{and} \quad G_C(u) = 2A\left[\frac{J_1(\pi Du)}{\pi Du}\right] \qquad (2.7)$$

where $A = \pi D^2/4$.

The same results can be derived by the use of the lambda functions. Spencer [2.1] indicated that Chu [2.2] introduced the lambda functions of the first kind into microwave theory, and Ramsay [2.3–2.5] subsequently continued and systematized its use into the field of antenna theory.

We shall outline their approach for the following reasons:

1. The resulting theoretical framework can be used to derive the essential characteristics of circular and rectangular apertures.
2. The lambda functions constitute a useful analysis/synthesis tool.

In Chapter 3, modern approaches to the illumination of phased arrays by the use of the Taylor or Bayliss functions will be outlined. The former functions resemble the cosine squared on a pedestal illumination function which we shall consider here.

The lambda functions are defined by the equation

$$\Lambda_\nu = \frac{\Gamma(\nu+1)}{(x/2)^\nu} J_\nu(x) \tag{2.8}$$

Equations (2.5) and (2.6) can be generalized in terms of the Lambda functions by $G_\nu(u)$ given by the equation

$$G_\nu(u) = \frac{2\pi^{\nu+1}}{\Gamma(\nu+1)} \int_0^{a/2} F(x)\Lambda_\nu(2\pi u x) x^{2\nu+1}\, dx$$

$$= \left[\frac{\pi a^2}{4}\right]^{\nu+1} \frac{\Lambda_{\nu+1}(\pi a u)}{\Gamma(\nu+2)} \tag{2.9}$$

if $F(x) = f(\rho) = 1$. In order to derive the far-field patterns for the line source and circular aperture, ν takes the values of $-1/2$ and zero, respectively, in Equation (2.9). If we recall that

$$\Lambda_{-1/2}(x) = \cos x, \qquad \Lambda_0(x) = J_0(x) \qquad \text{and}$$

$$\Lambda_1(x) = \frac{2\Gamma(2)}{x} \qquad J_1(x) = \frac{2J_1(x)}{x} \tag{2.10a}$$

$$\Lambda_{1/2}(x) = \left[\frac{\pi}{2x}\right]^{1/2} \qquad J_{1/2}(x) = \left[\frac{\pi}{2x}\right]^{1/2}\left[\frac{2}{\pi x}\right]^{1/2} \sin x = \frac{\sin x}{x} \tag{2.10b}$$

$$\Gamma(n+1) = n! \qquad \text{and} \qquad \Gamma\!\left[m + \frac{1}{2}\right] = \frac{1\cdot 3\cdot 5 \cdots (2m-1)}{2^m}\sqrt{\pi} \tag{2.10c}$$

Equation (2.9) can be reduced to

$$G_{-1/2}(u) = G_L(u) = \left[\frac{\pi a^2}{4}\right]^{1/2} \frac{\Lambda_{1/2}(\pi au)}{\Gamma(3/2)} = a\frac{\sin(\pi au)}{\pi au} \quad (2.11)$$

for a line source and to

$$G_0(u) = G_C(u) = \frac{\pi D^2}{4}\Lambda_1(\pi au) = \frac{\pi D^2}{4}\frac{2J_1(\pi Du)}{\pi Du} \quad (2.12)$$

for a circular aperture of $a = D$.

The Lommel property that both patterns belong to the same functional family distinguished only by the order, ν, has been verified [2.3].

Figure 2.2 shows the resulting far-field patterns for a line source and circular aperture when the illumination function is uniform. The resulting far-field patterns have the narrowest beamwidths possible and are therefore useful as benchmarks against which the resulting beamwidths derived when other illuminations are used, can be compared.

The HPBWs for a line source and circular illumination are $0.89\lambda/a$ and $1.02\lambda/D$, respectively. The first sidelobes for line source and a circular aperture are -13.3 and -17.6 dB, respectively.

2.1.1.2 Half-Power Beamwidths of Apertures

The HPBW, θ, of an aperture, of length L wavelengths, is usually obtained from the well-known equation $\theta = b\lambda/L$, where b is a beam broadening factor dependent on the illumination used. Usually the beamwidth of an aperture is measured after the radiation pattern has been drawn; an approximate expression of the beamwidth, however, has been attained [2.6].

The HPBW given by $\lambda/4\sigma_x$, yields a close approximation to the beamwidth obtained conventionally. Here σ_x, the second central moment of the aperture distribution is given by the equation

$$\sigma_x^2 = \frac{\int_L x^2 i(x)\,dx}{\int_L i(x)\,dx} \quad (2.13)$$

where $i(x)$ is the amplitude illumination along the x direction. The HPBW of an aperture can therefore be derived without first calculating the radiation pattern.

2.1.1.3 Other Illumination Functions

The search for illuminations resulting in lower sidelobe levels led designers to illuminations having a variety of shapes, such as inverted parabolas, bell-type, cosine, or cosine squared. We

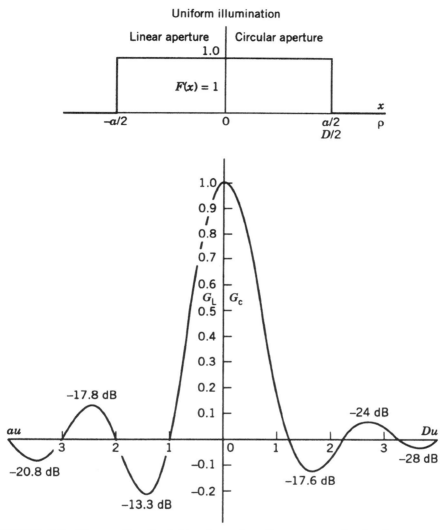

FIGURE 2.2 The resulting far-field patterns for a line source and a circular aperture when the illumination function is uniform, are shown on the lower LHS and RHS figures, respectively. (*Source*: [2.3].)

have listed some commonly used illuminations, and their resulting first sidelobe levels and HPBW b factors in Table 2.1. As the sidelobe levels decrease, the HPBWs increase.

Here we shall explore only two illumination functions and their resulting far field. When the line source and circular aperture illuminations are given

TABLE 2.1 The First Sidelobe Level and HPBW Factor b for Different Illumination Functions Corresponding to a Line Source/Circular Aperture [2.3, 2.4]

Illumination Function	First Sidelobe Level (dB)		HPBW Factor b	
	Line Source	Circular Aperture	Line Source	Circular Aperture
Uniform	−13.3	−17.6	0.89	1.02
Inverted parabolic	−21.3	−24.6	1.15	1.27
Inverted parabolic squared	−27.7	−30.6	1.38	1.48
Cosine/Bessel	−23	−27.5	1.19	1.32
Cosine squared/inflected Bessel	−31.7	−34.9	1.44	1.56
Cosine/Bessel on a 10-dB pedestal	−20	−24	1	1.14
Cosine squared/inflected Bessel on a 10-dB pedestal	−30	−28.4	1.06	1.15
Zero-order Sonine on a 14.8-dB pedestal	−24.7	−26.7	1.1	1.2
First-order Sonine	−28.9	−31.1	1.3	1.39

by a cosine and Bessel function given by the equations

$$F_L(x)|_{\cos} = \cos\left[\frac{\pi x}{a}\right]; \quad F_C(\rho)|_{\text{Bessel}} = J_0\left[\frac{4.81\rho}{D}\right] \quad (2.14)$$

and an inverted parabolic squared illumination function is given by

$$F_L(x)|_{(\text{parabolic})^2} = \left[1 - \left[\frac{2x}{a}\right]^2\right]^2; \quad F_C(\rho)|_{(\text{parabolic})^2} = \left[1 - \left[\frac{2\rho}{D}\right]^2\right]^2 \quad (2.15)$$

it has been shown [2.3] that the resulting far-field radiation patterns are given by the following equations:

$$G_L(u)|_{\cos} = \left[\frac{\pi}{2}\right]^2 \frac{\cos \pi a u}{(\pi/2)^2 - (\pi a u)^2} \quad (2.16)$$

$$G_C(u)|_{\text{Bessel}} = (2.405)^2 \frac{J_0(\pi D u)}{(2.408)^2 - (\pi D u)^2} \quad (2.17)$$

$$G_L(u)|_{(\text{parabolic})^2} = \Lambda_{5/2}(\pi au) = \frac{15}{(\pi au)^2}[\Lambda_{3/2}(\pi au) - \Lambda_{1/2}(\pi au)]$$

(2.18)

$$G_C(u)|_{(\text{parabolic})^2} = \Lambda_3(\pi au) = \frac{48 J_3(\pi Du)}{(\pi Du)}$$

(2.19)

respectively. Figures 2.3 and 2.4 illustrate the above-mentioned illumination functions and the resulting far-field radiation patterns.

It is clear that the cosine/Bessel illumination functions yield a good compromise between having a narrow beamwidth and low first sidelobe levels.

2.1.1.4 Difference Patterns

If separate halves of either a line source or a circular aperture are in relative antiphase, an antiphase lambda pattern will yield the far-field patterns of both types of antennas. The generalized antiphase pattern is given by [2.3]

$$G_\nu(u)|_{\text{antiphase}} = \frac{2\pi^{\nu+1}}{\Gamma(\nu+1)} \int_0^{a/2} F(x) \Lambda H_\nu(2\pi ux) x^{2\nu+1} \, dx \quad (2.20)$$

where $F(x)$ is an odd function on a linear radiator, or is odd in the plane of antiphase for a circular aperture that is symmetrically illuminated in intensity. The lambda–Struve function in the integrant is defined by the equation

$$\Lambda H_\nu(z) = \frac{\Gamma(\nu+1)}{(z/2)^\nu} H_\nu(z)$$

(2.21)

where $H_\nu(z)$ is the Struve function.

For an anti-phase uniform illumination, the far-field radiation pattern for the line source is given by the equation

$$G_\nu(u)|_{\text{antiphase}} = \left[\frac{\pi a^2}{4}\right]^{\nu+1} \left[\frac{\Lambda H_{\nu+1}(\pi au)}{\Gamma(\nu+2)}\right]$$

(2.22)

which reduces to

$$G_{-1/2}(u)|_{\text{antiphase}} = G_L(u)|_{\text{antiphase}} = a\Lambda H_{1/2}(\pi au) = a\frac{1 - \cos \pi au}{\pi au}$$

(2.23)

98 FROM FILLED APERTURES TO PHASED ARRAYS

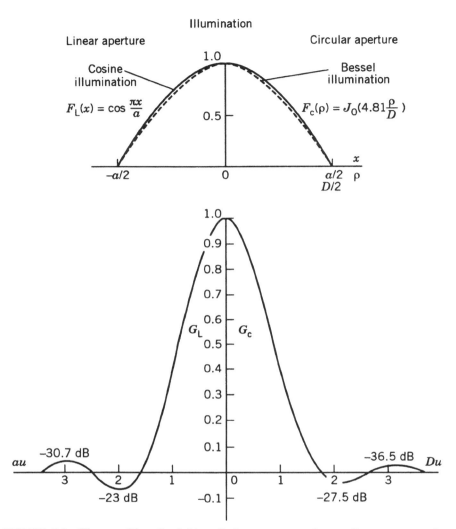

FIGURE 2.3 The resulting far-field radiation patterns when a line source and a circular aperture are illuminated by a cosine and Bessel function, respectively. The same convention used in Figure 2.2 applies. (*Source*: [2.3].)

$$G_0(u)|_{\text{antiphase}} = G_C(u)|_{\text{antiphase}} = \frac{\pi D^2}{4} \Lambda H_1(\pi Du) = \frac{\pi D^2}{4} \frac{2H_1(\pi Du)}{\pi Du}$$

(2.24)

for a circular aperture. The resulting far-field radiation pattern is shown in Figure 2.5.

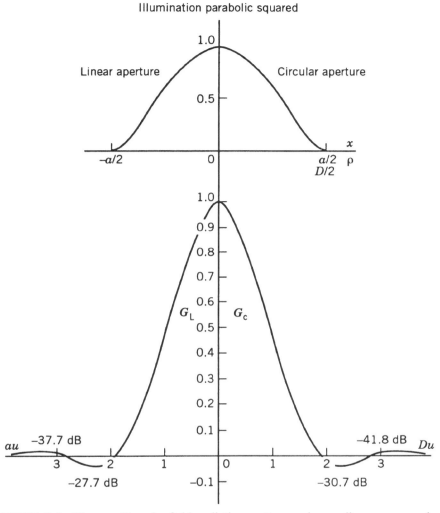

FIGURE 2.4 The resulting far-field radiation patterns when a line source and a circular aperture are illuminated by parabolic square functions. The same convention used in Figure 2.2 applies. (*Source*: [2.3].)

Ignoring constants, the far-field patterns for line sources having parabolic and parabolic squared illuminations are

$$G_{\text{L}}(u)|_{\text{antiphase}} \propto \Lambda H_{3/2}(\pi au) \tag{2.25}$$

$$G_{\text{L}}(u)|_{\text{antiphase}} \propto \Lambda H_{5/2}(\pi au) \tag{2.26}$$

100 FROM FILLED APERTURES TO PHASED ARRAYS

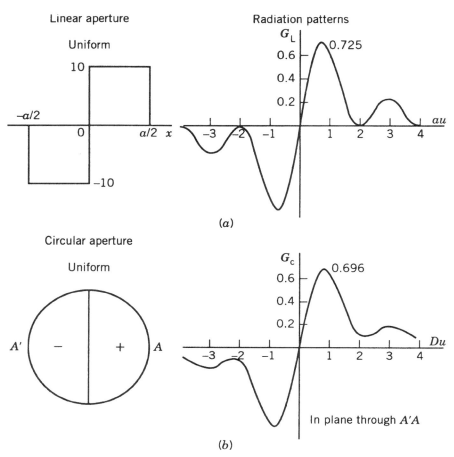

FIGURE 2.5 The resulting difference far-field radiation patterns when the illumination function is uniform: (*a*) for a linear aperture; (*b*) for a circular aperture. (*Source*: [2.3].)

respectively. Similarly, for a circular aperture

$$G_C(u)|_{\text{antiphase}} \propto \Lambda H_2(\pi Du) \tag{2.27}$$

$$G_C(u)|_{\text{antiphase}} \propto \Lambda H_3(\pi Du) \tag{2.28}$$

respectively.

Some of these functions approximate the radiation patterns of feed horns used to illuminate the apertures. In a phased array context the designer has more freedom to specify a greater variety of illumination functions. In

Section 3.3.2 we shall consider more appropriate illumination functions for phased arrays.

2.1.1.5 Conventional Measures of Aperture Efficiency In this section we shall briefly state the measures of aperture efficiency adopted by antenna engineers before we explore unconventional measures of efficiency usually adopted by systems-oriented engineers. Without any loss of generality, we shall consider the case where the aperture is a system of reflectors, a Cassegrain system, illustrated in Figure 2.6.

In Section 1.2, we defined η as the aperture efficiency of a reflector/lens; here we shall examine its many components, which are defined by the equation

$$\eta = \eta_s \ \eta_m \ \eta_i \ \eta_b \ \eta_r \tag{2.29}$$

where η_s = subreflector spillover efficiency
η_m = main reflector spillover efficiency
η_i = illumination efficiency

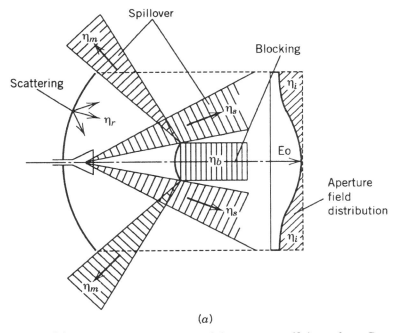

(a)

FIGURE 2.6 (a) The various components of the aperture efficiency for a Cassegrain system of reflectors. (*Source*: Miya [2.7], courtesy KDD.)

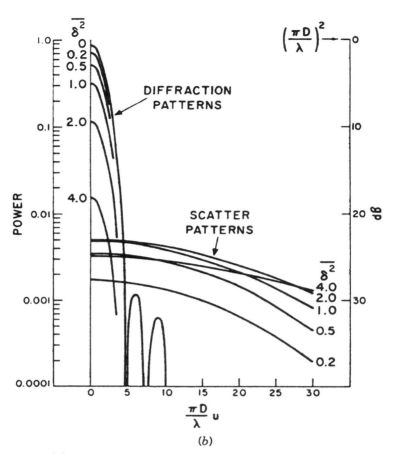

FIGURE 2.6 (*b*) Radiation patterns of a circular aperture having a 12 dB illumination taper. As the magnitude of the random phase errors increases, the main beam gain decreases and power is directed to the envelope of the sidelobes (scatter patterns); here $\bar{\delta}^2 = \left(\dfrac{4\pi\sigma}{\lambda}\right)^2$. (© 1966 IEEE, Ruze [2.9].)

η_b = blocking efficiency
η_r = the scattering efficiencies due to random and systematic phase errors

The first four efficiencies are given by the following well-known expressions [2.7]:

$$\eta_s = \frac{\text{power illuminating the subreflector}}{\text{total power radiated from the primary radiator}} \quad (2.30)$$

$$\eta_m = \frac{\text{power illuminating the main reflector}}{\text{total power scattered from the subreflector}} \quad (2.31)$$

$$\eta_i = \frac{1}{A} \frac{\left|\int F \, dA\right|^2}{\int |F|^2 \, dA} \quad (2.32)$$

where F is the illumination function across the aperture and A is the aperture area

$$\eta_b = \frac{\text{power blocked}}{\text{power radiated from the main reflector}} \quad (2.33)$$

η_r has two components—one, η_1, due to random phase errors and the other, η_2, due to systematic phase errors on the surface of the aperture.

2.1.1.6 Random Errors
Ruze [2.8, 2.9] examined how the gain of a reflector antenna is affected when phase errors, random in nature, exist on its surface; these errors are attributed to the roughness of the surface of a shallow paraboloid that is manufactured with an rms error σ. Additionally, the phase values are completely uncorrelated for distances larger than c, which is much smaller than the diameter of the aperture but larger than the wavelength of operation. The resulting gain is given by the equation

$$G = G_0 \exp\left[-\left(\frac{4\pi\sigma}{\lambda}\right)^2\right] \quad (2.34)$$

where

$$\eta_1 = \exp\left[-\left(\frac{4\pi\sigma}{\lambda}\right)^2\right] \quad (2.35)$$

Figure 2.6b illustrates the dependence of the main lobe and the envelop of the sidelobes of the reflector on σ; as σ increases, the main beam gain decreases and the envelop of its sidelobes increases. Power is therefore diverted from the main beam to the sidelobes.

Zarghamee [2.10] derived a small correction to the formula deduced by Ruze [2.8 and 2.9] by not assuming a uniform error distribution over the aperture.

2.1.1.7 Systematic Errors
In this section the effect of systematic errors on the antenna's surface are considered. Systematic errors having a correlation interval $c \sim D$ are usually due to (1) gravity loading [2.11], (2) thermal

distortions [2.12], and (3) initial rigging or adjustment bias on the structure of the antenna [2.13]. Assuming that the third cause can be eliminated by proper adjustments, the remaining causes affect the antenna gain decidedly.

At some elevation angle the opposite sides of a reflector would move near the focus while the other two quadrants would move farther away. Gravitational sag tends to produce a deformation of the type, as will thermal expansion in certain cases.

Following Cogdell and Davis [2.14], the decomposition of tolerance errors into errors having short correlation interval and those having interval comparable to the diameter of the antenna becomes simple and natural. If $\eta(\lambda)$ is the antenna efficiency at λ and $\eta(\infty)$ its efficiency at $\lambda \approx \infty$ (or the longest wavelength of operation), then

$$\frac{\eta(\lambda)}{\eta(\infty)} = \exp\left[-\left[\frac{4\pi\sigma}{\lambda}\right]^2\right] \frac{\eta(l)}{\eta(\infty)} \quad (2.36)$$

where $\eta(l)/\eta(\infty)$ is the tolerance loss resulting from errors having long correlation interval. If the main beam is a Gaussian function, it can be shown that the efficiency is inversely proportional to the product of its beamwidths θ_1 and θ_2 along the two principal directions:

$$\eta(\infty) \propto \frac{1}{\theta_1 \theta_2} \quad (2.37)$$

$$\eta(l) \propto \frac{1}{\theta_1'} \frac{1}{\theta_2'} \quad (2.38)$$

If we substitute Equations (2.37) and (2.38) into (2.36), we obtain

$$\frac{\eta(\lambda)}{\eta(\infty)} = \exp\left[-\left[\frac{4\pi\sigma}{\lambda}\right]^2\right] \frac{\theta_1 \theta_2}{\theta_1' \theta_2'} \quad (2.39a)$$

where

$$\eta_2 = \frac{\theta_1 \theta_2}{\theta_1' \theta_2'} \quad (2.39b)$$

Since all the parameters shown in the RHS of Equation (2.39b) are measurable, η_2 can be deduced.

Systematic errors ultimately define the shortest wavelength of operation and not random errors; in a typical example [2.14] the shortest wavelength of operation of the 16-ft radio telescope of the University of Texas was 1.76 mm, defined by systematic errors while the shortest wavelength of operation defined by random errors was 0.96 mm.

From the foregoing considerations it is easy to appreciate why most antennas operating at mm wavelengths are either enclosed in radomes with appropriate windows to the cosmos or else precautions are taken to ensure that the causes responsible for systematic surface errors are eliminated.

2.1.1.8 Maximum Diameters for Conventional Apertures The gravitational and thermal constrains defining the maximum diameter of reflector antennas have been examined by taking into account random and systematic errors. It was deduced that thermal constraints define the maximum diameter of a reflector antenna; if conventional materials such as steel are used, the best spatial resolution that can be attained is independent of the frequency of operation and is equal to about 12 arc seconds [2.15]. The maximum diameter as a function of λ_{min} (in mm) is given by the equation [2.15]

$$\lambda_{min} = 5 \text{ mm } \frac{D}{100} \qquad (2.40)$$

where D is in meters. Thus for operation at 1, 2, 3, or 10 mm, the maximum diameter of the aperture is 20, 40, 60, or 200 m.

Phased arrays are therefore used whenever the required spatial resolution is less than 12 arc seconds, the limiting resolution of conventional apertures utilizing steel members.

2.1.1.9 Unconventional Definition of Efficiency Although the above-mentioned efficiencies are useful in the antenna engineering context, these efficiencies are only the prerequisites for other more important efficiency measures. An interesting way of exploring other measures of efficiency is to ponder over what we would have if we optimize all these efficiencies. The aperture will yield one, and only one, antenna beam. This contrasts poorly with cheap cameras where an aperture (a lens) can yield many pixels, each corresponding to an independent antenna beam in antenna engineering terms. While there are important differences between the two systems, we should not be satisfied with apertures yielding one antenna beam.

We therefore have either one of two objectives:

1. The generation of as many independent and simultaneous antenna beams as possible from one aperture. This way the "effective" efficiency of an aperture operating at radio wavelengths will approximate that of an aperture operating at optical wavelengths; put differently, such apertures will tend to approximate the nearest equivalents to radio cameras depending on the number of independent and simultaneous antenna beams generated.
2. The generation of a radio image by electronically scanning the scene with one beam.

In what follows we should go along with the maximization of the conventional aperture efficiency measures before we explore how we can further increase the efficiency of our operations by generating more than one antenna beam.

2.2 THE QUEST FOR MORE EFFICIENT APERTURES

We have already defined the limiting spatial resolution of conventional apertures; in what follows we shall endeavor to define other parameters that delineate the regime of applications for apertures and phased arrays.

Of all the aperture efficiencies of reflectors, the blocking efficiency is intrinsically related to symmetrical apertures or reflectors having either a prime focus geometry or a Cassegrain/Gregorian geometry. Apart from a decrease in the overall aperture efficiency, the blockages raise the sidelobes of the aperture [2.16]. We would like to explore other shortcomings related to symmetrical reflectors and arrangements that render $\eta_b = 1$.

Finally we shall define the number of independent and simultaneous antenna beams a reflector can yield or the extent of the aperture's FOV. This is important for the definition of the domains of applications for apertures and phased arrays.

2.2.1 Offset Reflectors

Spencer [2.1] proposed a simple "off center" reflector illustrated in Figure 2.7a, as a way of minimizing the sidelobes of an antenna and his suggestion was taken up by several authors. With this arrangement the phase center of the antenna feed is placed at the focus, F of the paraboloid. Silver [2.17] pointed out that prime-focus offset paraboloids have lower reflection coefficients than their symmetrical counterparts. In Figure 2.7b and c we have illustrated Cassegrain versions of offset reflectors, where the prime focus of the main reflector is at F_1 while the secondary focus is at F_2. To the extent that equivalent parabolas can be derived for the Cassegrain systems [2.18], these systems (and the Gregorian systems that are considered in the next section) share the same characteristics of their prime-focus counterparts.

Dragone and Hogg [2.19] showed that the improvement in the reflection coefficient of offset systems (with respect to their symmetrical counterparts) can be more than 20 dB, a significant improvement especially for multibeam systems. The same authors considered the sidelobe levels corresponding to symmetrical and offset near-field Cassegrain systems of reflectors, illustrated in Figures 2.7d and e, respectively. They deduced that the sidelobe level corresponding to the offset system is about 10 dB below that corresponding to its symmetrical counterpart. These are significant improvements, which come with a penalty of high cross-polarization fields predicted by Silver [2.17] and calculated by Cook et al. [2.20].

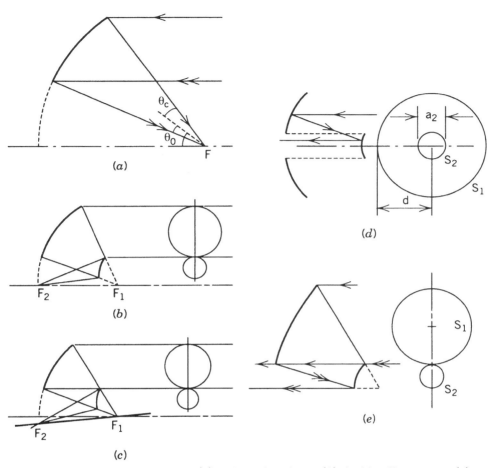

FIGURE 2.7 Reflector systems. (*a*) Offset prime focus; (*b*) double offset system; (*c*) optimized double offset system; (*d*) near-field symmetric Cassegrain; and (*e*) near-field bisected Cassegrainian.

Chu and Turrin [2.21] first published detailed graphical data of the maximum cross-polarization level of primary fed offset reflectors when the feed horns used to illuminate the reflectors are linearly polarized as a function of the offset angle θ_0 and the half-illumination angle θ_c. They have also shown that when the same reflectors are illuminated by circularly polarized feeds, no circular cross-polarization exists; the beams corresponding to the two polarizations are, however, displaced with respect to the plane of symmetry. The maximum cross-polarization corresponding to a prime-focus offset reflector is given by the equation

$$C(\theta, \phi)_{max} = -\theta_c \tan\frac{\theta_0}{2} \sqrt{\frac{10}{eT \ln 10}} \quad (2.41)$$

where T is the illumination taper (in power dB) and the angles θ_0 and θ_c are as shown in Figure 2.7a. Minimizing the angles θ_0 and θ_c minimizes the cross-polarization level for the case where linearly polarized illumination is used and the beam displacement for the case where circularly polarized illumination is used. Figure 2.8 illustrates the co- and cross-polarization

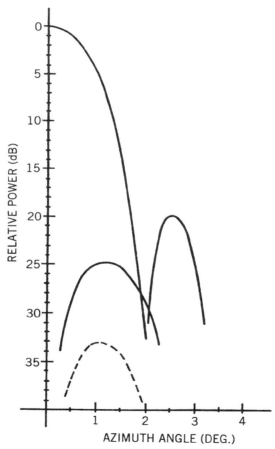

FIGURE 2.8 The theoretical co- and cross-polarization radiation patterns of two offset reflectors: (a) the case where $\theta_0 = 52°$ and $\theta_c = 20°$ (———); (b) the case where $\theta_0 = 40°$ and $\theta_c = 10°$ (-----).

radiation patterns of two offset antennas when (*a*) $\theta_0 = 52°$ and $\theta_c = 20°$ (———); and (*b*) $\theta_0 = 40°$ and $\theta_c = 10°$ (- - - -). As can be seen, the cross-polarization fields decrease if the angles θ_0 and θ_c decrease.

From the foregoing considerations it is clear that the cross-polarized fields result from the asymmetry of offset reflectors, that is, from the inherent geometric distortions. Thus, one can deduce that the introduction of another geometric distortion would render the resulting cross-polarization radiation negligible over a very wide band, a proposition that was corroborated by Graham [2.22]. Graham's paper constituted the first verification that the resulting cross-polarization can be minimized.

The cross-polarization fields under consideration are known as *geometric optic* (g.o.) cross-polarized fields to distinguish them from other sources of cross-polarization, such as the cross-polarization caused by diffraction effects originating at the subreflector. If the size of the subreflector is over 25λ, the latter cross-polarization fields are negligible. In what follows we shall assume that the latter cross-polarization fields are negligible.

Before we explore methods of canceling the g.o. cross-polarization fields in dual-offset systems of reflectors, it is worth considering the cross-polarization of prime-focus offset reflectors. If $\theta_0 = \theta_c = 32°$, the resulting maximum cross-polarization field is -25 dB; the corresponding F/D (focal distance/diameter) ratio is 0.4, which results in a mechanically manageable structure. One has to select a reflector having an F/D $= 0.8$ to attain a cross-polarization level of say -35 dB, an unattractive proposition from a mechanical engineering point of view. The designer therefore has no option but to use dual-offset systems of reflectors to either minimize the cross polarization fields or to cancel them altogether; while the first approach is straightforward, we shall consider the latter approach in some detail.

2.2.1.1 Cancellation of the G.O. Cross-Polarization of Dual-Offset Systems of Reflectors

Several researchers established a mathematical relationship that if satisfied, ensured the elimination of the g.o. cross-polarization of dual-offset reflectors; furthermore, this property was independent of frequency [2.23–2.25]. The underlying assumptions were (1) the feed horn used to illuminate the system of offset reflectors has negligible cross-polarization over the bandwidth of interest; and (2) the size of the subreflector used is such that diffraction effects can be neglected. Shore [2.26] formulated a simple derivation of the same basic formula.

A design procedure for dual-offset systems of reflectors having negligible cross-polarization has been outlined with the aid of several design equations that relate the defining parameters of the primary reflector to those of the secondary [2.27]. Only a brief outline of the proposed design equations is presented here, and the systems under consideration are the Cassegrain and Gregorian systems of reflectors; the methodology outlined here is, however, applicable to other more exotic systems of reflectors.

2.2.1.1.1 The General Equations

A Gregorian and a Cassegrain dual-offset system of reflectors are shown in Figure 2.9a and b respectively. $F_1 I$ is the axis of the primary reflector, a paraboloid, the focus of which is at F_1. The axis of the secondary is $F_1 F_2$, while the equivalent axes of the systems shown in Figures 2.9a and b are $IF_2 K$ and $F_2 IK$, respectively, and KR is the central ray of the primary reflector. Using the theory developed in Dragone [2.25], both systems of reflectors meet the negligible cross-polarization requirement because the points I, F_2, and K lie on a straight line. What follows is the design procedure outlined by Fourikis [2.27] for dual-offset systems of reflectors having negligible cross-polarization.

Given that R and M are on the main paraboloid reflector, the following equations, respectively, can be deduced:

$$\tan \frac{\theta_1}{2} = \frac{(d/2) + E}{2F} \tag{2.42}$$

$$\tan \frac{\theta_2}{2} = \frac{E}{2F} \tag{2.43}$$

Here F is the focal distance of the primary reflector.

Given that I and K are on the secondary reflector, the following equations can be derived:

$$\rho \tan \alpha = \tan \beta \tag{2.44}$$

$$\tan \alpha = \rho \tan \frac{\theta_1 + 2\beta}{2} \tag{2.45}$$

where

$$\rho = \frac{|e - 1|}{e + 1} \tag{2.46}$$

and e is the eccentricity of the secondary; for a Gregorian system $0 < e < 1$, while $e > 1$ for a Cassegrain system.

Combining Equations (2.44) and (2.45), one obtains

$$\tan \beta = \rho^2 \tan \frac{\theta_1 + 2\beta}{2} \tag{2.47}$$

It is now appropriate to study each system of reflectors, shown separately in Figures 2.9a and 2.9b.

2.2.1.1.2 The Gregorian System

We would like to explore the domains of definition of Equations (2.44), (2.45), and (2.47) for the Gregorian system of reflectors. To this end, and without any loss of generality, we shall set the

THE QUEST FOR MORE EFFICIENT APERTURES 111

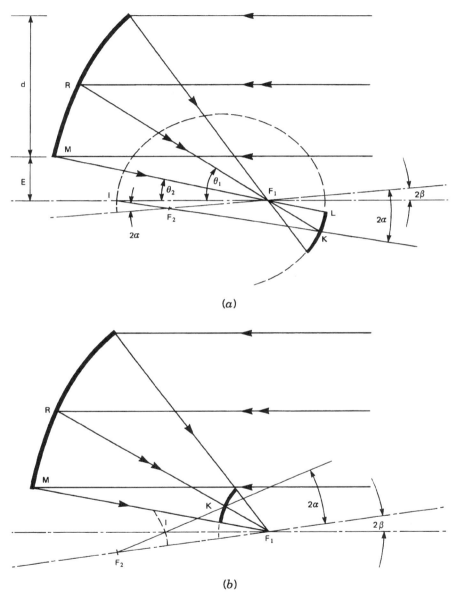

FIGURE 2.9 The dual-offset system of reflectors: (*a*) the Gregorian; (*b*) the Cassegrain. (*Source*: Fourikis [2.27], © 1988 IEEE.)

112 FROM FILLED APERTURES TO PHASED ARRAYS

angle θ_1 and F_1F_2 as constants and investigate how the quantity ρ and the angle α vary as the angle β varies.

If we differentiate Equation (2.47) with respect to β and set $\partial\rho/\partial\beta = 0$ to obtain

$$\sin 2\beta = \sin(\theta_1 + 2\beta) \tag{2.48}$$

we can deduce that

$$4\beta = 180° - \theta_1 \tag{2.49}$$

If we use Equation (2.49) in Equation (2.47), we obtain

$$\rho_{max} = \tan \beta_{max} \tag{2.50a}$$

or

$$\beta_{max} = \arctan \rho_{max} \tag{2.50b}$$

If we use Equation (2.50b) in conjunction with (2.44), we obtain

$$2\alpha = 90° \tag{2.51}$$

when ρ and β reach their maximum values. Equations (2.50a) and (2.49) yield

$$\rho_{max} = \tan\left(45 - \frac{\theta_1}{4}\right) \tag{2.52}$$

The derived equations have been used for a parametric study that defined a very large number of species of Gregorian systems having negligible cross polarization and Figure 2.10a illustrates the results of this study.

2.2.1.1.3 The Cassegrain System The Cassegrain system is illustrated in Figure 2.9b and the same approach is used here. The limiting values for ρ have been deduced to be [2.27]

$$\cot^2[\beta + (\theta/2)] \le \rho \le \tan\beta \tan\left(\beta + \frac{\theta_1}{2}\right) \tag{2.53}$$

The left-hand side (LHS) of inequality (2.53) can be written as

$$\beta + \frac{\theta_1}{2} < \arctan\left(\frac{1}{\rho}\right)^{1/2} \tag{2.54}$$

THE QUEST FOR MORE EFFICIENT APERTURES   113

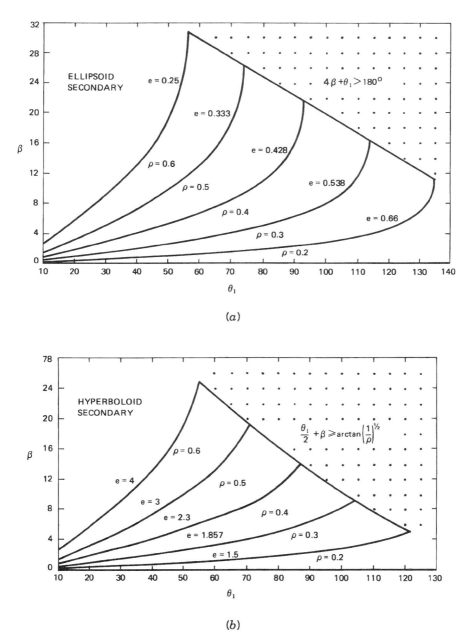

FIGURE 2.10 θ_1 versus β maps of systems of offset reflectors that meet the negligible cross-polarization requirement: (*a*) the Gregorian systems; (*b*) the Cassegrain systems. (*Source*: Fourikis [2.27], © 1988 IEEE.)

If Equation (2.54) is used in conjunction with (2.47) and (2.44), respectively, the following equations result:

$$\beta_{max} = \arctan(\rho)^{3/2} \qquad (2.55)$$

$$\alpha_{max} = \arctan(\rho)^{1/2} \qquad (2.56)$$

The derived equations have been used for a parametric study that defined a very large number of species of Cassegrain systems having negligible cross-polarization and Figure 2.10b illustrates the results of this study. The designer can therefore select volumetrically attractive systems, from this and the previous study.

It is clear that a very large number of species of these systems can be realized when the requirement for negligible cross-polarization is to be met. This result is a fortunate one because other requirements can be met without any conflict with the requirement for negligible cross-polarization. Alternatively, existing systems (where θ_1 is known) can be upgraded by using the derived equations and/or Figure 2.10. Equations (2.42) and (2.43) can be used to relate θ_1 to the F/D ratio of the systems considered. For some applications β and/or the quantity ρ might be known.

From the foregoing considerations it is clear that one can design a system of reflectors having a blocking efficiency of 1 and negligible cross-polarization radiation.

2.3 FOCAL PLANE IMAGING SYSTEMS

In this section we shall consider focal plane imaging systems before we consider hybrid systems that consist of a phased array placed in or near the focal plane of one or more reflectors or lenses.

With reference to Figure 2.11, let us consider the case where N antenna elements are used in the focal region of a parabolic reflector; each feed horn is connected to a receiver/transceiver, so that the resulting antenna beams, N_{tot}, are accommodated in as many receivers/transceivers. The paraboloid can have a prime-focus geometry or a Gregorian/Cassegrain system, in which case an equivalent parabola can be derived.

We would like to define N_{tot} and the key parameters that determine its value. The arrangement under consideration represents the radio equivalent of the optical camera in the sense that each pixel on a photographic film is substituted by a feed horn and receiver/transceiver combination. The beam scanning or imaging properties of different antennas have been considered in the literature [2.28–2.30].

While the main beam, accommodated in the feed horn positioned at the focus of the reflector, will have sidelobes determined by the theory outlined in Section 2.1.1, the antenna beams accommodated in the neighboring feeds

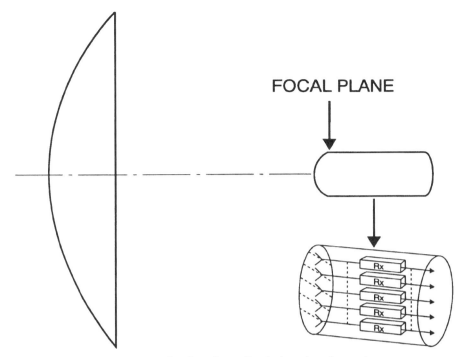

FIGURE 2.11 A prime-focus, focal plane imaging system.

directed toward $\theta_1, \theta_2, \theta_3$, and so on with respect to the boresight axis will have their first sidelobe, also known as the *coma lobe*, raised because Abbe's sine condition (modified for reflectors). is not satisfied. Apart from the raising of the coma lobe, the gain of the resulting beams decreases.

2.3.1 The Total Number of Antenna Beams

Ruze [2.28] concluded that the severity of the above effects is directly proportional to the feed displacement (or number of beams required) and the F/D ratio of the aperture. More specifically, if one accepts a maximum antenna gain loss of 1 dB, at which point the coma lobe is at about -10 dB relative to the main lobe, the total number of antenna beams is given by the equation [2.31]

$$N_{tot} = 1520\left(\frac{F}{D}\right)^4 \tag{2.57}$$

For prime-focus reflectors F/D is typically equal to 0.4, and N_{tot} is only 39. For Cassegrain and Gregorian systems, however, the F/D ratio is typically greater than one, so a substantial number of beams can be generated. While the derived N_{tot} is overgenerous [2.31] for high-quality imaging systems, it is indicative of what is theoretically possible. Given that N_{tot} depends so much on the F/D ratio of the reflector, other systems of reflectors can offer even wider beam scanning capabilities [2.31, 2.32].

2.3.2 Spacing of Antenna Elements at the Focal Plane

Full sampling of the focal plane for incoherent illumination of a scene and measurement of the intensity in the focal plane requires an element spacing, which is given by the equation [2.33]

$$\Delta x = \frac{F}{D}\frac{\lambda}{2} \tag{2.58}$$

If $F/D = 1$, the spacing between feeds ought to be $\lambda/2$. The minimum spacing between scalar feeds and feeds employing one hybrid mode is 4λ and 2.4λ, respectively [2.34]; these feeds were found to be very efficient illuminators of paraboloids and generate negligible cross-polarization fields. The same feeds can be used in conjunction with reflectors having larger F/D ratios.

The related issues are the cross-coupling between closely spaced antenna elements [2.35] and the resulting aperture efficiency [2.36].

2.3.3 Focal Plane Imaging Systems: Applications

We can discern two kinds of applications for focal plane imaging systems: the radio-astronomy and applied-science applications, which range over many unrelated areas of endeavor, some of which are of considerable benefit to the community and/or aviation safety and plasma research [2.33].

2.3.3.1 Radio-Astronomy Applications In the radio-astronomy context the requirement for focal plane imaging systems emanates from the need to map large regions of the cosmos in the shortest possible time. As the carbon monoxide molecule is abundant, the mapping of molecular clouds contributes to our knowledge of the cloud kinematics, which often shed light on cloud formation processes. The CO molecule has transitions at about 115.3 GHz and multiples thereof.

Given that the resolutions attainable by moderate-size (e.g., 14-m) antennas are of the order of 50 arc seconds at 115.3 GHz [2.33], the time taken to map a molecular cloud of angular extent of 2.5° by 3° would be far too long by a conventional single-beam aperture. Furthermore, the observation times for radio telescopes operating at millimeter-range wavelengths are often severely limited by meteorological conditions.

Goldsmith, in reference [2.31] collated reports of all known focal plane imaging systems. Of interest are the following arrays:

1. A 4 × 2 element array operating in the frequency band of 220–230 GHz in conjunction with the 12-m radio telescope operated by the NRAO (National Radio Astronomy Observatory). As the F/D ratio of the telescope is 13.8, it utilizes scalar feeds.
2. A 3 × 5 element array operating in the frequency band of 86–115 GHz in conjunction with the 14-m radio telescope operated by the FCRAO (Five Colleges Radio Astronomy Observatory) [2.37]. The array consists of suitably modified corrugated feed horns, and polarization interleaving is used.

Both systems utilize cooled front-end mixers to down-convert the incoming signals to a suitable IF. Other developments in progress at different observatories are also outlined in the same reference.

2.3.3.2 Applied-Science Applications Passive mm wave radiometers are ideal imagers for they are compact, offer adequate spatial resolution and can operate in the presence of fog, clouds, smoke, and sandstorms. These imagers promise to be an effective aid, or an alternative to existing technology for aircraft landing and surface operations under inclement weather conditions. The impact of this technology on air traffic can be significant in terms of reducing weather-related accidents and delays as well as affecting the economies of airlines, and parcel carriers [2.33]. Other applications of imagers include plasma diagnostics, which are essential for understanding the energy-loss processes [2.33] and for the detection of concealed weapons and explosives in airports [2.33].

While current systems typically have 64 antenna elements or less, future systems having 10^4 antenna elements are seriously being considered [2.33]. As space for commercial applications is limited, lenses rather than reflectors are typically used; even then their F/D ratio is typically ≤ 1.25 [2.33]. The other problem with lenses is their insertion loss, which is usually excessive at the short mm wavelengths. The use of Fresnel zone plates [2.38] should go some way towards overcoming this limitation.

Focal plane imagers have adequate sensitivity even when images are obtained at videolike frame rates—specifically, 10–30 frames per second. A focal plane imager having an 8 × 8 pixels in conjunction with a 60-cm lens aperture was realized at 94 GHz [2.39]. After the feasibility of concept was demonstrated, a 16 × 16-pixel imager was realized. If heterodyne systems are to be used when the number of pixels is 10^4, enough LO power is required to feed as many mixers. The whole system has to be simplified if this high number of pixels is to be derived [2.31]. Goldsmith [2.31] suggested direct low-noise amplification without any heterodyning as a means of realizing a simpler system.

Figure 2.12 shows two images: one obtained by optical means and the other using a 94-GHz novel imager. The system consisted of a LNA amplifier operating at 94 GHz and the incoming radiation was directly detected without using a heterodyne receiver. The receiver had a noise figure of 5.5 dB, a predetection bandwidth of 8 GHz, an integration time of 2 ms, and the minimum resolvable temperature was 0.3 K. The system represented a significant breakthrough for it simplified the conventional system, utilizing a mixer, down converter, local oscillator and IF amplifiers, considerably [2.39, 2.40].

In another realization a pushbroom one dimensional focal plane array imaging camera is at present operational on board an aircraft; it is being used

FIGURE 2.12 Images obtained by two imagers operating at (*a*) optical wavelengths; (*b*) 94 GHz. (*Courtesy*: TRW.)

for detection of oil spills when they are caused by oil tanker accidents in the ocean. Finally, in an effort to increase the spatial resolution of the imagers, Shoucri et al. [2.39] reported the establishment of a test bed for passive synthetic aperture imaging studies at millimeter waves.

2.4 HYBRID OR LIMITED-SCAN PHASED ARRAY SYSTEMS

Reference [2.41] is an up-to-date source of important papers related to both focal plane imaging systems and hybrid systems, while reference [2.42] is a thorough treatment of hybrid systems.

Hybrid systems are usually known as *limited-scan phased array systems* [2.43, 2.44]. Typically they consist of a system of offset reflectors and a phased array placed near or at the focal plane of the secondary reflector. Figure 2.13 illustrates a Gregorian system where the phased array is placed at the near field of the secondary. Other realizations consist of a combination of lenses/reflectors with a phased array [2.41]. These systems meet the following requirements:

1. The scan angle is at best not more than $\pm 10°$.
2. The systems have considerable collecting area.
3. The resulting systems are inexpensive, when compared to fully populated phased array systems occupying the same area.

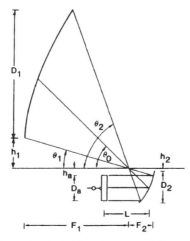

FIGURE 2.13 Illustration of a hybrid system consisting of a near-field Gregorian system and a phased array.

FIGURE 2.14 A hybrid system operating at 11 GHz with a 29-element feed array [2.45] (*Courtesy*: ERA.)

While the cost of the T/R (transmit/receive) modules used in phased arrays was high, these systems represented the best solutions available for applications where the scan angles were moderate.

2.4.1 Applications

Given that the angle subtended by a satellite located at the geostationary orbit is only 17.34°, it is natural to use hybrid systems for communication systems. The transmit antenna of the INTELSAT-VI (4/6 GHz) global system made use of a 3.2-m offset reflector fed by an array of 146 dual-polarized feed horns; groups of feeds are excited together to produce the required beam patterns and to cancel sidelobes [2.45]. Figure 2.14 shows a hybrid system designed for communication satellites; it had a 29-element feed array [2.45] and operated at 11 GHz. Studies of hybrid systems operating at 20 GHz have also been reported [2.46].

With future trends in satellite systems moving toward higher capacity, stringent demands have been placed on the antenna design to provide more reuse of the spectrum. These requirements have driven conventional parabolic reflector designs to their limits [2.47, 2.48]. The resulting mass and volumetric constraints constitute the limits for the preceding systems [2.48].

FIGURE 2.15 The AN/TPN-25 precision approach radar of the AN/TPN-19 system [2.43]. (*Courtesy*: Raytheon.)

The case for using phased arrays in on-board communication satellites has been put fairly succinctly by Sorbello et al. [2.48]:

> A number of advanced concepts employing active phased arrays are under investigation at COMSAT Laboratories [2.49]. Phased arrays have been given increased attention for satellite antenna applications due to their ability to form multiple beam, provide power-sharing among beams through distributed amplification and rapidly reconfigure and repoint beams. The drawbacks to phased arrays have been centered around concerns over mass, DC-to-RF efficiency and the uniformity of performance of a large number of elements. The application of MMICs to arrays has helped to eliminate many of these concerns.

An example of a hybrid radar system is the precision approach radar (PAR) of the ANTPN-19 system shown in Figure 2.15; the system operates at X-band and guides the aircraft during its final approach; it requires a limited coverage sector of 15° in elevation and 20° in azimuth [2.43, 2.44]. A less expensive version of the PAR is the AN/GPN-22 has a coverage of 8° in elevation only.

2.5 TOWARD PHASED ARRAYS

In this section we shall outline the long road from single aperture systems to phased arrays utilizing high-gain antenna elements. This is necessary before we move to the subsequent chapters of the book, which are devoted to phased arrays. More importantly, however, it is our intention to: (a) define the maximum spatial resolution attainable by single aperture systems utilizing modern techniques; and (b) explore the possibility of realizing phased arrays mounted on fully steerable structures.

2.5.1 Attempts to Overcome the Fundamental Limitations of Single-Aperture Systems

We have already deduced the resolution limit of single aperture systems in Section 2.1.1.8. If operation at 1 mm is contemplated, for instance, the theoretical maximum diameter of an antenna using conventional materials, such as steel, is 20 m. A proposal for an active surface reflector antenna 50 m in diameter and capable for operation at 1 mm has been put forward recently [2.50].

The proposal calls for a surface that consists of 126 hexagonal segments in a classic Cassegrain configuration, shown in Figure 2.16; the segments are actively controlled to reduce the effects of gravitation on and/or thermal gradients across the reflector's surface. The design was based on the successful design of the 10-m Keck optical telescope, the surface of which consists of 36 actively controlled hexagonal segments [2.51]. The *large millimeter tele-*

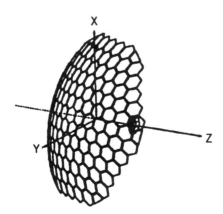

FIGURE 2.16 The segmented, actively controlled surface of the 50-m proposed large millimeter telescope. (*Source*: Cortez-Medellin and Goldsmith [2.50], © 1992 IEEE.)

scope, as it is known, will have a HPBW of 5 arc seconds and a gain of 102 dB, at 1 mm.

It is often less expensive to mount N reflectors, of diameter d, on a mechanically steerable structure to realize a reflector of diameter $d\sqrt{N}$ [2.52]. In the same reference a steerable structure supporting 2×4 reflectors, each 16 m in diameter was reported.

2.5.1.1 Phased Arrays Mounted on Fully Steerable Structures While phased arrays having many low-gain antenna elements, each followed by a T/R module, will be considered in Chapter 3, phased arrays mounted on fully steerable structures having high-gain antenna elements have been realized.

2.5.1.1.1 Satellite Mobile Communications Application An array designated for satellite mobile communications employing a small number of high-gain antenna elements has been investigated [2.53]. The aim was to measure the performance of such an array generating multiple beams at the IF.

Phased arrays consisting of 16, 18, and 19 antenna elements in the form of short backfire radiators, depicted in Figure 2.17, have been realized at L-band under contract from ESA (European Space Agency) [2.53]. The diameter of the array was 2.1 m, and each array element had a gain in excess of 15 dB. The array was deemed suitable for future generations of mobile communication satellites.

2.5.1.1.2 Annular Synthesis Antenna A circular array of antenna elements in the form offset reflectors, shown in Figure 2.18, has been reported [2.54, 2.55a, 2.55]. It consists of eight high-gain antenna elements or unit collecting areas arranged in circle, and all signals are processed in the central area of the circle. The array, the annular synthesis antenna, operates at mm wavelengths and the signals from all unit collecting areas are processed in one cryodyne.

The array can serve many purposes, some of which are:

1. When used in conjunction with a high-resolution radio-astronomy array, it can provide the short spacings (between antenna elements) often missing in such arrays. These spacings increase the quality of the resulting images.
2. It can be used as a power combiner [2.56].
3. As an imager having a FOV determined by the diameter d of the unit collecting areas and a resolution defined by the overall diameter D of the array.

124 FROM FILLED APERTURES TO PHASED ARRAYS

FIGURE 2.17 The L-band multibeam array model (MAM). The MAM, of diameter 2.1 m, served the purposes of demonstrating its capability for future mobile communication satellites. (*Courtesy*: Ericsson.)

With reference to Figure 2.18*b*, d is related to D by the equation

$$d = \frac{D \sin(\pi/N)}{1 + \sin(\pi/N)} \quad (2.59)$$

where N is the number of unit collecting areas used. A collecting area utilization factor k representing the ratio the areas of the annulus having a width of d, to the total area of a circle having a diameter D is given by the equation

$$k = \frac{4 \sin(\pi/N)}{[1 + \sin(\pi/N)]^2} \quad (2.60)$$

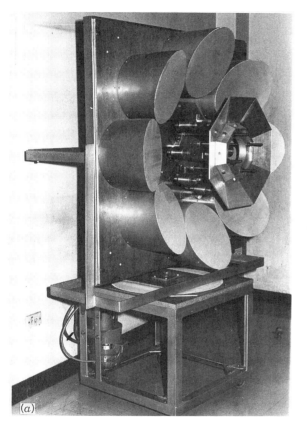
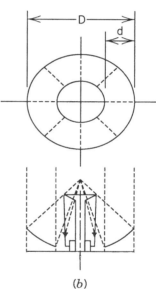

FIGURE 2.18 The annular synthesis antenna: (*a*) a model of one realization of an annular synthesis antenna that operated at 38 GHz—the overall diameter was 1.2 m, while the diameter of each collecting area was 0.33 m; (*b*) the antenna geometry. (*Source*: Fourikis [2.55a].)

As N increases, d decreases, and the unutilized area increases. Typically k is equal to 0.902 and 0.809 when N is equal to 6 and 8, respectively.

The number of the resulting antenna beams M is equal to the number of nonredundant baselines possible when N takes odd or even values. For the systems under consideration, M is given by the following equations:

$$M \approx \frac{N^2}{2} \quad \text{when } N \text{ is even} \qquad (2.61)$$

$$M \approx N^2 - N + 1 \quad \text{when } N \text{ is odd} \qquad (2.62)$$

If N is equal to 8 or 9, M is equal to 32 or 73; the savings in receivers (or T/R modules) is substantial. This is especially the case when the receivers are cryogenically cooled and/or operate at mm wavelengths. The systems under consideration have the following attractive characteristics:

1. The resulting beams sample the scene perfectly, if the Rayleigh criterion is used in forming the simultaneous and independent antenna beams.
2. Scalar feeds or hybrid mode feeds illuminate the unit collecting areas efficiently.
3. Mutual coupling between the feeds of the unit collecting areas can be neglected.
4. A conventional paraboloid can be converted to an annular synthesis antenna.
5. Thermal influences (expansion and contraction) affect all unit collecting areas equally.

2.5.1.1.3 Single-Aperture Reflectors versus Phased Arrays? Are single-aperture reflectors and phased arrays rivals or complementary systems? We have already seen that ensembles of single apertures form phased arrays; we also noted that the annular synthesis antennas can provide the short spacings (often missing) between antenna elements of a radio-astronomy phased array.

The elements of a radio-astronomy array, the single-aperture reflectors, define the FOV of the array. However, we have seen that single-aperture structures can yield several independent beams. In theory, these additional beams can be used to provide the FOV for adjacent images to that corresponding to the main beam. Even if a conservative number of independent antenna beams, say, 10 or 100, can be realized, the data rate emanating from this novel array will be phenomenal, but the returns substantial for the observation time of a given field will be reduced dramatically. This is an area worth considering for further research.

2.6 IDEAL FEED HORNS

It is useful to end this chapter by considering ideal but realizable feed horns, for their characteristics will form a useful benchmark to assess other feed horns or antenna elements of a phased array.

Ideal feeds serve the purpose of defining what is required from a realizable feed horn. Ludwig [2.57] assumed circularly symmetrical feeds and deduced the following conditions that render the cross-polarization component to be zero:

1. The difference between the E- and H-plane amplitude patterns should be zero.
2. The difference between the E- and H-plane phase patterns should be zero.

Both hybrid-mode and multimode feeds approximately meet these requirements.

Figure 2.19a shows a hybrid mode horn that has corrugations perpendicular to the horn's axis; the cone angle is typically less than 40°. The rationale on which these horns were realized was that if the same boundary conditions apply inside the horn in the E- and H-planes, the field would have the same pattern in both planes. A metallic surface made up of many closely spaced corrugations presents a reactive boundary condition for TM and TE modes at glazing incidence. If the corrugations are deep enough (viz., about $\lambda/4$, where λ is the free-space wavelength), the surface reactance is capacitive and surface waves cannot be supported.

The resulting horns have almost identical E- and H-plane radiation patterns; the phase centers for the E- and H-planes coincide, and as a consequence, the cross-polarization radiation is negligible over an octave. This is an important property for polarimetric radiometers and radars. For the scalar feeds, having larger cone angles and corrugations that are perpendicular to the flared surface of the cone, the position of the phase center is independent of frequency and the useful frequency band can extend over two octaves.

A complete set of papers dealing with the design and realization of scalar and hybrid mode feeds is collated in Love [2.58]. The protagonists in this field are B. MacA Thomas and P. J. B. Clarricoats and their collaborators.

Procedures to design hybrid mode feeds operating at mm-wavelengths have been outlined in Fourikis [2.55], and Figures 2.19b and 2.19c show the copolarization E- and H-plane radiation patterns of a hybrid mode horn and its cross-polarization pattern, respectively, taken at 38 GHz. The horn was designed to have a cross-polarization level of -35 dB, and horns having lower cross-polarization levels of, say, -40 dB are realizable. As can be seen, the E- and H-planes are almost identical; hence the resulting cross-polarization is low.

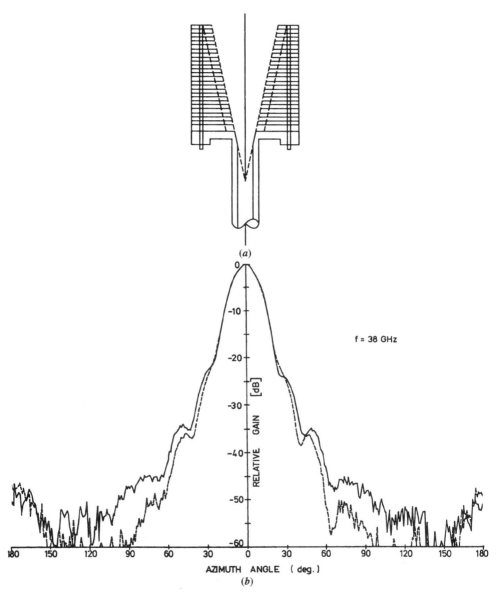

FIGURE 2.19 A hybrid mode feed horn designed to operate at 38 GHz: (*a*) the horn is made up of thick and thin plates held together by long bolts; (*b*) the copolarization E- and H-plane radiation patterns. (*Source*: Fourikis [2.55, 2.55a].)

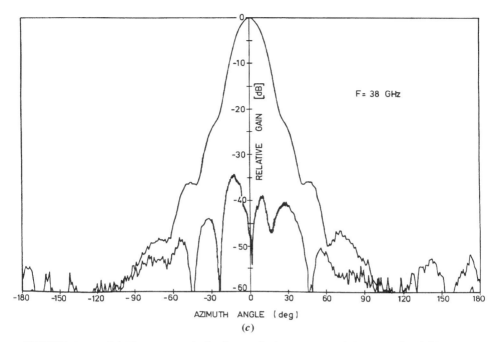

FIGURE 2.19 (*c*) The cross-polarization radiation pattern of the same feed. (*Source*: Fourikis [2.55, 2.55a].)

2.7 CONCLUDING REMARKS

In this chapter we have traced the path from filled apertures to phased arrays mounted on fully steerable structures, through focal plane imaging systems and hybrid phased arrays. Inter alia we have:

1. Outlined a theoretical framework that can be used to derive the essential characteristics of circular and rectangular apertures.
2. Defined the consequences of random and systematic errors present on the surface of reflectors as the frequency of operation increases. Systematic errors ultimately define the highest frequency of operation of an aperture and data from several operational radiotelescopes support these conclusions.
3. Derived the resolution limit attained by filled apertures utilizing conventional materials such as steel and filled apertures supporting active surfaces.

Findings 1 and 2 established the regimes of applications for phased arrays and filled apertures when the criterion is spatial resolution.

When spatial resolution is not of prime importance, the definition of the maximum number of independent and simultaneous antenna beams a filled aperture can yield, is important. A plethora of applied science and radioastronomical applications utilizing systems that consist of a filled apertures capable of imaging a scene have been described. We have also outlined the regime of applications for hybrid antenna systems that typically use offset systems of reflectors and proposed a design procedure for dual-offset reflectors which have negligible G.O. cross-polarization radiation.

The road to phased arrays naturally led us to phased arrays mounted on fully steerable structures and their properties and applications were described.

The chapter concludes with the presentation of measurements of the E- and H-plane radiation patterns as well as the cross-polarization pattern of ideal (hybrid mode) horns. These horns are used to feed reflectors and lenses but can be used as elements of phased arrays. More importantly however these measurements serve the purpose of defining the cross-polarization levels achievable by realizable ideal antennas.

We have already seen that phased array-based radiotelescopes utilize several filled aperture reflectors; phased arrays mounted on fully steerable mounts can be used to provide the short spacings (often missing) between antenna elements of a radiotelescope. As filled apertures can have N independent beams which in turn define N fields of view, a phased array-based radiotelescope, can in principle derive images of a celestial body corresponding to the N adjacent fields of view. Although highly desirable, the realization of such systems will depend on how much data parallel processors can handle in real-time and the space available in the focal plane of conventional filled aperture reflectors to accommodate the N receiving systems; MMIC-based receivers can be a solution. Perhaps the next generation of phased array-based radiotelescopes will utilize offset reflectors. This is an area worth considering for further research.

CHAPTER THREE

Phased Arrays: Canonical and Wideband

...Go to encounter for the millionth time the reality of experience...

James Joyce (1882–1941)

Canonical phased arrays, which operate over narrow instantaneous bandwidths, have been in use for a long time. In the context of this chapter, the fractional bandwidth of canonical arrays is 10% or less and the spacing between antenna elements is equal. The fractional bandwidth of a system is the ratio of the instantaneous bandwidth of operation over the center frequency of the system. This figure for the fractional bandwidth was not derived from some esoteric and/or theoretical considerations but from real-world considerations, in the sense that the derived systems met the system requirements at reasonable costs and development times.

By contrast, wideband phased arrays can operate over wide bandwidths (e.g., 2–3 octaves), and their realization had to wait for the availability of wideband components and T/R modules. In Chapter 1 we made a case for multifunction wideband phased array systems. We also distinguished systems that have an instantaneous bandwidth of 2–3 octaves from the systems that can operate over 2–3 octaves while their instantaneous fractional bandwidth is about 10% or less.

In this chapter we shall explore the generic characteristics of phased arrays and their architectures before we outline the domains of application for passive and active phased arrays and the in-between variants. *Active phased arrays* are defined here as arrays where minimum losses are incurred between the antenna elements and the IF signal processing subsystems.

The potential of active phased arrays has been recognized for a long time, but costs associated with their T/R modules prevented their widespread use. Recently all the significant indicators point to the realizability of affordable

T/R modules. Furthermore, array designs particularly suitable to active phased arrays have evolved. This has resulted in a plethora of approaches being placed at the disposal of the array designer.

3.1 INTRODUCTORY BACKGROUND

The early radio-astronomy and radar phased arrays had many similarities. Both types of arrays were truly canonical, and it was very expensive to change their frequency of operation, which was derived from either astronomical considerations or the required radar functions (surveillance or tracking). The effective bandwidth of operation of their antenna elements (dipoles, monopoles, horns, or paraboloid antennas fed by dipoles), and active circuits, simply matched the system requirements.

The field of view (FOV) of phased arrays is normally defined by the beamwidth of the unit collecting areas or antenna elements and the array scanning capabilities; for both applications the FOV of the unit collecting areas either matched the required FOV or was much wider. Optimum equivalent noise temperatures, derived by the use of discrete circuits and vacuum tubes or transistors operating over narrow bandwidths, enabled radio-astronomy systems and radars to have increased sensitivities. Similarly, power sources and their associated circuitry used in radar systems were optimized to yield maximum output powers over a narrow bandwidth.

The striking differences between radar and radio-astronomy arrays are:

1. The radar arrays are relatively small in size and yield beamwidths of the order of a few degrees, while radio-astronomy arrays are large and yield beamwidths of the order a few arc seconds or narrower.
2. Typically the radar arrays are fully populated, while the radio-astronomy arrays are sparsely populated.
3. The radar arrays operate in the transmit/receive mode, while the radio-astronomy arrays typically operate in the receive mode only.
4. For radar phased arrays the total FOV (or surveillance volume) is usually electronically scanned at high speeds while radio-astronomy arrays form an image of a celestial source that extends over the instantaneous FOV of the array, which is usually much smaller than the total FOV (or the radiotelescope's horizon).
5. The instantaneous FOV corresponding to monolithic radar arrays, is equal to the total FOV and is fixed in space, while the instantaneous FOV corresponding to radio astronomy arrays is not; more explicitly all unit collecting areas can be pointed to any celestial source within the radiotelescope's horizon.
6. While the total FOV of a radar phased array is usually $\pm 45°$ or $\pm 60°$ in azimuth and $45°$ in elevation, the instantaneous FOV corresponding to

a radio astronomy phased array is usually of the order a few degrees or less and its horizon is limited by the shadowing effects between adjacent unit collecting areas.

While the trend for radar arrays is to approach a hemispherical total FOV (by using one or more phased arrays), radio-astronomy arrays utilizing unit collecting areas of ever-increasing diameters such as those used in the VLA, have instantaneous FOVs well below 1° at the shorter wavelengths of operation. As can be expected their horizon is not quite hemispherical.

With the passage of time radio-astronomy phased arrays having ever-increasing spatial resolutions, ever-increasing total bandwidths and dual-polarization capabilities, have been realized; their instantaneous fractional bandwidth, however, is 10% or less. Over the horizon radars, OTHRs have had similar bandwidth capabilities and the trend for future radar phased arrays is similar to current radio astronomy arrays. The following are indicative of probable developments:

1. Future radar phased arrays have an instantaneous (or total) bandwidth of 2–3 octaves.
2. The spatial resolutions of radar phased arrays will not be as high as those used in radio astronomy; however possibility of using radio-astronomy arrays to perform communications or radar functions cannot be discarded.

3.1.1 Phased Array-Based Radar, EW, and Communication Systems

Radar arrays are designed to yield maximum power aperture products and adequate spatial resolution, which determines the size of the array aperture. How the aperture is populated depends on the availability of funds as a function of time. A designer for instance might realize an aperture that has the required beamwidth and fully populate it prior to the commissioning stage, if adequate funds are available. Alternatively the same aperture is partly populated at the commissioning stage and fully populated at a later stage as funds become available. In the latter case, however, the different stages have to be preprogrammed at the design stage to ensure the smooth implementation of the different stages.

If the inter- (antenna) element spacing is regular and too wide, grating lobes will cause problems. It is recalled here, from optics, that grating lobes resemble the main array beam and move toward the main beam and away from it when the interelement spacing is increased and decreased, respectively. When the array is in the transmit mode, power will be wasted in the directions of the grating lobes; in the receive mode the array will receive signals from several directions and ambiguities will result. In both cases the beam efficiency will decrease. The management of the grating lobes of all arrays is therefore one of the main concerns for the array designer. As the array is required to operate over ever-increasing bandwidths the management of the array grating lobes will become more difficult. The phased arrays used in conjunction with OTHRs, for instance, represent an extreme case for the designer because they have to operate over a 30:1 frequency range and be free of grating lobes.

As the designer decreases the inter-element spacing to avoid grating lobes and maximize the power aperture product, at some spacings the array radiation pattern will have nulls toward certain directions within the array total FOV, if precautions are not taken. These nulls occurring in directions where the array experiences scan blindness, will therefore limit the array total FOV. The causes of scan blindness and approaches to avoid them will constitute another important task for the array designer. While we are able to notionally consider the above two problems separately this is not always the case.

Given the above problems, the designer will seek optimum ways of populating the available area, so that the resulting array is efficient and its total FOV matches the requirement over the frequency band of operation.

In broad terms, the early phased array-based radars performed a limited number of radar functions and were venerable to ECM, because they operated over a narrow band of frequencies. With the passage of time, dual-polarization, truly multifunction phased arrays operating over several octaves and capable of performing a variety of radar-related functions such as the EW functions and communication functions on a time-sharing mode will be realized. The identification (or imaging) function will be added to the many functions several current systems perform on a piecemeal fashion. The integration of all these functions will in turn result in an overall increased efficiency of operations.

The integration of another set of functions related to airport traffic management will also result in an overall increased efficiency and safety of operations in civilian airports and air force bases.

These developments and the adoption of phased arrays in communication systems utilizing geostationary and Earth-orbiting satellites stem from the availability of affordable T/R modules.

3.1.2 Astronomy, Geodesic, and Remote Sensing Systems

It is recalled from Chapter 1 that dual-polarization, high resolution images of celestial sources taken at several spot frequencies were required before one could make any statements about the nature of the emission mechanisms associated with celestial sources. The FOV of radio astronomy arrays is solely determined by the radiation pattern of the unit collecting antennas, which is narrow, so the resulting grating lobes, when multiplied by the radiation pattern of the unit collecting areas are heavily attenuated—see Section 3.2.1.

The realization of several modern high resolution, phased arrays operating at several spot frequencies are based on the above considerations.

The quest for high resolution and high sensitivity images of celestial objects of interest has driven radio astronomers to VLBI an OVLBI systems, which are used for radio-astronomical observations and geodesic applications. We have traced parallel developments for high spatial resolution optical astronomy systems, in Chapter 1.

Airborne remote sensing systems that are either operational or in the proposal stage utilize the techniques developed by radio-astronomers.

From the foregoing considerations it is clear that the many diverse applications of phased arrays share the same theory, which we are about to explore.

3.2 THEORETICAL CONSIDERATIONS

Let us consider a system that requires a spatial resolution, which is attainable by the continuous aperture, shown in the center of Figure 3.1. The continuous aperture is then sampled at certain points on its surface chosen according to the criterion that the sampled points are equi-distant in both the X and Y directions. We then substitute the monolithic aperture with smaller apertures, unit collecting areas, placed at the sampling points we have chosen. A phased array has resulted, shown in Figure 3.1a that has the same resolution as the continuous aperture we started with. In many cases the continuous aperture is not realizable while the phased array is. The sampling points can form a cross, a circle or a Y as illustrated in Figure 3.1b, 3.1c, and 3.1e respectively. In Figure 3.1d the distribution of the sampling points is random.

The above considerations apply to phased arrays required for radio-astronomical applications and similar considerations apply to phased arrays for radar applications that require one or more inertialess antenna beams; for the latter case the corresponding continuous surface on which the phased array is placed is of modest dimensions.

3.2.1 The Far-Field Radiation Pattern of a Linear Phased Array

In a manner similar to the approach taken in Section 2.1.1, we shall consider the resulting array pattern of a planar phased along two linear dimensions separately. Figure 3.2 illustrates N line sources of length d, separated by a distance s, along the z axis. The resulting far-field pattern of the array $P(u)$, is the product of the far-field pattern corresponding to one line source, $G(u)$, and the array geometric factor (pattern) (AGF), which constitutes the sum of the contributions attributable to the line sources:

$$P(u) = G(u)\text{AGF} \qquad (3.1)$$

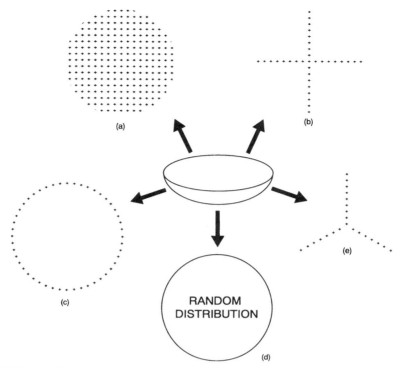

FIGURE 3.1 If a continuous aperture shown in the center of the figure is sampled at points indicated by (+) where unit collecting areas are placed, a set of phased arrays results. In many cases the phased array is realizable but the continuous reflector is not. Cases (a) and (d) represent typical radar phased arrays while cases (b), (c) and (e) represent typical radioastronomy phased arrays.

Both $G(u)$ and AGF are power patterns. In Chapter 2 we derived the resulting $G(u)$ patterns when the line sources have different illuminations; here, therefore, we shall proceed with the derivation of the AGF.

The phase difference between consecutive unit collecting areas ψ is given by the equation

$$\psi = ks \sin\theta + \beta = \frac{2\pi s \sin\theta}{\lambda} + \beta \qquad (3.2a)$$

or

$$\psi = ks \cos\vartheta + \beta = \frac{2\pi s \cos\vartheta}{\lambda} + \beta \qquad (3.2b)$$

where β the difference in phase excitation between elements and ϑ is the complementary angle of θ.

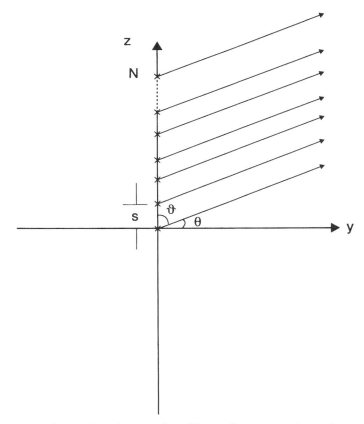

FIGURE 3.2 A linear phased array of equidistant line sources located at points (×), is illuminated uniformly.

If we assume uniform illumination for the line sources of the array, the sum of all voltages attributable to the consecutive line sources, or AGF, is given by the following equations:

$$\text{AGF} = f(u) = 1 + \exp j\psi + \exp j2\psi + \exp j3\psi + \cdots + \exp j(N-1)\psi$$

$$= \sum_{1}^{n=N} \exp j(n-1)\psi \quad (3.3)$$

which can be reduced to

$$f(u) = \frac{\exp jN\psi - 1}{\exp j\psi - 1} \quad (3.4)$$

If the reference point is the physical center of the array, Equation (3.4) is further reduced to

$$f(u) = \frac{\sin(N\psi/2)}{\sin(\psi/2)} \qquad (3.5)$$

At an angle θ_0, the main beam is formed and θ_0 is in the range $0° \leq \theta_0 \leq 180°$. To accomplish this, the phase excitation β between elements must be adjusted so that

$$\psi = ks \sin \theta_0 + \beta|_{\theta=\theta_0} = ks \sin \theta_0 + \beta = 0 \quad \text{or} \quad \beta = -ks \sin \theta_0 \quad (3.6)$$

Thus by changing β, the progressive phase change between antenna elements, we can direct the maximum radiation pattern toward any direction to form a scanning array. Recalling that $\sin \theta/\lambda = u$, $P(u)$ when the array beam is along the y axis is given by the normalized equation

$$P(u) = \left[\frac{\sin(d\pi u)}{d\pi u}\right]^2 \left[\frac{\sin(Ns\pi u)}{N \sin(s\pi u)}\right]^2 \qquad (3.7)$$

Figures 3.3a–3.3c depict the far-field radiation patterns of the line source, the array geometric pattern, and the resultant pattern $P(u)_x$, respectively. For this example, $s = 2d$ and $N = 20$. The grating lobes are clearly seen and the sidelobes are close to the main beam and the grating lobes.

In Figure 3.4 the power patterns $E(u)$, $F(u)$, and $P(u)$ are illustrated when s equal to $3d$; as the separation between unit collecting areas increases, the magnitude of the grating increases and the lobes move closer to the main beam.

If a linear phased array of omnidirectional antenna elements has an east–west orientation, the resulting main beam and grating lobes will extend from north to south. In Figure 1.3 a a linear phased array along the E–W axis is illustrated together with the resulting main-beam and grating lobes, within the FOV defined by the radiation pattern of the unit collecting areas, which is not omnidirectional.

Let us compare linear arrays and planar arrays; linear arrays can have one fan beam which is scanned from east to west in our example. The position of a target or a source can therefore be defined in one coordinate. For monostatic radar applications, the target's range defines a sphere. Linear arrays "see" a large volume of space where many sources/targets/jammers might be located. By contrast planar arrays are capable of locating a source or a target in two coordinates. However the costs and complexity associated with the latter arrays are considerable. Before we consider planar arrays it is worth examining other important properties of linear phased arrays.

Let us compare the far-field radiation patterns of a line source and a phased array when both have the same length; here we have assumed that

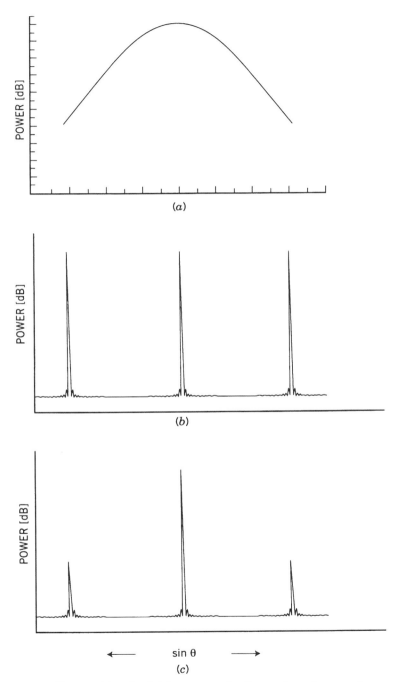

FIGURE 3.3 The resulting far-field patterns of a linear phased array where $s = 2d$, $d = \lambda/2$, and $N = 20$. The following patterns are shown: (a) one line source—$G(u)_x$, (b) the geometric array pattern—$F(u)_x$, and (c) the array pattern—$P(u)_x$.

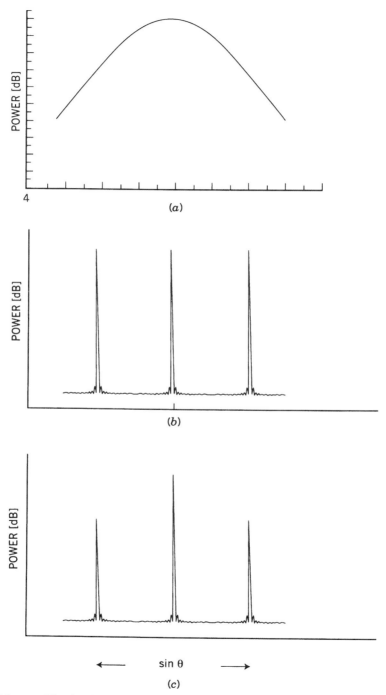

FIGURE 3.4 The far-field patterns of a linear phased array are similar to the one illustrated in Figure 3.3 but for the condition that $s = 3d$.

$s = d = \lambda/2$ and $N = 32$; consequently L, the length of the line source, is equal to 16λ. For a line source and a phased array the respective far-field radiation patterns are given by the equations

$$P(u)|_{\text{line source}} = \left[\frac{\sin(L\pi u)}{L\pi u}\right]^2 \tag{3.8}$$

$$P(u)|_{\text{phased array}} = \left[\frac{\sin(Ns\pi u)}{N\sin(s\pi u)}\right]^2 \tag{3.9}$$

We have plotted the resulting far-field radiation patterns in Figure 3.5 together with the difference

$$\Delta = P(u)|_{\text{line source}} - P(u)|_{\text{phased array}} \tag{3.10}$$

drawn as a solid line close to the sin θ axis. The array length is taken to be Ns when N is a large number [if N is not large, the array length L is equal

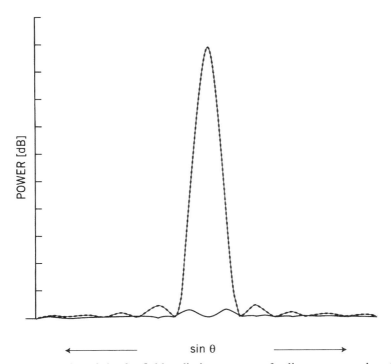

FIGURE 3.5 A plot of the far-field radiation pattern of a line source and a phased array when both have the same total length and $d = s = \lambda/2$. The case where $N = 32$ is illustrated. The difference Δ between the two far-field patterns for each case is also shown, in solid lines near the sin θ axis.

to $(N-1)s$]. The solid line corresponds to the line source, while dashed lines correspond to the phased array. All three radiation patterns coincide around the main beam and the positions of the nulls coincide for all patterns. Δ is seen to decrease as θ increases.

The theoretical framework on the equivalence between continuous and discrete arrays is outlined in Krienski [3.1].

From the foregoing considerations, we can rewrite Equation (3.9) as

$$P(u)|_{\text{phased array}} = \left[\frac{\sin(Ns\pi u)}{Ns\pi u}\right]^2 \quad (3.11)$$

This is an important result, for one can utilize Equation (3.11) to derive the far-field radiation pattern of a phased array instead of the more complex Equation (3.9). More importantly, there is a substantial body of knowledge derived from studies related to conventional reflectors, outlined in Chapter 2, which is directly applicable to phased arrays.

3.2.2 Array Grating Lobes

In Figures 3.3–3.4, we have illustrated the main beam, the sidelobes, and grating lobes of a couple of phased arrays. We know how to vary the beamwidth and the sidelobes of the array. We aim to derive the means by which the position and amplitude of the grating lobes can be varied.

The array geometric factor can be generalized to reflect the situation where the array is scanned to θ_0; the equation for the resulting AGF is

$$f(u) = \frac{\sin[Ns\pi(\sin\theta - \sin\theta_0)/\lambda]}{N\sin[s\pi(\sin\theta - \sin\theta_0)/\lambda]} \quad (3.12)$$

We can deduce that the grating lobes occur whenever the denominator of the RHS of Equation (3.12) vanishes, that is, whenever

$$\frac{s\pi[\sin\theta - \sin\theta_0]}{\lambda} = \pm\pi, \pm 2\pi, \pm 3\pi \ldots \quad (3.13)$$

For the first grading lobe $(\pm\pi)$ to be positioned at the horizon, that is, when $\theta = 90°$ or $-90°$, we can deduce that

$$\frac{s}{\lambda} = \frac{1}{1 + |\sin\theta_0|} \quad (3.14)$$

Thus the condition that the spacing between antenna elements is $\lambda/2$ is enough to place the grating lobes of the phased array at the horizon.

Microstrip dipoles are commonly used as antenna elements for phased arrays and a spacing of about $\lambda/2$ between antenna elements is adequate to meet the above condition. Appropriate design guidelines, however, relating the substrate thickness, its relative permittivity and the interelement spacing of the array, have to be followed so that the array is scan-blindness free. Similarly the interactions between the antenna elements (microstrip or printed board dipoles) and their feed structures have to be optimized—see Chapter 5. The far-field radiation pattern of dipoles is almost omni-directional, so the designer has to rely entirely on the AGF for the grating lobe management. It is not therefore surprising that some authors advocate a spacing of $\lambda/2$ or less for canonical phased arrays [3.2]. As the bandwidth of operation of an array increases, the designer encounters difficulties meeting the above grating lobe criterion.

3.2.3 Phased Array Beamwidth and Bandwidth

We seek to define simple expressions for the beamwidth and bandwidth of phased arrays; in Section 2.1.1 we outlined a method to derive the half-power beamwidth (HPBW) of filled apertures. The same method can be used when the phased array illumination is known. A simpler method has been derived for the beamwidth and bandwidth of phased arrays when the illumination function is of standard form [3.3]. Here we shall follow the approach taken by Hemmi [3.3].

Let us recall Equation (3.12) and substitute λ for c/f, furthermore, we introduce f_0 to represent the frequency used to determine the phase gradient. Equation (3.12) can therefore be rewritten as

$$f(\theta, f) = \frac{\sin[N\pi s(f\sin\theta - f_0\sin\theta_0)/c]}{N\sin[\pi s(f\sin\theta - f_0\sin\theta_0)/c]} \tag{3.15}$$

It is easy to verify that the array geometric factor reaches maximum at θ_0 and f_0. At any other frequency, the maximum is reached at an angle given by the equation

$$\sin\theta = \frac{f\sin\theta_0}{f_0} \tag{3.16}$$

The -3 dB bandwidth is defined as $f(\theta, f) = 0.707$, if its maximum is equal to unity; if we assume that $\sin(Nx)/[N\sin(x)] \approx \sin(Nx)/Nx$, the half-power points of Equation (3.15) are given by the equation

$$\frac{N\pi s}{c}(f \sin \theta - f_0 \sin \theta_0) = \pm \frac{b}{2}\pi \quad (3.17)$$

where b is the beam broadening factor, which depends on the illumination. For uniform illumination, $b = 0.886$.

The beamwidth is determined by letting $f = f_0$ and using the approximation

$$\sin \theta - \sin \theta_0 = \frac{\theta - \theta_0}{\cos \theta_0} \quad (3.18)$$

which is applicable to large arrays near the main beam. The HPBW is therefore given by the equation

$$\text{HPBW} = \frac{0.886c}{Nsf_0 \cos \theta_0}$$

$$= \frac{0.886c}{Lf_0 \cos \theta_0} \quad (3.19)$$

where the array length L was taken to be equal to Ns. If the array length and frequency of operation are set, the HPBW increases as θ_0 increases. For this reason the maximum scanning angle of phased arrays very seldom exceeds $\pm 60°$.

Similarly, the 3-dB bandwidth can be deduced by letting $\sin \theta = \sin \theta_0$ and solving for $f = f_0$. Hence

$$\text{Bandwidth } (Hz) = 2(f_{3\,\text{dB}} - f_0) = \frac{0.886c}{Ns \sin \theta_0}$$

$$= \frac{0.886c}{L \sin \theta_0} \quad (3.20)$$

Although the bandwidth is independent of the frequency, the HPBW is frequency- and angle-dependent. For other illuminations the beam broadening factors tabulated in Table 2.1 can be used to derive the HPBW and bandwidth of a phased array.

For long phased arrays, and/or substantial scanning angles, the array bandwidth is necessarily narrow. If wideband operation is required, switchable time delays have to substitute the switchable phase shifters.

3.2.4 Directivity of a Linear Phased Array

The beamwidth and directivity of large scanning arrays have been deduced by Elliott in a key paper having two parts [3.4a and 3.4b]. If the antenna elements are isotropic, the directivity D is solely defined by the geometric array factor [3.42] or by the equation

$$D = \frac{\text{p.d. in the direction of the main beam maximum}}{\text{average p.d. from the array}} \quad (3.21)$$

where p.d. stands for power density.

When the spacing between antenna elements is $\lambda/2$ and the origin of the coordinate system coincides with the central array element Equation (3.21) simplifies to [3.4a]

$$D = \frac{\left[\sum_{-n_z}^{n_z} A_n\right]^2}{\sum_{-n_z}^{n_z} A_n^2} \quad (3.22)$$

where A_n are the amplitudes of the contributions from the antenna elements.

Directivity, therefore, is a measure of how well the array directs energy in a particular direction. The numerator is proportional to the total coherent field squared, whereas the denominator is proportional to the sums of the squares of the individual fields from each element.

As can be seen, the directivity is not dependent on the scan angle; as the array is scanned toward endfire, its beamwidth broadens but the cone of the resulting conical beam occupies a smaller solid angle in space, an effect that just cancels the beam broadening.

For a linear array of length L_x using an aperture distribution which has uniform progressive phase and an element spacing of $\lambda/2$, $D = 2L_x/\lambda$.

3.2.5 The Half-Power Beamwidth × Directivity Product for a Linear Phased Array

When a linear array is uniformly illuminated, the product of its HPBW and directivity at broadside is constant and is given by the equation

$$D \approx \frac{101.5}{\Theta} \quad (3.23)$$

where Θ the HPBW at broadside is expressed in degrees.

3.3 LINEAR ARRAY OF EQUISPACED LINE SOURCES OF NONUNIFORM AMPLITUDE

We seek to derive the array geometric factor (AGF) of a linear array having an even number of equispaced line sources of nonuniform amplitude, when the total number of line sources is $2N$, where N is an integer. The additional constraint is that the illumination is symmetrical with respect to the center of the array.

The AGF in this case is given by the equation

$$\begin{aligned}
\text{AGF}|_{2N,\text{even}} &= A_1 \exp j\frac{ks \sin \theta}{2} + A_1 \exp -j\frac{ks \sin \theta}{2} \\
&\quad + A_2 \exp j\frac{3ks \sin \theta}{2} + A_2 \exp -j\frac{3ks \sin \theta}{2} \\
&\quad + \cdots + A_N \exp j\frac{(2N-1)ks \sin \theta}{2} \\
&\quad + A_N \exp -j\frac{(2N-1)ks \sin \theta}{2} \\
&= 2 \sum_{n=1}^{N} A_n \cos\left[\frac{(2n-1)}{2} ks \sin \theta\right]
\end{aligned} \quad (3.24)$$

In its normalized form the factor of 2 will be omitted in Equation (3.24). If the array has an odd number of line sources, $2N + 1$, where N is again an integer, the normalized AGF is given by the equation

$$\text{AGF}|_{2N+1,\text{odd}} = \sum_{n=1}^{N+1} A_n \cos[(n-1)ks \sin \theta] \quad (3.25)$$

By assigning different values for the amplitudes A_n, a variety of array geometric factors can be attained.

3.3.1 The Binomial Array

The coefficients A_n of a binomial array [3.5] are determined from the binomial expansion

$$(1 + x)^{m-1} = 1 + (m-1)x + \frac{(m-1)(m-2)}{2!} x^2 + \frac{(m-1)(m-2)(m-3)}{3!} x^3 + \cdots \quad (3.26)$$

where m represents the number of elements of the array. The resulting coefficients take the form of Pascal's [B. Pascal, 1623–1662] triangle.

For 2, 3, 4, and 5 elements, $2M = 2$, $2M + 1 = 3$, $2M = 4$, and $2M + 1 = 5$, respectively:

```
m = 1                    1
m = 2                 1     1
m = 3              1     2     1
m = 4           1     3     3     1
m = 5        1     4     6     4     1
```

An interesting characteristic of the binomial arrays is that they have no sidelobes when the spacing between elements is $\lambda/4$ or $\lambda/2$ [3.6]. In general, binomial arrays exhibit low-level sidelobes and wider beamwidths, compared to other arrays, such as the Dolph–Chebyshev arrays, which we shall consider in Section 3.3.2.1. One drawback of the binomial arrays is the large variation in current amplitudes, which usually contributes to problems related to mutual coupling. When $m = 11$, the variation of the amplitudes is $1:352$; for larger arrays the variation is even greater.

3.3.2 Array Synthesis Procedures

So far we have analyzed arrays in the sense that we deduced their AGF when the A_n coefficients (or illumination) were known; in what follows we shall explore the synthesis of linear arrays when the AGF is specified and the A_n coefficients are deduced. The fundamental assumption made in these approaches is that the designer has the freedom to vary the above-mentioned coefficients of the array to any value derived from the synthesis procedure. While this is fundamentally true, other factors centered around the cost of T/R modules, have to be considered.

In Section 3.9 it will be shown that active arrays (arrays where each antenna element is followed by a T/R module) have many attractive features. In that context it is more convenient and less costly if several ensembles of T/R modules have the same characteristics, such as the same gain and output power. Attention is therefore focused on these approaches while an outline of the fundamentals is given here of the well-established synthesis procedures. Furthermore, it will be demonstrated that the well-established synthesis procedures form the basis for the modern synthesis methods.

3.3.2.1 The Synthesis of Dolph–Chebyshev Arrays   The Dolph-Chebyshev (also transliterated as Chebychev, Tschebyscheff, or Tchebysceff) arrays were first proposed by Dolph [3.7] and further investigated by others [3.8–3.11].

The beamwidth of these arrays are a compromise between the uniformly illuminated arrays and binomial arrays. Similarly, their sidelobe levels are lower than those corresponding to arrays having a uniform illumination but higher than those corresponding to binomial arrays. For a given sidelobe ratio the Dolph–Chebyshev synthesis yield arrays having the highest directivity and narrowest beamwidth.

From Equations (3.24) and (3.25) we can deduce that the AGF is nothing but the sum of N or $N+1$ cosine terms. The largest harmonic of these cosine terms is one less than the total number of elements of the array. Each cosine term, the argument of which is an integer times a fundamental frequency, can be rewritten as a series of cosine functions with fundamental frequency as the argument. More explicitly

when $n = 0$ $\cos(nu') = 1$

$n = 1$ $\cos(u') = \cos u' = \cos z$

$n = 2$ $\cos(2u') = 2\cos^2 u' - 1 = 2z^2 - 1$

$n = 3$ $\cos(3u') = 4\cos^3 u' - 3\cos u' = 4z^3 - 3z$

$n = 4$ $\cos(4u') = 8\cos^4 u' - 8\cos^2 u' + 1 = 8z^4 - 8z^2 + 1$

(3.27)

where

$$u' = \frac{\pi s \sin \theta}{\lambda} = \frac{\pi s \cos \vartheta}{\lambda}$$

Equations (3.27) were deduced from Euler's [L. Euler, 1707–1783] formula

$$[\exp ju']''' = \exp jmu' = \cos(mu') + j\sin(mu') \quad (3.28)$$

the identity $\sin^2 u' = 1 - \cos^2 u'$ and the equation $\cos u' = z$.

Equations (3.27) bear a remarkable similarity to the Chebyshev functions

$$T_m(z) = \cos[m \arccos(z)] \qquad -1 \le z \le 1 \quad (3.29)$$

$$T_m(z) = \cosh[n \arccos h(z)] \qquad z < -1, z > 1 \quad (3.30)$$

which can be written as

$$T_0(z) = \cos nu' = 1 \quad \text{when} \quad n = 0$$

$$T_1(z) = \cos u' = z \quad n = 1$$

$$T_2(z) = \cos 2u' = 2z^2 - 1 \quad n = 2$$

$$T_3(z) = \cos 3u' = 4z^3 - 3z \quad n = 3$$

$$T_4(z) = \cos 4u' = 8z^4 - 8z^2 + 1 \quad n = 4 \quad (3.31)$$

These relations between the Chebyshev polynomials and the cosine functions are valid only in the $-1 \leq z \leq 1$ range. Within that range

$$|T_n(z)| \leq 1 \quad \text{and} \quad |\cos(nu)| \leq 1 \quad (3.32)$$

For $|z| > 1$, the Chebyshev polynomials are related to the hyperbolic cosine functions. Figure 3.6 illustrates the Chebyshev polynomials $T_0(z)$ to $T_4(z)$. The following characteristics of the Chebyshev polynomials are of interest:

1. All polynomials pass through the point (1.1).
2. Within the range $-1 \leq z \leq 1$, the polynomials have values within -1 to 1.
3. All roots occur within the range $-1 \leq z \leq 1$, and all maxima and minima have values of 1 and -1, respectively.

The unknown coefficients of the array factor can be determined by equating the series representing the cosine terms of the array geometric factor to the appropriate Chebyshev polynomial. The order of the polynomial should be one less than the total number of elements of the array.

The requirement that an array of N elements has a sidelobe level R_0 dB below the maximum of the main lobe can be met by the procedure outlined in Balanis [3.6], the equations derived in reference [3.8] and the relationship

$$R_0 = 20 \log R'_0 \quad (3.33)$$

where R'_0 is the ratio of the maximum value attained by the main beam to the maximum value attained by the sidelobe nearest to the abscissa $z = 1$.

The beam broadening factor b of the array is given by the equation [3.4a]

$$b = 1 + 0.636 \left\{ \frac{2}{R'_0} \cos h \left[\sqrt{(\operatorname{arccos} h R'_0)^2 - \pi^2} \right] \right\}^2 \quad (3.34)$$

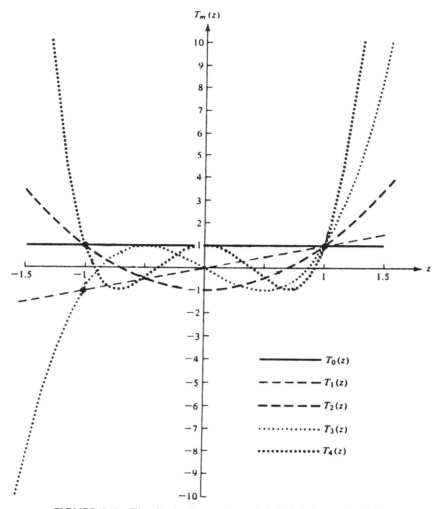

FIGURE 3.6 The Chebyshev polynomials $T_0(u)$ through $T_4(u)$.

while its directivity D_0 in terms of the beam broadening factor is

$$D_0 = \frac{2R_0'^2}{1 + (R_0'^2 - 1)b(\lambda/L)} \qquad (3.35)$$

where L is the length of the array.

The conditions that apply for the beam broadening factor and directivity are that the array is not scanned too close to endfire and that the sidelobe levels are within the range of -20 to -60 dB. The beam broadening factor and directivity of the Chebyshev arrays are illustrated in Figures 3.7a and 3.7b, respectively [3.4].

In Figure 3.8 the beam broadening factors of two arrays are illustrated as a function of sidelobe levels attained [3.4]. One array has a Chebyshev distribution, while the other has a cosine on a pedestal distribution. It is clear that the sidelobe level of the Chebyshev array has a lower beam broadening factor than the array having a cosine on a pedestal distribution.

When the array is used in an endfire mode, the Hansen–Woodyard conditions for maximum directivity apply [3.6, 3.12].

3.3.2.2 The Schelkunoff Circle

We have already considered the cases where the linear array illumination is nonuniform, the number of elements were odd or even and the amplitude distribution was symmetrical in Section 3.3. Additionally equations (3.24) and (3.25) were derived for the resulting AGF. Here we seek to apply the same concepts to the array illustrated in Figure 3.2. The AGF for this case can be deduced to be

$$\text{AGF} = A_1 + A_2 \exp j(ks \sin \theta + \beta) + A_3 \exp 2j(ks \sin \theta + \beta) \\ + \cdots + A_N \exp(N-1)j(ks \sin \theta + \beta) \quad (3.36)$$

If we introduce the following substitution

$$z = \exp j\psi = \exp j(ks \sin \theta + \beta) \quad (3.37)$$

into Equation (3.36), we obtain

$$\text{AGF} = \sum_{n=1}^{N} A_n z^{n-i} = A_1 + A_2 z + A_3 z^2 + \cdots + A_N z^{N-1} \quad (3.38)$$

which is a polynomial of the variable z and of the degree $(N-1)$; thus we can rewrite the RHS of Equation (3.38) as [3.13]

$$\text{AGF} = A_n (z - z_0)(z - z_1)(z - z_3) \cdots (z - z_{N-1}) \\ = A_n \prod_{n=1}^{N} (z - z_n) \quad (3.39)$$

where $z_0, z_1, z_2, z_3, \ldots, z_{N-1}$ are the roots of the polynomial on the right-hand side of Equation (3.39). Similarly, the magnitude of the AGF is given by the equation

$$|\text{AGF}| = |A_n||z - z_1||z - z_2||z - z_3| \cdots |z - z_{N-1}| \quad (3.40)$$

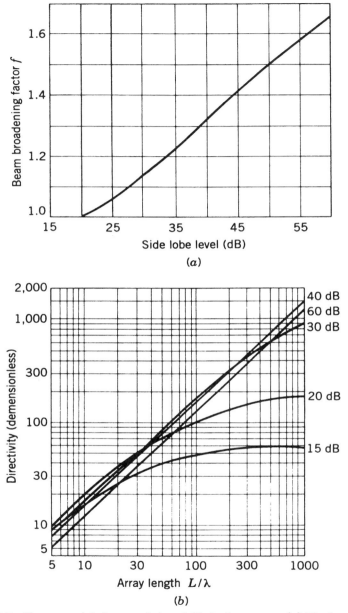

FIGURE 3.7 Two essential characteristics of Chebyshev arrays. (*a*) The beam broadening factor of Chebyshev arrays as a function of sidelobe level; and (*b*) the directivity of Chebyshev arrays as a function of array length and when the sidelobe level takes the values shown. (*Source*: Elliott [3.4a], © 1963 *Microwave J.*)

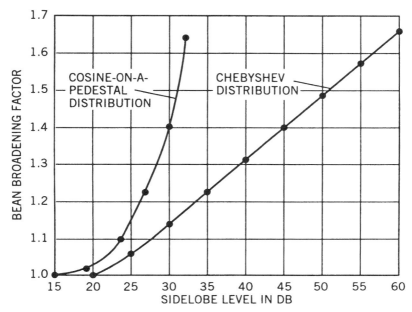

FIGURE 3.8 A graph of the beam broadening versus sidelobe level for two arrays having a Chebyshev distribution and a cosine on a pedestal distribution. (*Source*: Elliott [3.4a], © 1963 *Microwave J.*)

Given that

$$z = |z|\exp j\psi = |z|\angle \psi = 1\angle \psi \quad \text{and} \quad \psi = \frac{2\pi}{\lambda}s\sin\theta + \beta \quad (3.41)$$

the magnitude of z will lie on a unit circle, usually referred to as the *Schelkunoff circle* [3.13], irrespective of what the values of s, θ, and β are; its phase, however, will depend on the values of s, θ, and β.

3.3.2.3 Elliott's Synthesis Procedure Elliot's contributions in the field of array synthesis based on the generalized concepts just outlined are significant. While reference [3.14] serves as an excellent introduction to his synthesis procedure, reference [3.15] outlines in considerable detail his improved pattern synthesis procedure applicable to equispaced linear arrays. Here we shall outline the key features of linear arrays synthesized by his procedure.

Figures 3.9a, 3.9b, and 3.9c show the AGFs of arrays having a sum, difference and shaped beam pattern, respectively, together with their corresponding roots on, in or outside the Schelkunoff circle. As θ takes different values within the range $0° < \theta < 180°$ (defining the visible region), the position of z, the angle ψ and the distances $|z - z_n|$ vary. The AGF is equal

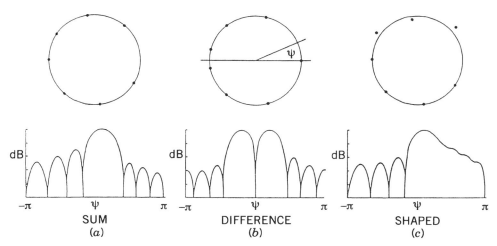

FIGURE 3.9 The roots and array geometric factors for (*a*) a sum pattern, (*b*) a difference pattern; and (*c*) a shaped pattern. (*Source*: Elliott [3.14], © 1988 IEEE.)

to zero whenever z coincides with one of its roots located on the unit circle; a null will therefore result in the AGF. As can be seen in Figure 3.9*b*, the difference pattern is generated by locating a pole at the point $1\angle 0°$. Similarly whenever z comes near a root located in or outside the unit circle a dip in the AGF will appear. The depth of the dip will depend on how close the position of the root is to z.

From the foregoing considerations it is clear that the locations of the zeros not only determine the width of the main lobe but dictate the sidelobe level also. If we for example, require low sidelobes in region A of the AGF, and can tolerate slightly higher sidelobe levels in region B of the AGF, then the number of zeros per unit arc in the region A ought to be higher than that corresponding to region B.

In Figures 3.10*a* and 3.10*b* the AGFs of two arrays having eight elements are illustrated [3.15]. The former was designed to have a sum pattern and sidelobe levels of -15 dB on one side of the main lobe and -25 dB sidelobe level on the other side of the main lobe. The latter AGF was designed to have a difference pattern with a uniform -20 dB sidelobe level.

The design procedure begins with the placement of the seven zeros appropriate for the required pattern, for example, by placing a zero at the point $1\angle 0°$ for a difference pattern and by not placing a zero at the same point for a sum pattern. After several iterations that essentially minimize the differences between the specified sidelobe level and the level resulting from that iteration, the required pattern is attained. The resultant normalized coefficients A_n are tabulated in Table 3.1.

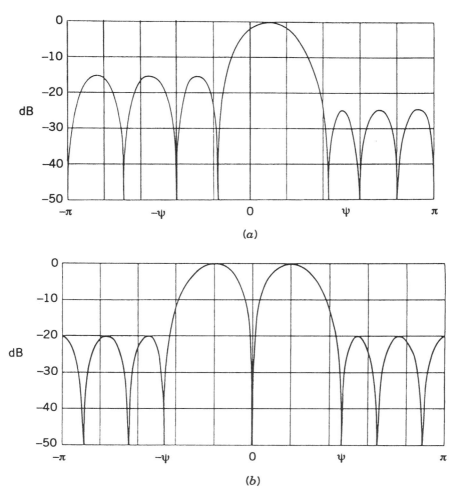

FIGURE 3.10 The array geometric factors for two arrays having eight antenna elements: (*a*) sum pattern having sidelobe levels of -15 and -25 dB on either side of the main beam; and (*b*) difference pattern having a sidelobe level of -20 dB. (*Source*: Elliott [3.14], © 1988 IEEE.)

The amplitudes of the A_n coefficients in Table 3.1 clearly illustrate the requirement to have T/R modules operating efficiently over a wide range of powers (transmit mode). The other observation is that the ratio of the amplitude coefficients is not great, a requirement that decreases the effects of mutual coupling.

In what follows we shall briefly outline the pertinent characteristics and parameters of the many well-established synthesis techniques that yield the

TABLE 3.1 The A_n Coefficients of the AGFs Illustrated in Figures 3.12a and 3.12b

A_n	1	2	3	4	5	6	7	8
Σ	0.622	0.732	0.855	1	1	0.855	0.732	0.622
Δ	−0.671	−1	−0.988	−0.419	0.429	0.988	1	0.671

required one- or two-dimensional distributions; while the treatment is necessarily brief, the references cited will expose the reader to a fuller treatment.

3.3.2.4 Taylor's One-Parameter Synthesis Procedure What follows are the most popular array synthesis procedures and their popularity is based on their simplicity and robustness. Hansen summarized these procedures succinctly [3.16] in 1992.

Although these procedures were derived for linear and planar apertures, one can sample the derived distributions to provide the excitation for large arrays. The robustness of these procedures is related to the fact that the resulting distributions have sidelobe level envelopes approximating $1/u_1$ [3.17 and 3.18], where $u_1 = (d/\lambda)\sin\theta$.

Taylor [3.18] outlined a synthesis design procedure for line sources; the procedure is based on the observation that the first few sidelobes of a uniformly illuminated line source are high while the envelope of the remaining sidelobes decays as $1/u_1$, and can be acceptable. Figure 3.11a illustrates the point made; specifically if a sidelobe level of −20 dB is acceptable, only the first two sidelobes are above the required level. The rearrangement of the zeros around that region will therefore bring the level of all sidelobes within the specification. Figure 3.11b illustrates the resultant AGF after the modification [3.14]. In this and the remaining sections we shall follow Hansen's [3.16] outline of these important procedures.

The positions of the close-in zeros are now given by the equation

$$u_1 = \sqrt{n^2 + B^2} \tag{3.42}$$

where B is a single parameter which defines all the parameters of the array. The resulting pattern is a modified sinc function of πu_1 and is expressed in two forms:

$$F(u_1) = \frac{\sin \pi\sqrt{u_1^2 - B^2}}{\pi\sqrt{u_1^2 - B^2}} \qquad u \geq B \tag{3.43}$$

$$F(u_1) = \frac{\sinh \pi\sqrt{B^2 - u_1^2}}{\pi\sqrt{B^2 - u_1^2}} \qquad u \leq B \tag{3.44}$$

LINEAR ARRAY OF EQUISPACED LINE SOURCES 157

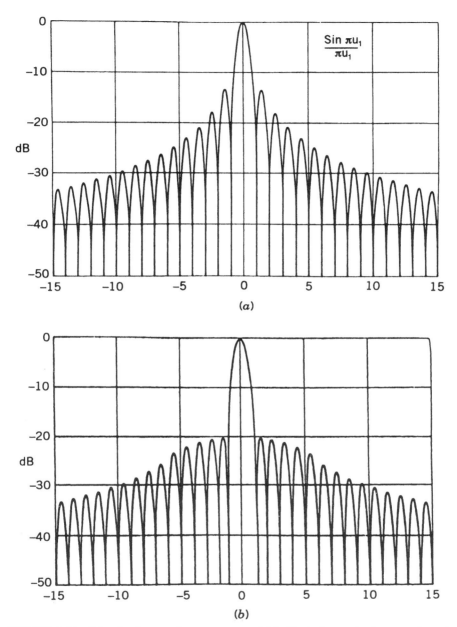

FIGURE 3.11 The fundamental concept on which Taylor's synthesis procedure is based: (*a*) the array geometric factor of a uniformly illuminated array; and (*b*) the array geometric factor of the previous array after a rearrangement of the zeros around the region corresponding to the first two sidelobes. In the case illustrated, a sidelobe level of -20 dB is acceptable. (*Source*: Elliott [3.14], © 1988 IEEE.)

The transition from the sinc pattern to the hyperbolic form occurs at $u_1 = B$ on the side of the beam. The sidelobe level (SL) in decibels and the aperture distribution are given by the equations

$$\text{SL}|_{\text{dB}} = 13.26 + 20 \log \frac{\sinh \pi B}{\pi B} \tag{3.45}$$

$$g(p) = I_0\left(\pi B \sqrt{1 - p^2}\right) \tag{3.46}$$

respectively, where p is the distance from the center of the aperture to either end and I_0 is the modified Bessel function of the first kind and order zero.

In Table 3.2 all the pertinent parameters of the distribution have been tabulated where η and η_b are the illumination and beam efficiencies, respectively [3.16]. The latter efficiency is defined as the fraction of the power in the main beam, null-to-null.

In summary then the designer selects the required sidelobe level, which in turn defines the value of B from Table 3.2 (or from Equation (3.45)). Given B, Equations (3.43) and (3.44) are used to derive $F(u)$.

The application of Taylor's approach to linear phased arrays requires the sampling of Taylor's distribution at the appropriate intervals. A more accurate approach has been outlined by Elliott [3.19] and it involves null-matching.

3.3.2.5 The Taylor $\bar{n}$ Distribution The Taylor $\bar{n}$ distribution [3.17] is a compromise between the Dolph–Chebyshev and the Taylor one-parameter distribution; additionally, it offers a modest improvement in efficiency over the one-parameter Taylor distribution. Again, the proposed method modifies the level of the first sidelobes only, while the remaining sidelobes follow the $1/u_1$ envelope. The distribution is based on two variables, A and $\bar{n}$.

TABLE 3.2 The Pertinent Parameters of the Taylor One-Parameter Distribution [3.16]

SLR (dB)	B	HPBW/2 (radians)	η	η_b
13.26	0	0.4429	1	0.9028
20	0.7386	0.5119	0.9333	0.982
30	1.2762	0.6002	0.8014	0.9986
35	1.5136	0.6391	0.7509	0.9996
40	1.7415	0.6752	0.7090	0.9999
45	1.9628	0.7091	0.674	1
50	2.1793	0.7411	0.6451	1

LINEAR ARRAY OF EQUISPACED LINE SOURCES

The pattern is given by the equation

$$F(u_1) = \sum_{n=-\bar{n}+1}^{\bar{n}-1} F(n, A, \bar{n}) \operatorname{sinc} \pi(u_1 + n) \qquad (3.47)$$

where

$$F(n, A, \bar{n}) = \frac{[(\bar{n}-1)!]^2}{(\bar{n}-1+n)!(\bar{n}-1-n)!} \sum_{m=1}^{\bar{n}-1} \left(1 - \frac{n^2}{z_m^2}\right) \qquad (3.48)$$

The zeros are given by the equation

$$z_n = \pm \sigma \sqrt{A^2 + (n - \tfrac{1}{2})^2} \qquad 1 \le n \le \bar{n}$$

$$z_n = \pm n \qquad n \ge \bar{n} \qquad (3.49)$$

where σ is the dilation factor, used to modify the first few zeros and is given by the equation

$$\sigma = \frac{\bar{n}}{\sqrt{A^2 + (\bar{n} - \tfrac{1}{2})^2}} \qquad (3.50)$$

Typical $\bar{n}$ values and the corresponding A values are tabulated in Table 3.3.

The efficiencies of the Taylor distributions that utilize the largest values of $\bar{n}$ that allow a monotonic distribution are listed in Table 3.4 together with the values of $\bar{n}$ that produce maximum efficiency. Given that the difference in efficiencies is not that significant, the monotonic cases are usually preferred [3.16]. Guidelines for the selection of $\bar{n}$ are given in the same reference. Both the Taylor distributions are widely used.

TABLE 3.3 Pertinent Parameters of the Taylor $\bar{n}$ Distribution [3.16]

SLR (dB)	A	HPBW/2 (radians)	σ				
			$\bar{n} = 2$	4	6	8	10
20	0.9528	0.4465	1.1255	1.1027	1.0749	1.0582	1.0474
25	1.1366	0.4890		1.0870	1.0683	1.0546	1.0452
30	1.3200	0.5284		1.0693	1.0608	1.0505	1.0426
35	1.5032	0.5653			1.0523	1.0459	1.0397
40	1.6865	0.6000			1.0430	1.0407	1.0364
45	1.8697	0.6328				1.0350	1.0328
50	2.0530	0.6639					1.0289

TABLE 3.4 Typical Aperture Efficiencies for the Taylor $\bar{n}$ Distributions (Maximum η, Monotonic $\bar{n}$) [3.16]

SLR (dB)	Max η Values $\bar{n}$		η $\bar{n}$	Monotonic $\bar{n}$
25	12	0.9252	5	0.9105
30	23	0.8787	7	0.8619
35	44	0.8326	9	0.8151
40	81	0.7899	11	0.7729

3.3.2.6 The R. C. Hansen Synthesis Procedure for Circular Distributions

Hansen [3.20] outlined a synthesis procedure similar in approach to that proposed by Taylor but applicable to circular distributions. Here, again, the starting point is the observation that for a uniform illumination only the first sidelobes of the resulting pattern, having the form of $J_1(u_1)/u_1$, need lowering.

The two forms of the resulting patterns are

$$F(u_1) = \frac{2I_1\left(\pi\sqrt{H^2 - u_1^2}\right)}{\pi\sqrt{H^2 - u_1^2}} \qquad u_1 \leq H \tag{3.51}$$

$$F(u_1) = \frac{2J_1\left(\pi\sqrt{u_1^2 - H^2}\right)}{\pi\sqrt{u_1^2 - H^2}} \qquad u_1 \geq H \tag{3.52}$$

where J_1 is the Bessel function of the first kind and order 1, I_1 is the modified Bessel function of the first kind and order 1 and H is the one parameter that defines all the pertinent parameters of the distribution given in Table 3.5. The SL in decibels and the aperture distribution are given by the equations

$$\mathrm{SL}|_{\mathrm{dB}} = 17.57 + 20\log\frac{2I_1(\pi H)}{\pi H} \tag{3.53}$$

$$g(p) = I_0\left(\pi H\sqrt{1 - p^2}\right) \tag{3.54}$$

respectively, when the aperture extends from $p = \pm 1$.

Given the required sidelobe level, H can be read from Table 3.5 or derived from Equation (3.53). $F(u_1)$ in turn can be derived from Equations (3.51) and (3.52).

LINEAR ARRAY OF EQUISPACED LINE SOURCES 161

TABLE 3.5 Pertinent Parameters of the Hansen One-Parameter Distribution [3.16]

SLR (dB)	H	HPBW/2 (radians)	η	η_b
17.57	0	0.5145	1	0.8378
25	0.8899	0.5869	0.8711	0.9745
30	1.1977	0.6304	0.7595	0.993
35	1.4708	0.6701	0.6683	0.9981
40	1.7254	0.7070	0.5964	0.9994
45	1.9681	0.7413	0.539	0.9998
50	2.2026	0.7737	0.4923	1

3.3.2.7 The Taylor $\bar{n}$ Circular Distribution The Taylor $\bar{n}$ circular source distribution [3.21] offers a modest improvement in efficiency and beamwidth over the Hansen one-parameter distribution just as observed for the linear distributions. Again, the starting point is the $2J_1(\pi u_1)/(\pi u_1)$ function, which corresponds to uniform illumination. The positions of $\bar{n}$ zeros on either side of the main beam are modified to produce the desired sidelobe level.

The radiation pattern is given by the equation

$$F(u_1) = \frac{2J_1(\pi u_1)}{\pi u_1} \prod_{n=1}^{\bar{n}-1} \frac{1 - (u_1^2/z_n^2)}{1 - (u_1^2/\mu_n^2)} \tag{3.55}$$

where μ_n are the zeros of $J_1(\pi u)$. The pattern zeros are given by the equation

$$z_n = \pm \sigma \sqrt{A^2 + \left(n - \tfrac{1}{2}\right)^2} \qquad 1 \leq n \leq \bar{n} \tag{3.56}$$

and the dilation factor is given by the equation

$$\sigma = \frac{\bar{n}}{\sqrt{A^2 + \left(\bar{n} - \tfrac{1}{2}\right)^2}} \tag{3.57}$$

The reader will find a more detailed derivation of the above equations and extensive tables that are readily usable in references [3.22, 3.23].

In Table 3.6 the values of the dilation factor and the A parameter are tabulated for various combinations of $\bar{n}$ and sidelobe levels; typical efficiencies for sets of values of $\bar{n}$ and sidelobe level are tabulated in Table 3.7. As can be observed, the value of $\bar{n}$ that yields maximum efficiency increases as the required sidelobe level increases.

3.3.2.8 The Bayliss (Difference) Distributions In Chapter 2 we considered some difference far-field radiation patterns; these patterns are in principle

TABLE 3.6 Some Pertinent Parameters of the Taylor Circular $\bar{n}$ Distribution [3.16]

SLR (dB)	A	HPBW/2	σ $\bar{n}=3$	4	5	6	7	8	9	10
20	0.9528	0.4465	1.2104	1.1692	1.1398	1.1186	1.1028	1.0996	1.0910	1.0732
25	1.1366	0.4890	1.1790	1.1525	1.1296	1.1118	1.0979	1.0870	1.0782	1.0708
30	1.3200	0.5384	1.1455	1.1338	1.1180	1.1039	1.0923	1.0827	1.0749	1.0683
35	1.5032	0.5653		1.1134	1.1050	1.0951	1.0859	1.0779	1.0711	1.0653
40	1.6865	0.6000		1.0916	1.0910	1.0854	1.0789	1.0726	1.0560	1.0620

required to track targets. Although the illumination functions were straight forward, the resulting far-field radiation patterns had high sibelobe, which in turn resulted in poor tracking capabilities.

Bayliss [3.24] proposed a difference pattern that is generated by the following process:

1. The difference pattern is derived from the Taylor sum pattern by differentiation.
2. The resulting pattern has high sidelobes, which are corrected by modifying the first four zeros to yield the typical Taylor distribution, that is, the first sidelobes are at the specified sidelobe level while the level of the remaining sidelobes decreases monotonically.

The first four zeros are given in Table 3.8 as a function of the sidelobe ratio (SLR), while the zeros where $n > 4$ are given by the equation

$$z_n = \pm\sqrt{A^2 + n^2} \tag{3.58}$$

TABLE 3.7 Typical Aperture Efficiencies for the Taylor Circular $\bar{n}$ Distribution [3.16]

SLR (dB)	$\bar{n}=4$	5	6	8	10
20	0.9723	0.9356	0.8808	0.7506	0.6238
25	0.9324	0.9404	0.9379	0.9064	0.8536
30	0.8482	0.8623	0.8735	0.8838	0.8804
35	0.7708	0.7779	0.7880	0.8048	0.8153
40	0.7056	0.7063	0.7119	0.7252	0.7365

TABLE 3.8 Some Pertinent Parameters of the Bayliss Distribution [3.16]

SLR dB	15	20	25	30	35	40
A	1.00790	1.22472	1.43546	1.64126	1.84308	2.04154
z_1	1.51240	1.69626	1.88266	2.07086	2.26025	2.45039
z_2	2.25610	2.36980	2.49432	2.62754	2.76748	2.91234
z_3	3.16932	3.24729	3.33506	3.43144	3.53521	3.64518
z_4	4.12639	4.18544	4.25273	4.32738	4.40934	4.49734
u_0	0.66291	0.71194	0.75693	0.79884	0.83847	0.87649

The same Table also lists the parameters A and u_0, the difference peak lobe location. The resulting radiation pattern is given by

$$F(u_1) = \sum_{n=0}^{\bar{n}-1} B_n \frac{(-1)^n u_1 \cos \pi u_1}{(n+1/2)^2 - u_1^2} \qquad (3.59)$$

where the coefficient B_n is defined in reference [3.16]. Figure 3.12 illustrates a typical Bayliss distribution where the SLR is 25 dB and $\bar{n} = 5$ and the resulting sidelobe level has a typical Taylor distribution.

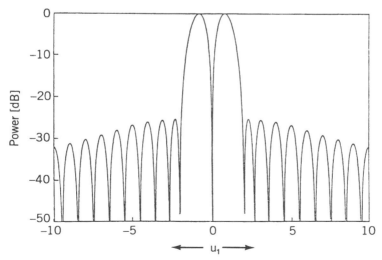

FIGURE 3.12 A typical array geometric factor corresponding to an array having a Bayliss distribution having SLR = 25 dB and $\bar{n}$ = 5. (*Source*: Hansen [3.16], © 1992 IEEE.)

164   PHASED ARRAYS: CANONICAL AND WIDEBAND

3.3.2.9   Other Synthesis Procedures   Orchard et al. [3.25], Kim and Elliott [3.26], and Powers [3.27] outlined other synthesis approaches. The synthesis procedures outlined by Taylor and Bayliss constitute the radar industry standards.

The review paper by Hansen [3.16] and his other contributions in references [3.22 and 3.23] constitute valuable guides to the general theme of array pattern control and synthesis. The Taylor illumination is somewhat similar to the cosine squared on a pedestal, which we have considered in Chapter 2 and is readily implemented, especially to phased arrays.

3.4   FAR-FIELD RADIATION PATTERN OF PLANAR ARRAYS

We seek to derive the array geometric factor of a planar array, having M (along the x direction) $\times N$ (along the y direction) antenna elements, as illustrated in Figure 3.13; to this end it is more convenient to use the usual

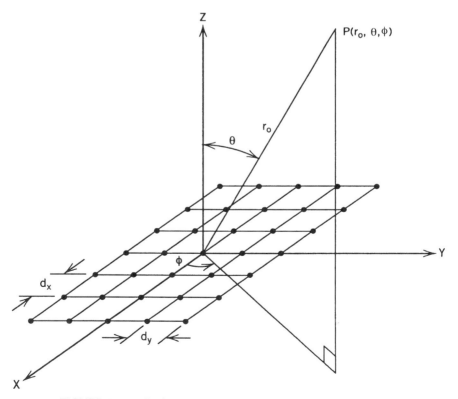

FIGURE 3.13   A planar array having $M \times N$ antenna elements.

spherical coordinates. The array factor for a linear array along the x axis can be deduced from Equation (3.3) to be

$$\text{AGF}_x = \sum_{m=1}^{M} I_m \exp j(m-1)(kd_x \sin\theta \cos\phi + \beta_x) \quad (3.60)$$

where d_x and β_x are the spacing between antenna elements and the progressive phase shift between antenna elements, respectively. Similarly, the AGF corresponding to a linear array along the y axis is

$$\text{AGF}_y = \sum_{n=1}^{N} I_n \exp j(n-1)(kd_y \sin\theta \sin\phi + \beta_y) \quad (3.61)$$

where d_y and β_y are the spacing between antenna elements and the progressive phase shift between antenna elements, respectively. The required AGF is therefore

$$\text{AGF} = [\text{AGF}_x][\text{AGF}_y] \quad (3.62)$$

If all the amplitude excitations I_n are equal to I_m, then

$$\text{AGF} = \frac{1}{M} \frac{\sin[M(\psi_x/2)]}{\sin(\psi_x/2)} \frac{1}{N} \frac{\sin[N(\psi_y/2)]}{\sin(\psi_y/2)} \quad (3.63)$$

where

$$\psi_x = kd_x \sin\theta \cos\phi + \beta_x$$
$$\psi_y = kd_y \sin\theta \sin\phi + \beta_y$$

It is often required that the main beam of the planar array be formed at $\theta = \theta_0$ and $\phi = \phi_0$, in which case

$$\beta_x = -kd_x \sin\theta_0 \cos\phi_0 \quad (3.64a)$$
$$\beta_y = -kd_y \sin\theta_0 \sin\phi_0 \quad (3.64b)$$

For the planar array under consideration, the main lobe and grating lobes occur at

$$kd_x \sin\theta \cos\phi + \beta_x = \pm 2m\pi \qquad m = 0, 1, 2\ldots \quad (3.65a)$$
$$kd_y \sin\theta \sin\phi + \beta_y = \pm 2n\pi \qquad n = 0, 1, 2\ldots \quad (3.65b)$$

If we insert Equations (3.64a) and (3.64b) into Equations (3.65a) and (3.65b), respectively, we obtain the positions of the main lobe and grating lobes from the equations

$$\sin\theta\cos\phi - \sin\theta_0\cos\phi_0 = \pm\frac{m\lambda}{d_x} \qquad (3.66a)$$

$$\sin\theta\sin\phi - \sin\theta_0\sin\phi_0 = \pm\frac{n\lambda}{d_y} \qquad (3.66b)$$

When $m = n = 0$, the position of the main beam is defined.

Von Aulock [3.28] proposed a useful way of visualizing the movement of the grating lobes as the array is scanned in two dimensions and his proposal gained wide acceptance [3.29]. With reference to Figure 3.14a, the resulting beam of a phased array located at the center of a hemisphere of unity radius is steered toward a direction θ and ϕ. A is the intersection of the hemisphere with the direction of the beam and A' is its projection on to a

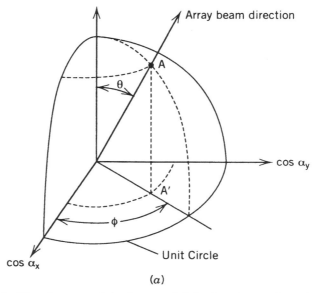

FIGURE 3.14 The elements of sin θ space. (a) A phased array is at the center of a hemisphere and the array beam is directed toward θ and ϕ. The x and y coordinates of the hemisphere are the direction cosines cos α_x and cos α_y, respectively.

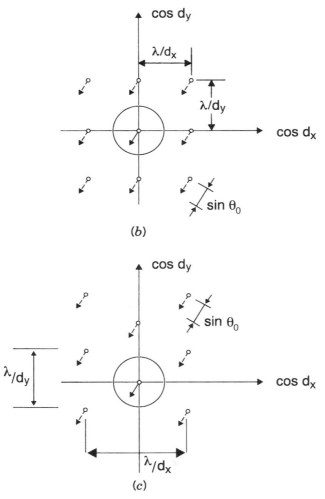

FIGURE 3.14 (b) The real-space circle or unit circle within which grating lobes are visible is shown. Outside the circle grating lobes are not visible. The case where $\lambda/d_x = \lambda/d_y$ and a square pattern results, is shown. (c) The case where $\lambda/d_x \neq \lambda/d_y$ and a triangular pattern results, is shown. For both (b) and (c) cases the array is scanned to θ_0.

coordinate system with axes $\cos \alpha_x$ and $\cos \alpha_y$, which are also called direction cosines. For any direction on the hemisphere the direction cosines are

$$\cos \alpha_x = \sin \theta \cos \phi \qquad (3.67a)$$

$$\cos \alpha_y = \sin \theta \sin \phi \qquad (3.67b)$$

and the region inside the circle is defined as

$$\cos^2 \alpha_x + \cos^2 \alpha_y \leq 1 \tag{3.68}$$

The direction of scan is defined by the direction cosines $\cos \alpha_{xs}$ and $\cos \alpha_{ys}$. The plane of scan is defined by the angle ϕ measured counterclockwise from the $\cos \alpha_x$ axis and is given by the equation

$$\phi = \arctan \frac{\cos \alpha_{ys}}{\sin \alpha_{xs}} \tag{3.69}$$

The angle of scan θ is defined by the distance of the point, $A(\cos \alpha_{xs}, \cos \alpha_{ys})$ from the origin and is equal to $\sin \theta$. This form of representation is therefore termed "sin θ space." An important feature of sin θ space is that the antenna pattern shape is invariant to the scan direction. As the beam is scanned, every point on the plot is translated in the same direction and by the same distance as is the main beam.

Figure 3.14b illustrates the unit circle that defines the visual or real space, while the space outside the circle is the imaginary space. Any grating lobes within the circle are visible, whereas grating lobes outside the circle are not.

The most common element lattices have either a rectangular or a triangular grid. The triangular grid may be thought as a rectangular grid where every other element is missing [3.30]. Equation (3.66) corresponding to the rectangular and triangular lattices are therefore

$$\cos \alpha_{xs} - \cos \alpha_x = \pm \frac{m\lambda}{d_x} \tag{3.70a}$$

$$\cos \alpha_{ys} - \cos \alpha_y = \pm \frac{n\lambda}{d_y} \tag{3.70b}$$

$$\cos \alpha_{xs} - \cos \alpha_x = \pm \frac{m\lambda}{2d_x} \tag{3.71a}$$

$$\cos \alpha_{ys} - \cos \alpha_y = \pm \frac{n\lambda}{2d_y} \tag{3.71b}$$

where $m + n$ is even.

As the array is scanned away from broadside, each grating lobe in sin θ space will move a distance equal to the sine of the angle of the scan and in the direction determined by the plane of the scan. Figures 3.14b and 3.14c illustrate the movement of the grating lobes for the cases where a rectangular and a triangular lattice geometry of the array elements is adopted, respectively.

For example, if a maximum scan angle of 60° from broadside is required, no grating lobes should exist within a circle of radius $1 + \sin 60° = 1.866$ [3.29]. If the lattice of the antenna elements is square the following conditions can be deduced:

$$\frac{\lambda}{d_x} = \frac{\lambda}{d_y} = 1.866 \quad \text{hence} \quad d_x = d_y = 0.536\lambda \quad (3.72a)$$

and the area per element is

$$d_x d_y = 0.287\lambda^2 \quad (3.72b)$$

If the lattice of the antenna elements is an equilateral triangle, we can deduce

$$\frac{\lambda}{d_y} = \frac{\lambda}{\sqrt{3}\,d_x} = 1.866 \quad \text{hence} \quad d_y = 0.53 \quad \text{and} \quad d_x = 0.309\lambda$$

$$(3.73a)$$

and the area per element is

$$2 d_x d_y = 0.332\,\lambda^2 \quad (3.73b)$$

If the elements are placed at positions forming equilateral triangles, the number of antenna elements that fully populate the array is decreased by approximately 16% compared to the case where the elements are placed on a square grid [3.29]. For both cases the condition that the grating lobe not enter the visible space at the maximum scan angle applies.

While the triangular lattice of the antenna elements is more economical (less T/R modules are required), the square lattice offers maximum power aperture products at increased cost and more serious heat-dissipation problems.

3.4.1 Grating/Side Lobes: System Considerations

Another aspect worth considering is whether the grating lobes of a phased array can be tolerated in some circumstances such as for receive-only applications, wide-frequency band operation, or in the radar context where the radiation patterns of the receive and transmit mode are multiplied, see Section 3.12.7. Here we are cognizant of the fact that grating lobes decrease the array beam efficiency but there are system tradeoffs to be considered.

We have already considered the multielement crossed radio telescope realized by Christiansen and his collaborators in Chapter 1 [1.30]. For this telescope the first order grating lobes either side of the main beam were present. Given that the Sun is much brighter than any other celestial source in close proximity to it, the effects of the resulting ambiguities (due to the presence of grating lobes) introduced negligible errors to the observations.

For non-solar observations the grating lobes of one of the arms of the cross can be eliminated if it is converted into a compound interferometer. The conversion is implemented by placing one additional fully steerable reflector at the end of the EW arm and multiplying the combined output of the linear array with the output of the additional reflector. The diameter of the reflector, usually greater than the spacing between the elements of the cross, ensures that the grating lobes are negligible at the output of the multiplier. Compound interferometers were first proposed by Covington [3.32] and the modification of the EW arm of the Christiansen cross to a compound interferometer was described by Labrum et al. [3.33]. Similar considerations hold for the other arm of the cross. The added advantage of compound interferometers is that the spatial resolution of the resulting beam is twice that corresponding to the linear phased array used.

The AGF of large radio astronomy arrays have grating lobes; their amplitudes however are negligible because the FOV of the phased array is solely defined by their unit collecting areas; more explicitly the grating lobes of the AGF are placed at or farther from the nulls the radiation pattern of the unit collecting areas.

From the foregoing considerations it is clear that grating lobes of phased arrays have to be considered in the context of specific applications. For some applications (e.g., radar), extraordinary measures have to be taken to decrease their amplitude. For other applications the effects caused by array grating lobes can be negligible.

In a radar context the minimization of the grating and side lobes of a phased array decreases the probability of false alarm and provides some ECCM protection for the radar. We have already considered the case where jamming power enters the high sidelobes of an array and disables it. How low should the sidelobes be? The lower the better of course but what are the levels that are considered adequate? The AWACS have (one way) sidelobes ranging between -40 and -50 dB and these systems are considered as having ultralow sidelobes. A distinction has to be drawn here between the design and measured sidelobe level of phased arrays. As the required sidelobe level decreases, the designer has to control a multitude of parameters such as mechanical tolerances, mutual coupling effects (which we shall not consider in this book), and the tolerances on the amplitude and phase of the T/R modules used. The lower the required sidelobe levels the tighter the required tolerances and the higher the costs.

As this topic is of considerable import, we shall consider a variety of issues related to sidelobe minimization of phased arrays in Sections 3.6 to 3.8.

3.4.2 HPBW and Directivity of Planar Phased Arrays

We are interested in defining the HPBW of a planar phased array, of $(2N_x + 1) \times (2N_y + 1)$ antenna elements, along the x and y axes, respectively, when the resulting beam is pointed towards the direction, θ_0, ϕ_0. To this end we shall define two beamwidths Θ_1 and Θ_2 illustrated in Figure 3.15, [3.4] having HPBWs defined as

$$\Theta_1 = \theta_2' - \theta_1' \quad \text{and} \quad \Theta_2 = (\phi_2' - \phi_1')\sin\theta_0 \quad (3.74a)$$

where θ_2', θ_1', ϕ_2', and ϕ_1' are two sets of values that satisfy the equations

$$A(\theta', \phi_0) = 0.707 A(\theta_0, \phi_0) \quad \text{and} \quad A(\theta_0, \phi') = 0.707 A(\theta_0, \phi_0) \quad (3.74b)$$

After considerable manipulations, the two HPBWs have been deduced to be [3.4b]

$$\Theta_1^{-2} = \cos^2\theta_0 \left[\Theta_{xo}^{-2} \cos^2\phi_0 + \Theta_{yo}^{-2} \sin^2\phi_0 \right] \quad (3.75a)$$

and

$$\Theta_2^{-2} = \Theta_{xo}^{-2} \sin^2\phi_0 + \Theta_{yo}^{-2} \cos^2\phi_0 \quad (3.75b)$$

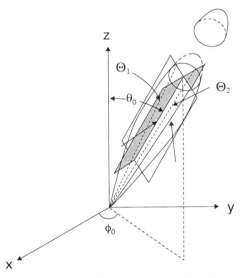

FIGURE 3.15 The HPBWs Θ_1 and Θ_2 of a planar phased array when its beam is pointed toward θ_0, ϕ_0.

where Θ_{xo} and Θ_{yo} are the beamwidths corresponding to the case where $\theta_0 = 0$ and the arrays are linear having $2N_x + 1$ and $2N_y + 1$ antenna elements along the x and y axes, respectively. Inspection of equation (3.75a) suggests that for a scan in a ϕ constant plane, $\Theta_1 \sim \sec \theta_0$, which is consistent with the notion that the projected aperture in the pointing direction is reduced by the factor $\cos \theta_0$, suggesting the broadening of the beam. For a square array ($\Theta_{xo} = \Theta_{yo}$) we can deduce from equation (3.75a) that $\Theta_1 = \Theta_{xo} \sec \theta_0$, indicating that Θ_1 is independent of ϕ_0 and that it broadens with scan in accordance with the projected aperture. Θ_2 by contrast does not depend on θ_0; for square arrays it is also independent of ϕ_0. For any scan direction, Θ_2 is determined with the aid of equation (3.75b) and a knowledge of Θ_{xo} and Θ_{yo}, which can be derived from figures similar to Figure 3.8.

A beamwidth B can be defined as $\Theta_1 \Theta_2$ and is equal to

$$B = \frac{\Theta_{xo} \Theta_{yo} \sec \theta_0}{\left[\sin^2 \phi_0 + \frac{\Theta_{yo}^2}{\Theta_{xo}^2} \cos^2 \phi_0\right]^{1/2} \left[\sin^2 \phi_0 + \frac{\Phi_{xo}^2}{\Theta_{yo}^2} \cos^2 \phi_0\right]^{1/2}} \quad (3.76)$$

From Equation (3.76) it can be seen that the beamwidth B is dependent on both θ_0 and ϕ_0.

The directivity D of a planar phased array is defined by the equation

$$D = \frac{4\pi A(\theta_0, \phi_0) A^*(\theta_0, \phi_0)}{\int_0^{\pi/2} \int_0^{2\pi} A(\theta, \phi) A^*(\theta, \phi) \sin \theta \, d\theta \, d\phi} \quad (3.77a)$$

which reduces to

$$D = \pi D_x D_y \cos \theta_0 \quad (3.77b)$$

where D_x and D_y are the directivities of the two linear arrays. From Equation (3.77b) it can be seen that the directivity of a planar array is dependent on θ_0 and independent of ϕ_0.

3.4.3 The Half-Power Beamwidth × Directivity Product for Planar Arrays

For a uniform illumination, the product of HPBW times the directivity for a planar phased array is given by the equation [3.4b]

$$D = \frac{32,400}{B_0} \quad (3.78)$$

where B_0 is expressed in square degrees.

3.4.4 Input/Output SNR of Arrays

We wish to derive expressions relating the input to the output SNR of a phased array having N elements. To this end let us assume that the input signal level, proportional to the envelope of the RF signal voltage in the antenna elements, is equal to S_i; if all signals are cophased by a set of phase shifters, the output signal voltage S_o will be equal to NS_i. The output signal power will therefore be equal to $N^2 S_i$.

If the input noise power is N_i at each antenna element, the output noise power N_o will be equal to NN_i. We have here assumed that the noise from element to element is independent.

The output $(SNR)_o$ will then be given by the equation

$$(SNR)_o = \frac{S_o}{N_o} = \frac{N^2 S_i}{NN_i} = N(SNR)_i \qquad (3.79)$$

The array SNR gain, G_{SNR} is therefore equal to N.

The product of the array G_{SNR} and array beamwidth (at boresight) for a long linear array having an interelement spacing of $\lambda/2$ is invariant and equal to 2. The assumptions made here are that the array illumination is uniform and that the length of the long array is $N(\lambda/2)$. Similarly for a planar array, the same product under the same assumptions is equal to 4.

3.5 CYLINDRICAL AND CIRCULAR PHASED ARRAYS

To meet the requirement for a 360° azimuth surveillance volume, several planar arrays are used, each covering a 90° or a 120° segment. For some applications a cylindrical phased array is an attractive competitor. The additional attraction of these arrays is that the interelement mutual coupling is less severe.

The coordination of the actions of an entire group of battleships presented an important application where cylindrical or circular phased arrays are used [3.34, 3.35]. The requirements formulated by the Cooperative Engagement Capability (CEC) program were to:

1. Share radar measurement data from battleship group radars in near real time so that a composite air picture is formed.
2. Provide cooperative engagements where different units support one another's missile operations.

A secure communications system with an unprecedented response time, ECM resistance, high data rate, and terminal track accuracy to support gridlock alignment was required to meet these requirements. A directive antenna, positioned at a relatively high antenna location on a ship's mast, able to point its beam rapidly to any azimuth angle, was needed to meet the system-level requirement for data latency and directive point-to-point networking scheduling. The cylindrical or circular phased array meets this requirement without the azimuthal beam broadening and gain loss associated with conventional wide-angle scanning planar arrays.

The prototype cylindrical array antenna is divided into several columns of radiating elements (microstrip antenna elements) spaced evenly around the cylinder. Only one-fourth of the antenna elements are active at any time to create an aperture in the desired azimuth direction, but the aperture is not fixed and can be electronically switched around the cylinder with the aid of double-pole four-throw switches and the amplitude commutator. The phase switches are used for elevation beam steering, phase compensation of the curved aperture, and fine azimuth beam steering.

In the radio-astronomy context circular phased arrays are not as popular as planar arrays of different geometries, such as crossed, Y and T arrays. However, circular arrays have one major technological advantage over their popular planar counterparts—all connections from the antenna elements to the central processing station have equal lengths. Given that the sidelobe level of circular arrays is relatively high, a synthesis procedure is required to minimize it.

3.5.1 The J. P. Wild Procedure to Minimize the Sidelobes of Circular Phased Arrays

With reference to Figure 3.16, the far-field radiation pattern of a circular array having N uniformly illuminated antenna elements is very complex. The figure illustrates the case where $N = 96$, and it consists of a main beam, several concentric rings of ever-decreasing amplitudes, namely, the sidelobes and grating lobes of a complex structure. While the main beam and the sidelobes are independent of the azimuth angle, the grating lobes are azimuth-angle-dependent.

Before we arrive at the AGF corresponding to the circular array having N antenna elements, it is worth recalling that the radiation pattern of a thin annulus of radius a wavelengths is

$$F_0(r) \propto [J_0(2\pi a r)]^2 \qquad (3.80)$$

where r is the angular displacement in radians from the boresight axis of the array. Similarly, the AGF, $F(u, \phi)$, of a circular array having N antenna elements, is given by the equation [3.36]

$$F(u, \phi) = N[J_0(p) \pm 2J_N(p)\cos(N\phi) + 2J_{2N}(p)\cos(2N\phi) \pm \cdots]^2 \qquad (3.81)$$

CYLINDRICAL AND CIRCULAR PHASED ARRAYS   175

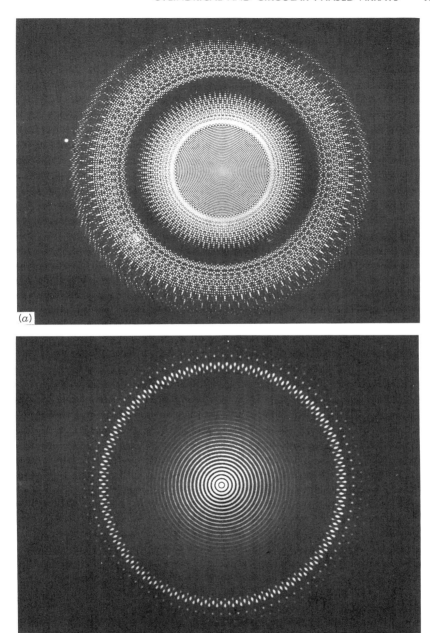

FIGURE 3.16 The far-field radiation pattern of a circular array having 96 antenna elements. (*a*) The main beam, sidelobes and two orders of grating lobes are shown; and (*b*) the main beam, sidelobes and the first order grating lobes are shown in some detail. (*Courtesy*: CSIRO Division of Radiophysics.)

where $p = \pi Du$, $u = \sin\theta/\lambda$, D is the diameter of the circular array, and the angles θ and ϕ are the zenith and azimuth angles respectively. J_N are the Bessel functions of the first kind of order N and the positive signs on the RHS of Equation (3.81) are taken when $N/2$ is an even number, while the negative signs are taken when $N/2$ is odd. The first term of the RHS of Equation (3.81) defines the main beam and sidelobes of the array, while the subsequent terms define its grating lobes, which are azimuth-dependent.

Given that the Bessel functions $J_N(p)$ and $J_{2N}(p)$ are zero when $0 < p < N/2$, only the $J_0(p)$ term on the RHS of Equation (3.79), defining the main beam and sidelobes, yields nonzero values. As $\sin\theta \to \lambda N/\pi D$ (or $p \to N$), the term $J_N(p)$ rises steadily so that the terms

$$4J_N^2(p)\cos^2(N\phi) \pm 4J_0(p)J_N(p)\cos(N\phi)$$

in the expansion of the RHS of Equation (3.79) give rise to a set of lobes with the form of modulated rings, which are further modulated, in azimuth, by the $\cos^2(N\phi)$ and $\cos(N\phi)$ terms; the grating lobes appear at angular distances r_1 where

$$r_1 = \pm q\frac{N}{2\pi a} \quad \text{and} \quad q = 1,2,3\ldots \quad (3.82)$$

Considering that $2\pi a = 2\pi r/\lambda$ is the circumference of array of radius r, Equation (3.82) is comparable to Equation (3.66). The designer aims at eliminating the grating lobes of a circular array with the aid of the radiation pattern of the antenna elements, so here we will examine the AGF around the main beam and the sidelobe area.

The profile of the radiation pattern $P(r)$ of the main beam is deduced, using Equation (3.79), to be

$$P(r) = \text{AGF}|_{\text{circular}}^2 = J_0^2(2\pi ar)$$

Given that the first sidelobe of the circular array is about 8 dB below the main lobe, a reduction of the sidelobes of the circular array is essential for many applications. While a reduction of the sidelobe level for linear arrays is attained by an amplitude taper in the illumination of the array, for circular arrays an appropriate phase taper is required to decrease their sidelobe level.

The method is based on the following observations:

1. If all antenna elements are connected to a central point with the aid of equal cables, the resulting radiation pattern will have the form of $J_0(2\pi ar)$ [Eq. (3.79)].
2. Circularly symmetrical patterns may be generated by inserting a phase shift, which increases uniformly around the circle, in each antenna channel. Any antenna element can be a reference antenna where the

CYLINDRICAL AND CIRCULAR PHASED ARRAYS   177

phase shift is zero and the total phase shift around the complete circle is $2\pi k$, where k is an integer.

The resulting far-field radiation pattern is [3.36, 3.37] is

$$P(r)_k = J_k^2(2\pi ar) \tag{3.83}$$

The implementation of the J_k^2 synthesis procedure requires the generation of a suitable set of $J_k^2(\pi Du)$ patterns that, when combined with the $J_0^2(\pi Du)$ pattern, yield an array geometry pattern that has sidelobes well within the required specifications. The derived array geometry pattern is then given by the equation

$$F(u) = \sum_{k=0}^{\infty} t_k J_k^2(2\pi ar) \tag{3.84}$$

where t_k are the weighting factors, which are determined from the integral equation [3.37]

$$t_k = 2^*(2\pi)^2(-1)^k \int_0^\infty r\,dr \int_0^{2a} F(r) J_0(2\pi\rho r) \cos\left[\frac{2k \arccos \rho}{2a}\right] \rho\,d\rho \tag{3.85}$$

where the star designates that the factor 2 is omitted in the case $k = 0$. If we adopt

$$F(r) = \frac{J_1(4\pi ar)}{2\pi ar} \tag{3.86}$$

to attain the resolution limit, then the t_k values obtained from Equation (3.85) are

$$t_k = 1, 0, -\tfrac{2}{3}, 0, -\tfrac{2}{15}, 0, -\tfrac{2}{35}, 0, -\tfrac{2}{63} \ldots \quad \text{when} \quad k = 0,1,2,3,4,\ldots \tag{3.87}$$

Although the series of the RHS of Equation (3.87) is theoretically infinite, it has been shown [3.37] that the series can be truncated within the range defined by $N/2 > k > 0$. It has been found by trial and error that a good compromise between resolution and low sidelobes is afforded by the pattern:

$$F(r) = 0.3\Lambda_1(4\pi ar) + 0.7\Lambda_3(4\pi ar) \tag{3.88}$$

in which case

$$t_k = 1, 0.7, -0.48, -0.42, -0.24, -1\ldots \tag{3.89}$$

Figure 3.17 illustrates the J_k^2 synthesis procedure outlined by Wild; Figure 3.17a shows the J_k^2 terms where $k = 0, 1, 2, 3,$ and 4 as a function of the

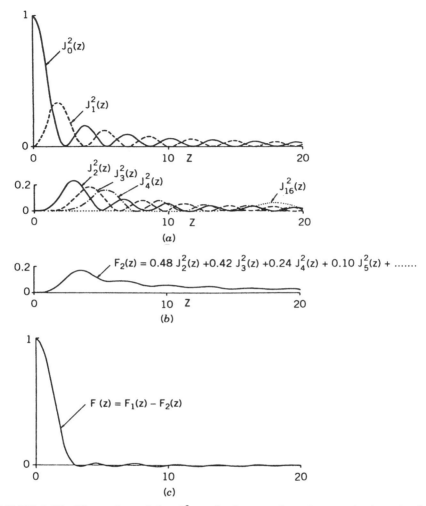

FIGURE 3.17 Illustration of the J_k^2 synthesis procedure that results in a far-field pattern given by the equation $0.3\Lambda_1(x) + 0.7\Lambda_3(x)$: (a) several $J_k(x)$ functions; (b) the $F_2(z)$ sidelobe corrective function; and (c) the resulting corrected far-field pattern (from [3.37]).

radial distance z. While $F_1(z) = J_0^2(z) + 0.7J_1^2(z)$, Figure 3.17b shows $F_2(z)$. Lastly the desired far-field radiation pattern $F(z) = F_1(z) - F_2(z) = 0.3\Lambda_1(4\pi ar) + 0.7\Lambda_3(4\pi ar)$ is shown in Figure 3.17c.

The synthesis procedure just outlined was used to lower the sidelobes of the Culgoora Radioheliograph, a circular radio telescope 3 km in diameter which produced dual-polarization radio pictures of the Sun every second at 80 MHz and utilized 96 fully steerable paraboloids 13m in diameter.

A variant method for the reduction of the sidelobes of circular arrays by adjusting the phases of the array has been described in Goto and Tsunada [3.38] and the references cited therein.

3.6 MODERN ARRAY SYNTHESIS PROCEDURES

Modern array synthesis procedures have evolved not because the conventional synthesis procedures needed refinement but because of fundamental changes in the array architectures. This said, the conventional synthesis techniques constitute necessary design tools.

3.6.1 The J. J. Lee Synthesis Procedure

Active phased arrays—where a T/R module follows every antenna element—have significant advantages over passive phased arrays—see Section 3.9. For a long time costs related to active phased arrays prevented the widespread use of active phased arrays; the costs were essentially related to the costs of T/R modules. This being the case, J. J. Lee examined the problem of controlling the sidelobe levels of active phased arrays [3.39] and his approach was based on the following observations:

1. It is too expensive to produce T/R modules that are identical in all respects, including amplitude and phase tracking, but have gains or output powers that can be varied continuously to meet the requirements for an amplitude taper across the active phased array.
2. An amplitude taper across the active phased array can be approximated by the use an N step staircase approximation to the taper required.
3. In the interest of keeping the costs of T/R modules down, the N steps should be kept to a minimum (e.g., five).
4. It is economical to manufacture five sets of T/R modules, each of which have identical characteristics.

With a five-step approximation, Lee reported a peak sidelobe level of -36 dB in all azimuth planes, a significant achievement.

With reference to Figure 3.18, Lee partitioned the array into five zones; each zone is then populated by T/R modules having the same power output. His method consisted in varying (1) the power of the modules located in one zone and (2) the geometric extent of each zone area to minimize the resultant sidelobe level. More explicitly, the task was to find a global minimum for the array sidelobe level using a gradient method by varying the values of E_n and b_n.

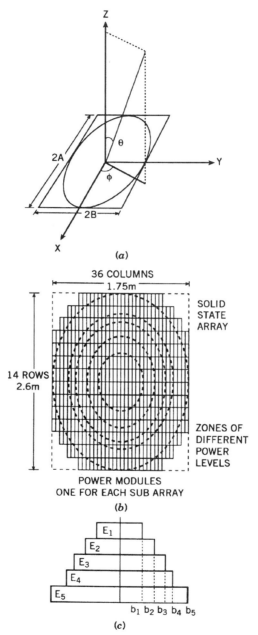

FIGURE 3.18 The J. J. Lee method of partitioning the array area into five zones; each zone is populated by T/R modules having the same transmitted power output: (*a*) the coordinate system used; (*b*) the five zones of the array; and (*c*) the stepped amplitude distribution of the array. (*Source*: Lee [3.39], © 1988 IEEE.)

MODERN ARRAY SYNTHESIS PROCEDURES   181

If the five zones were circular, the far-field pattern in spherical coordinates would be as follows [3.39, 3.40]:

$$F(\theta, \phi) = [\hat{x} f(\theta, \phi)]^2 \tag{3.90}$$

where

$$\hat{x} = \hat{a}_\theta \cos \phi - \hat{a}_\phi \sin \phi \cos \theta \tag{3.91}$$

and

$$f(\theta, \phi) = \sum_{i=1}^{5} 2 A_i E_i \frac{J_1(u_i)}{u_i} \tag{3.92}$$

where A is the area of the circular zones and E_i is the corresponding voltage amplitude of the ith layer of the aperture. If the zones are ellipses where the major and minor axes are a_i and b_i, respectively, then

$$A_i = \pi a_i b_i \tag{3.93}$$

$$u_i = k_o a_i \sin \theta \sqrt{\cos^2 \phi + \left[\frac{b_i}{a_i}\right]^2 \sin^2 \phi} \tag{3.94}$$

$f(\theta, \phi)$ is again characterized by the familiar $J_1(u_i)/u_i$ function except that the pattern is now compressed along a direction perpendicular to the minor axis of the ellipse.

In the computer program organized by Lee a starting set of b_n, E_n and an acceptable level of sidelobes SLL were stipulated. One variable was changed at a time and the resulting sidelobe level SLL_1 was recorded; if $SLL_1 >$ SLL, SLL_1 was declared unacceptable and another iteration was undertaken. The starting values of b_n and E_n can be such that an approximation to the Taylor 35-dB illumination is attained; convergence is easily reached with the aid of a gradient method. If SLL_1 is set unrealistically low—say, -45 dB, convergence is not possible. There has been no theoretical work to establish a relationship between the attainable global minimum sidelobe level and the number of zones used.

The strength of Lee's method lies in the observation that the attained minimum sidelobe level is not exceeded in any azimuth angle. In Table 3.9 we

TABLE 3.9 The E_i and b_i Values for Lee's Synthesis Procedure for a -36-dB Peak Sidelobe Level

E_1/b_1	E_2/b_2	E_3/b_3	E_4/b_4	E_5/b_5
1/0.34	0.74/0.52	0.52/0.68	0.36/0.78	0.2/1

have tabulated the E_i and b_i values calculated by Lee which resulted in the peak sidelobe level of -36 dB.

It is interesting to note that the power changes from one level to the next are approximately 3 dB. An additional stage of amplification can therefore accommodate the derived requirement.

The importance of the Lee synthesis procedure and other derivatives will depend entirely on the additional cost the designer will have to budget for T/R modules that have different power outputs. It is conceivable that some time in the future the additional cost of T/R modules having a range of preset power outputs (in contradistinction to continuously variable output powers) will not be high; until that time the Lee synthesis procedure and its derivatives will be widely used.

3.6.2 Spatial Tapers: The R. E. Willey Approach

Willey [3.41] recognized, in 1962, that the conventional amplitude tapers used to minimize the sidelobe levels of phased arrays would be impractical to implement for active phased arrays. He therefore proposed the use of active arrays where the antenna elements are placed in such a way as to simulate a space taper instead of an amplitude taper used by conventional arrays. In the former arrays each antenna element is connected to a T/R module and all T/R modules are the same. The two arrays have sidelobe levels that are below a specified level. In Willey's terminology the conventional phased array utilizing an amplitude taper is the reference array.

An essential quantity that defines many parameters of the arrays utilizing spatial tapers is the number of elements R that the arrays use. As can be seen from Figure 3.19 the array sidelobe level tends to rise as θ increases, above the first sidelobe level when R is too low or when the specified sidelobe is too low.

R for large arrays utilizing many antenna elements has been determined in terms of the "optimum sidelobe level," SL_{dB}, which is defined as the peak sidelobe level of the spatially thinned array. For linear and planar arrays the $SL_{dB}|_L$ and $SL_{dB}|_P$ are given by the equations

$$SL_{dB}|_L \approx -10\log\frac{R}{2} + 10\log\left[1 - k\frac{R}{U}\right] \qquad (3.95)$$

and

$$SL_{dB}|_P \approx -10\log\frac{R}{4} + 10\log\left[1 - k\frac{R}{U}\right] \qquad (3.96)$$

respectively, where U is the number of antenna elements of the reference array and k is related to the interelement spacing of the reference array by the relationship $kd = \lambda/2$; usually $k = 1$ and R/U is a measure of how

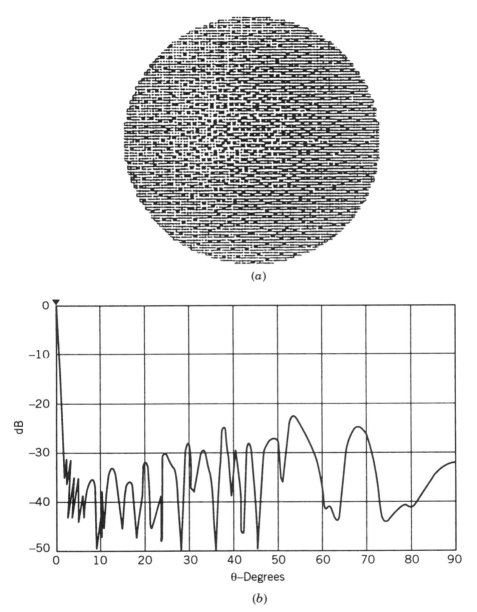

FIGURE 3.19 The sidelobe level of a thinned array rises at angular distances away from the main beam because either the number of antenna elements used is too low or the required sidelobe level is set too low. (*a*) The array topology; and (*b*) the resulting far-field radiation pattern. (*Source*: Willey [3.41], © 1962 IEEE.)

thinly or densely the array is populated. Having determined the required number of elements, the designer has to place them on the linear/planar array.

For a linear array the area under the illumination function $f(x)$ is divided into a number of equal segments and equal current sources, or elements, are assigned to each segment; by following this procedure we will have an approximation to the $f(x)$ function.

For circular arrays the array is divided into annular rings, with the width of each ring equal to the element spacing of the reference array. Next the illumination function $f(\rho)$ is integrated over each annular ring and over the total aperture. Elements are then assigned to each annular ring; the number of elements per ring is defined by the equation

$$\frac{\text{Integral over ring}}{\text{Integral over aperture}} = \frac{\text{number of elements in ring}}{\text{total number of elements}} \quad (3.97)$$

In an example cited, a planar array had a maximum sidelobe of 25 dB (Taylor distribution), when $k = 1$ and $R_P/U = 0.199$; this configuration is illustrated in Figure 3.20a.

For a planar thinned array, the resulting gain G_{thinned} is given by the equation [3.42]

$$G_{\text{thinned}} = (\text{element gain})\frac{R_P^2}{U}$$

$$= \frac{4\pi A}{\lambda^2 U}\frac{R_P^2}{U} = \frac{4\pi A}{\lambda^2}\frac{R_P^2}{U^2} \quad (3.98)$$

regardless of the positions of the antenna elements; here A is the array area.

3.6.2.1 Spatial Tapers: Applications The Cobra Dane (AN/FPS-108) [3.43] is a phased array-based radar having a diameter of 28.956 m or 95 ft and transmits 15.4 MW peak power and 0.92 MW average power at L-band. It was deployed in 1977 on one of the Aleutian islands, and its functions were intelligence gathering on Soviet missile systems undergoing test firing, space track support, and intercontinental ballistic missile (ICBM) early warning.

An essential feature of this radar is its ability to provide high-resolution observations of targets at long ranges, of the order of 1853 km. Although its aperture can accommodate 34,769 antenna elements, only 19,403 antenna elements are used; thus $R/U = 0.558$ and the remaining positions are occupied by antenna elements terminated into dummy loads. However, the method of populating the array is not known.

MODERN ARRAY SYNTHESIS PROCEDURES   185

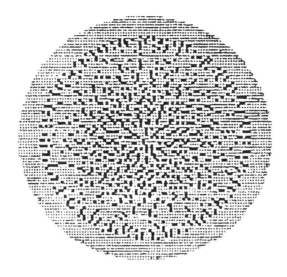

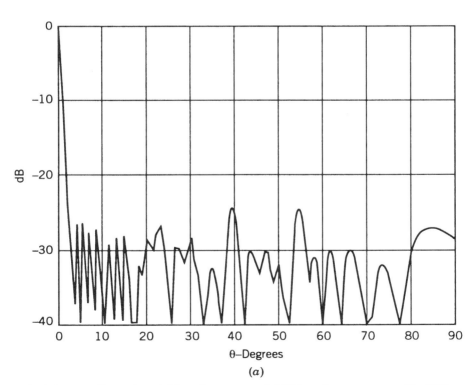

FIGURE 3.20 Examples of thinned apertures. (*a*) Thinned array proposed by Willey [3.41].

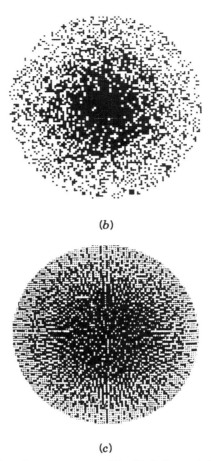

FIGURE 3.20 (*b*) Thinned array proposed by Skolnik et al. (*Source*: [3.44], © 1964 IEEE); and (*c*) thinned transmit and receive arrays occupying the same aperture [3.41].

Skolnik et al. [3.44] reported a maximum sidelobe level of −35 dB when an aperture had a Taylor distribution and the resulting aperture distribution is illustrated in Figure 3.20*b*.

When an aperture is thinned, it is possible to separate the receive from the transmit function in such a way as to fully utilize the aperture [3.41]. A combined transmit/receive array is illustrated in Figure 3.20*c*.

With this configuration the conventional T/R module is split into a receive and transmit portion; thus the front-end switch (or circulator) normally used to switch the antenna element to the transmitter/receiver is not required. For arrays operating at L-band and at lower frequencies this is a considerable saving in weight and space; the costs for the above arrays operating at any frequency are lower. Additionally less heat generated by the final power amplifier migrates to the front-end low-noise amplifier.

3.6.2.2 The Mailloux–Cohen Approach In an effort to minimize sidelobes for affordable active phased arrays, the Mailloux–Cohen [3.45] approach acknowledges that the designer could have a finite number of sets of T/R modules, in accord with the approach taken by Lee [3.39]. The required illumination function is therefore approximated by a staircase approximation of P sets of T/R modules, where P is, say, smaller than 10.

The major difference between the Mailloux–Cohen approach and Lee's is that in the former approach the transition from one quantized level to the next is smoothed by using one of the proposed three methods of statistical thinning.

Mailloux–Cohen used a three-level staircase approximation to a -50 dB sidelobe Taylor amplitude distribution and explored a range of values of radii ρ_p, which minimized the resulting average array sidelobe level; the selection of a particular value of ρ_p, corresponded to a particular value on the -50 dB sidelobe Taylor distribution. The further minimization of the resulting array sidelobe level was then sought by the implementation of statistical thinning which followed the criteria set by the three methods explored in the their paper.

The average sidelobe level for arrays thinned by methods 1, 2, and 3 were -42.5, -47, and -45 dB, respectively. As can be seen, thinning smooths the transition from one level of the staircase to the next and lowers the resulting sidelobe levels. Although these sidelobe levels are impressive, one can theoretically attain lower sidelobe levels by increasing the number of steps of the staircase approximation from three to four, five or six [3.45].

3.6.2.3 The Frank–Coffman Synthesis Procedure or Active Hybrid Arrays For a long time designers considered phased arrays that were either active or passive and these terms have been defined already. The term "active hybrid array" may cause some confusion. Although Frank–Coffman [3.34, 3.46] define it as an array in which the central part of its aperture is an active array while its outer annulus is a passive array, there is an element of confusion. Other authors and readers may use the term hybrid array for an array populated by T/R modules that are hybrid in the sense that the receive portion of the module is active while the transmit portion is passive (or vice versa). We shall keep this distinction in mind throughout this book.

The central thrust of the Frank–Coffman proposal is to realize affordable phased arrays when the cost of T/R modules is high. With reference to Figure 3.21 each antenna element is connected to a T/R module in the central portion of the array while in the outer annulus, each T/R module is connected to eight antenna elements. With reference to Table 3.10 the authors consider an array having 10,000 antenna elements but utilizes only 3,440 T/R modules—a considerable saving in costs.

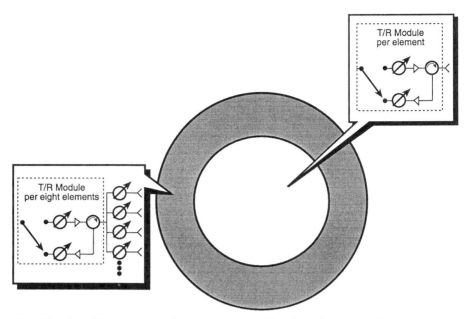

FIGURE 3.21 The array topology proposed by Frank and Coffman (*Source*: [3.46], © 1994 IEEE.)

TABLE 3.10 Antenna Elements and T/R Modules Distribution in a Typical Hybrid Array [3.46]

Annulus Area	Central Area	Total
Antenna elements 7500	Antenna elements 2500	Total antenna elements 10,000
Phase shifters 7500		
T/R modules ~ 940	T/R modules 2500	Total T/R modules 3440
1:8 power dividers ~ 940		

Only a quarter of the T/R modules feed subarrays in the outer annulus. Assuming a 2-dB loss for the phase shifter and power divider, the hybrid array would experience only a 0.5-dB loss since only the outer annulus (25% of the power) is subject to this loss.

The illumination pattern across the array is shown in Figure 3.22; the amplitude taper at the annulus is at the −13-dB level, while the pedestal is at the −22-dB level; in the proposed scheme it is not clear how the required taper is implemented.

For a given number of T/R modules, this technique provides higher ERP and gain along with narrower beams than the conventional one-T/R module

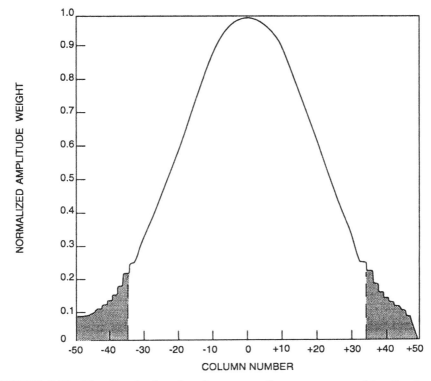

FIGURE 3.22 The illumination function across the array proposed by Frank and Coffman [3.46]. The shaded area represents the outer annulus.

per antenna element approach. It is assumed here that the greater geometric area needed to accommodate the extra antenna elements is available; for airborne platforms, this is not always the case.

With reference to Figure 3.23 the amplitude taper was chosen to yield −50-dB sidelobes, and the resulting maximum amplitude quantization (AQ) lobe level was about −40 dB. The AQ lobes occur because the amplitude control of the aperture illumination is at the subarray level rather than at the antenna element level. By analogy with conventional array theory, grating lobes of the array factor of subarrays for a rectangular subarray lattice will occur in sine space at locations given by [3.46]

$$u_p = u_0 \pm p \frac{\lambda}{d_{sx}}$$

$$v_q = v_0 \pm q \frac{\lambda}{d_{sy}} \qquad p, q = 0, 1, 2, \ldots \qquad (3.99)$$

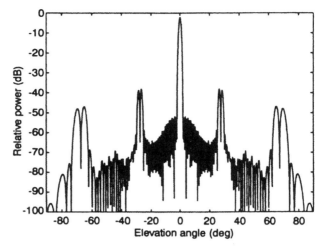

FIGURE 3.23 The radiation pattern of the array proposed by Frank and Coffman [3.46] illustrating the main beam, the sidelobes, and the amplitude quantization lobes. (*Source*: IEEE, © 1994 IEEE.)

and for a triangular subarray lattice at locations given by

$$u_p = u_0 \pm p \frac{\lambda}{2d_{sx}} \quad p, q = 0, 1, 2 \ldots$$

$$v_q = v_0 \pm q \frac{\lambda}{2d_{sy}} \quad p + q \text{ is even} \quad (3.100)$$

Here d_{sx} and d_{sy} are the intersubarray spacings in the x and y directions, rather than the interelement spacing more typically associated with these equations. AQ lobes are formed at these subarray induced grating lobe positions.

The nulls of the radiation pattern corresponding to the subarray of dimensions, say, $L \times L$, are at angular distances $\pm p(\lambda/L)$; for contiguous subarrays $L = d_{sx}$ or $L = d_{sy}$, so the position of the nulls coincide with the position of the grating lobes. When all subarrays are uniformly illuminated, the resulting main beam and grating lobes are narrow, so when the subarray radiation pattern is multiplied by the AGF, the grating lobes have negligible power levels. However, when the subarray illumination has a suitable taper to lower the resulting sidelobes, the main beam and grating lobes are broadened so that the resulting array radiation pattern has split grating lobes resembling the monopulse radiation patterns. The positions of the nulls of the grating lobes naturally correspond to the position of nulls of the subarray radiation pattern. In Figure 3.23 the monopulselike lobes closest to the main beam are seen always at a slightly higher level than those farther away from the main beam. The resulting AQ lobes are therefore consistent with the multiplication of the AGF and the radiation pattern of the subarray.

If the array has m sub-arrays of n elements, then the peak power of the AQ lobes, $G_{GL}|_p$, is given by the equation [3.47]

$$G_{GL}|_p = \frac{B^2}{m^2 n^2 \sin^2(\pi p/n)} \qquad (3.101)$$

where $p = \pm 1, \pm 2, \pm 3, \ldots$, and B is the ratio of the HPBW of the broadened beam to that corresponding to an array uniformly illuminated. For a known illumination, B can be deduced from Table 2.1 or Tables 3.2, 3.3, 3.5, and 3.6 where the HPBW/2 is given. Barton and Ward [3.48] also provide measures of B for a variety of illuminations.

The designer has the freedom to tailor the subarray lattice to avoid the formation of AQ lobes in critical pattern regions.

The major advantages of this method are:

1. The illumination taper can be naturally approximated by subarraying eight antenna elements; T/R modules used in the central region can be used for the annulus region.
2. There is merit in subarraying at the annulus of the array, where the taper is rather severe for the power in the AQ lobes is diminished.
3. It meets the criteria of utilizing fewer T/R modules and attaining maximum ERP.

For certain applications the four AQ lobes can be objectionable, even at the −39-dB level; this potential drawback, however, is not intrinsic to the proposed method and is related only to the particular illustrative example given in Frank and Coffman [3.46].

3.6.2.4 Derivative Approaches A compromise between using the minimal number of T/R modules and attaining low-AQ sidelobes can be reached if the subarray consists of four antenna elements (2 × 2). This design approach will displace the AQ lobes over the edge of the array FOV—see Equation (3.99). In Chapter 5 we shall see that subarrays consisting of 2 × 2 patch antennas have excellent cross-polarization characteristics. Additionally, the position of the resulting AQ lobes will be equidistant (in the angular space) from the main lobe. In order to further decrease the AQ lobes, a number of different subarray lattices can be used; this approach decreases the maximum level of the AQ lobes—see Section 3.12.2.

As can be seen, these approaches offer a plethora of choices to the designer who often has to meet a diverse set of requirements.

3.7 ADDITIONAL QUANTIZATION ERRORS

We have already explored the consequences of having amplitude quantization errors in the illumination patterns of phased arrays in the previous section. Here we shall consider the consequences of phase and time-delay quantization errors normally encountered in phased arrays; and in the subsequent sections we shall deal with random errors due to the imperfect manufacture of the many diverse phase array components and subsystems.

In broad terms quantization errors can be likened to the systematic errors of filled apertures (only because of their severity), while random errors give rise to increased sidelobe levels, broaden the main beam and decrease the array gain. Our treatment, however, will be brief because these topics have been explored in some detail in many journal and symposium articles and in a recent book [3.2].

3.7.1 Phase Quantization

Modern phased arrays utilize digital phase shifters to steer the beam anywhere within the array FOV and the least significant bit of digital N-bit phase shifters ϕ_0 is given by the equation

$$\phi_0 = \frac{2\pi}{2^N} \qquad (3.102)$$

Figure 3.24 shows the desired phase between element $-m$ and element m of a phased array as a diagonal straight line and the actual phase shift implemented by digital phase-shifters as the staircase approximation of the desired phase shift. The net phase error is also shown as a sawtooth along the horizontal axis of the graph. To a first approximation the average sidelobe level due to the quantization error alone is given as [3.47, 3.49]

$$\overline{\sigma^2} = \frac{1}{3N} \frac{\pi^2}{2^{2N}} \qquad (3.103)$$

Figure 3.25 illustrates the rms sidelobe level in decibels of a uniformly illuminated phased array which is 2 dB lower than that defined by Equation (3.102) is a function of N and the number of array elements.

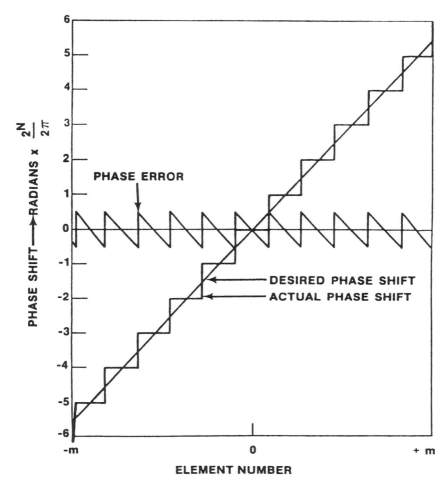

FIGURE 3.24 The desired phase-shift of an array across its elements ($-m$ to m) is shown by a straight line; the actual phase shift is shown as a staircase approximation the straight line. The net phase error is also shown as a sawtooth along the horizontal axis of the graph (from [3.49]).

More detailed considerations led Mailloux [3.47] to conclude that these quantization errors give rise to grating lobes, similar in shape as those due to AQ errors without the characteristic split of the latter lobes.

Let us consider that an array required to point toward the direction cosine u_0 points toward the u_s because the digital phase shifters can approximate only the required phased shifts.

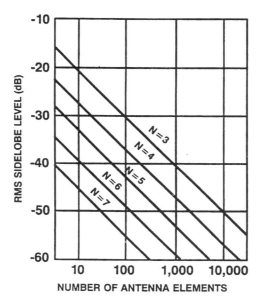

FIGURE 3.25 The rms sidelobe level of a uniformly illuminated phase array as a function of the N-bit phase shifters used and the number of antenna elements (from [3.49]).

The worst-case scenario occurs when the error buildup is entirely periodic, in which case the array is divided into virtual subarrays and each subarray has the same phase error gradient. This occurs when the distance between subarray centers is given by the equation [3.47]

$$|M\Delta\phi - M\Delta\phi_s| = \frac{2\pi d}{\lambda} M|u_0 - u_s| = \frac{2\pi}{2^N} \quad (3.104)$$

where M is the number of antenna elements in the virtual subarray and d is the interelement spacing. If $u_0 - u_s$ and N are known, Equation (3.104) can be used to deduce M. Subarraying would induce grating lobes to form, and the power in each lobe $P_{\Phi,\mathrm{GL}}$ is [3.47]

$$P_{\Phi,\mathrm{GL}} = \left[\frac{\pi/2^N}{M\sin(p'\pi/M)}\right]^2 \quad (3.105)$$

where $p' = p + 1/2^N$ and $p = \pm 1, \pm 2, \pm 3, \ldots$. In Table 3.11 we have tabulated the power in the first grating lobe as a function of M and N; as can

ADDITIONAL QUANTIZATION ERRORS

TABLE 3.11 The First Grating Lobe Power [in −dB (Reciprocal Decibels)] as a Function of N and M

N	M = 2	3	4	5	6	7	8	9	10
3	13.97	16.99	17.92	18.34	18.57	18.71	18.80	18.86	18.90
4	20.11	22.73	23.57	23.95	24.15	24.27	24.35	24.40	24.45
5	26.17	28.61	29.39	29.75	29.94	30.06	30.13	30.18	30.44
6	32.20	34.55	35.31	35.66	35.84	36.00	36.03	36.07	36.11

be seen, it can reach a high level of approximately -14 dB, when $M = 2$ and $N = 3$. As N is increased, however, the grating lobe power decreases dramatically. The power in the second grating lobe will be even lower.

Miller [3.49] also proposed methods to break up the periodicity of the quantization error and derived an expression for the beam deviation caused by periodic phase-shifter quantization. Again for a uniformly illuminated phased array the beam deviation δ as a function of the array beamwidth is given by [3.49]

$$\delta = \frac{\pi}{4} \frac{1}{2^N} \quad \text{beamwidths} \tag{3.106}$$

3.7.2 Time Delay Quantization

In a large array operating over a small fractional bandwidth, it may be advantageous and convenient to use time delays to steer the resulting beam at the subarray level and phase shifters at the element level; while convenient and economical, this arrangement again gives rise to grating lobes and the resulting grating lobe power, $P_D|_{GP}$, is given by the equation [3.47]

$$P_D|_{GP} = \frac{\pi^2 X^2}{\sin^2 \pi [X + p/M]} \tag{3.107}$$

where

$$X = \frac{u_0 d}{\lambda_0} \frac{\Delta f}{f_0} \tag{3.108}$$

Typical values of $\Delta f/f_0$, are 0.1, or less; the array fractional bandwidth has to be very small for the grating lobes to be negligible.

3.7.3 Applications: The Utilization of Phase and Time Delay in Large Phased Arrays

The COBRA DANE phased array-based radar uses time delays at the subarray level and phase shifters at the antenna elements [3.43]. The radar operates at L-band and its 95-ft aperture is divided into 96 subarrays, each having 160 antenna elements.

If the array is pointed 22° in elevation off the aperture boresight, the top subarray receives a 36-ns time delay with respect to the bottom subarray. A 200-MHz waveform can therefore provide a 2.5-ft range resolution [3.43].

3.8 RANDOM ERRORS

We have already considered the effects of random phase errors on the surface of filled apertures; as these errors increase (or the frequency of operation increases) the main beam broadens, the antenna gain decreases while its sidelobe level increases. Would you be surprised to read that a parallel situation exists for phased arrays? Indeed not, only the nature and causes of the random errors are different.

For filled apertures the phase errors are due to the manufacture of the reflector's surface and its supports, while for phased arrays we encounter phase and amplitude errors associated with the antenna feeds, the T/R modules, and the beamformers. It is well-known that it is too expensive to realize thousands of components and subsystems with negligible phase and amplitude variations.

The designer therefore launched a two pronged attack to realize affordable phased arrays:

1. The production of high yield T/R modules and subsystems.
2. The modules and subsystems are manufactured to tolerances that decrease monotonically with the passage of time.

These two aims are simultaneously satisfied by MMICs (monolithic microwave integrated circuits) which we shall consider in the next chapter.

Extensive studies were undertaken by Ruze [3.50], Elliott [3.51], Allen [3.52] Skolnik [3.53], Moody [3.54], and Cheston [3.55] to derive the impact of random array errors on sidelobe level, pointing error and directivity of phased arrays. Furthermore Hsiao [3.56, 3.57] and Kaplan [3.58] derived convenient curves of peak sidelobe level as a function of array parameters.

Random errors associated with active and passive components and subsystems add a random component to the theoretically designed sidelobe level. The "noise-like" addition is assumed to be Gaussian and to have a uniform variance over scan-angle in the planes of interest.

Phase and amplitude errors occurring at different parts of a phased array, direct a fraction of the available energy from the main beam and distribute it to the sidelobes. For small independent random errors, this small fraction σ_T is given by the equation

$$\sigma_T^2 = \sigma_\Phi^2|_{net} + \sigma_\alpha^2|_{net} \tag{3.109}$$

where $\sigma_\Phi|_{net}$ and $\sigma_\alpha|_{net}$ the net rms phase and amplitude errors, in radians and in ratios, respectively, are given by

$$\sigma_\Phi|_{net} = \left[\sigma_{\Phi_1}^2 + \sigma_{\Phi_2}^2 + \sigma_{\Phi_3}^2 + \cdots\right]^{1/2} \tag{3.110}$$

and

$$\sigma_\alpha|_{net} = \left[\sigma_{\alpha_1}^2 + \sigma_{\alpha_2}^2 + \sigma_{\alpha_3}^2 + \cdots\right]^{1/2} \tag{3.111}$$

here $\sigma_{\Phi_1}, \sigma_{\Phi_2}, \sigma_{\Phi_3}, \ldots, \sigma_{\alpha_1}, \sigma_{\alpha_2}, \sigma_{\alpha_3} \ldots$ are the rms phase and amplitude errors, respectively, occurring in different parts of the phased array such as the antenna feeds, the T/R modules, or the beamforming networks.

The energy due to amplitude and phase errors is radiated into the far field with the gain of the antenna element pattern. Phase and amplitude errors in different subsystems of a phased array are often expressed as $\pm X$ degrees and $\pm Y$ dB; suitable conversions are therefore needed before one can use Equations (3.109) to (3.111).

It has been shown that for a uniform error density function spread over an interval I, the rms error is $I/(12)^{1/2}$. So if a subsystem of a phase array, for example, the T/R modules, has phase errors defined as $\pm 2°$, the range is $4°$ or 0.0698132 radians or $0.0698132/(12)^{1/2} = 0.02$ rms radians. Similarly we require to convert a range of say $\pm Y$ dB to σ_α with aid of equation

$$\sigma_\alpha = \frac{10^{\alpha_+/20} - 10^{\alpha_-/20}}{\sqrt{12}} \tag{3.112}$$

where α_+ and α_- are the upper and lower values of the amplitude tolerance in decibels. Thus a tolerance of ± 1 dB will convert into an rms ratio of 0.0666.

To determine the mean-squared-sidelobe-level, MSLL, it is necessary to compare the energy due to errors with the peak of the pattern of the array of N antenna elements. More explicitly the MSSL is given as [3.55]

$$\text{MSSL} = \frac{\sigma_T^2}{\eta N(1 - \sigma_T^2)} \tag{3.113}$$

where η is the aperture efficiency of the array. The array will therefore have a mean floor of random sidelobes which on the average is $10\log(\mathrm{MSSL})$ dB below the peak of the beam.

If we assign the probability of a failed antenna element as $(1 - P)$, the MSSL is modified to take the form

$$\mathrm{MSSL} = \frac{(1 - P) + \sigma_\alpha^2|_{\mathrm{net}} + P\sigma_\Phi^2|_{\mathrm{net}}}{\eta PN} \quad (3.114)$$

If, $P = 1$ (no failed elements), Equation (3.114) reverts to (3.113), but for the bracketed term in Equation (3.113), which is not significant for low sidelobe arrays.

3.8.1 Realistic Examples of the Impact of Random Errors on MSSL

In this section we shall explore the impact of realistic phase and amplitude errors on the MSSL. Here we shall assume that amplitude and phase errors apportioned to current and future T/R modules are ± 1 dB and $\pm 5°$ and ± 0.25 dB and $\pm 3°$, respectively [3.58 and 3.59]. We also assigned the same errors to the beamformer [3.58] for illustrative purposes. The resulting MSSL levels, for the following cases have been computed with the aid of Equation (3.114):

1. Errors due to current T/R modules and beamformers are taken into consideration; and
2. Only errors due to current T/R modules are taken into consideration.
3. Same as above (1 and 2), but errors due to future T/R modules are considered; we also assumed the same tolerances for future beamformers.
4. The above calculations are performed for the cases where $N = 100$, 400, 1000, and 4000.

An aperture efficiency of 0.7 has been assumed for all calculations.

The results are tabulated in Table 3.12 as a function of antenna elements N; as can be seen the MSSL varies from -37 dB to -63.7. If more errors occur in other parts of the array the resulting MSSL values will further increase. These examples illustrate the importance of keeping errors in the many parts of the array as low as possible. MMICs hold the promise of meeting these challenges.

The other point Table 3.12 illustrates is the dependence of the MSSL level on the number of the array antenna elements N. When N is low, tolerances have to be kept tight if the required MSSL is low.

TABLE 3.12 The Derived MSSL as a Function of Antenna Elements and Random Amplitude and Phase Errors

N	$\alpha_1 \pm$	$\alpha_2 \pm$	$\phi_1 \pm$	$\phi_2 \pm$	MSSL {dB}
100	1	1	5	5	−37
	1	0	5	0	−40
	0.25	0.25	3	3	−44.7
	0.25	0	3	0	−47.7
400	1	1	5	5	−43
	1	0	5	0	−46
	0.25	0.25	3	3	−50.7
	0.25	0	3	0	−53.7
1000	1	1	5	5	−47
	1	0	5	0	−50
	0.25	0.25	3	3	−54.7
	0.25	0	3	0	−57.7
4000	1	1	5	5	−53
	1	0	5	0	−56
	0.25	0.25	3	3	−60.7
	0.25	0	3	0	−63.7

3.8.2 The Statistical Prediction of Peak Sidelobes

Let us revisit an ideal phased array that has no random errors and is illuminated by a symmetrical function; its far-field radiation is a sum of cosine functions. We can then consider a realistic array that has random phase and amplitude errors. These errors will be responsible for raising a mean floor of random sidelobes having a uniform average power in angular space when considered in correlation intervals of approximately the array HPBW; this component can be likened to Gaussian noise.

The radiation pattern of the ideal array can be considered as the "signal," which consists of a sum of cosines and is "unfluctuating." At the array nulls, the uniform average power due to phase and amplitude errors can be seen.

This situation is depicted in Figure 3.26, where the array residual, designed and specified (threshold) levels are shown. Thus sidelobes in all angular space have two components: one mathematically described, and a residue that is random in nature.

Hsiao [3.56] utilized the theoretical framework established by Rice [3.60] to describe statistically sidelobes that pop-up above the specified or threshold level; the key equation is [3.58]

$$p\left[\frac{E}{R}\right] = \frac{E}{R^2} \exp\left[-\frac{(E^2 + P^2)}{2R^2}\right] I_0\left(\frac{EP}{R^2}\right) \quad (3.115)$$

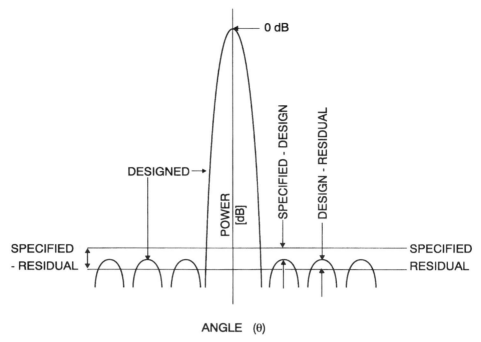

FIGURE 3.26 The radiation pattern of a phased array where the residual, designed, and specified (threshold) levels are shown.

where E is the ensemble of design plus random variable, R is the rms random component, P is the peak design component, and I_0 is the modified Bessel function of zero order.

A useful graph due to Kaplan, reproduced in Figure 3.27, illustrates the options designers and manufacturers have. The graph illustrates the relationship between the specified : design sidelobes ratio and the specified : residual sidelobes ratio as a function of probabilities ranging from 0.85 to 0.95.

It can be seen that one can attain the same probability of popups by overdesigning to attain a high specified : residual sidelobes ratio (and a low specified : design ratio) or by overdesigning to attain a high specified : design ratio (and a relatively low specified : residual ratio). In the case of 0.9 probability the respective ratios are 15, 2.5 dB and 11, 8 dB. In between these two extremes there are many tradeoffs that the manufacturer in consultation with the designer can consider.

The impact of errors on the array directivity and beam pointing error have been considered by Skolnik [3.53] and Steinberg [3.61]; for large phased arrays the impact of random errors on the directivity and pointing errors is not significant.

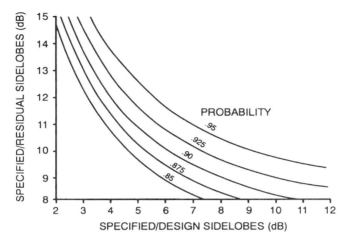

FIGURE 3.27 The relationship between the specified: designed sidelobe ratio and the specified: residual sidelobe ratio as a function of probabilities ranging from 0.85 to 0.95. (*Source*: Kaplan [3.58], © 1986 *Microwave J.*)

3.9 ACTIVE AND PASSIVE PHASED ARRAYS

In the interest of keeping the transmission losses to a minimum, the most important active circuits of an array, such as the LNA (low-noise amplifier) and the FPA (final power amplifier), should be positioned as close as possible to the antenna elements of the array. In a typical system, which consists of the subsystems shown in Figure 3.28, the system noise figure F_s can be deduced from the equation

$$F_s = F_A + \frac{L_M L_A F_R - 1}{G_A} \qquad (3.116)$$

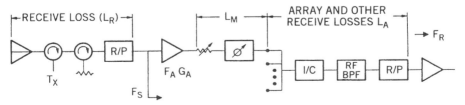

FIGURE 3.28 The block diagram of a typical front-end of a phased array T/R module. (*Source*: [3.62], © 1991 IEEE, McQuiddy et al.)

where F_A and G_A, respectively, are the noise figure and gain of the LNA and L_M represents the losses incurred between the output of the LNA (manifold losses) and the input of the remaining parts of the receiver; L_A shown in the figure includes array and other losses and F_R represents the noise figure of the remaining parts of the receiver. For a typical module operating at X-band where $L_M L_A = 11$ dB, $F_R = F_A = L_R = 2$ dB, and $G_A = 17$ dB, one can deduce that $T_S = 2.93$ dB [3.62].

If the gain of the LNA G_A is in the range of 20–30 dB, the noise figure of the LNA defines the system noise figure, provided F_R is not too high. Similarly, any losses between the antenna elements and the FPA will directly attenuate the signals transmitted.

If costs can be ignored, the fully active phased array would have been the most popular choice among designers because a variety of signal processing functions can be performed after the LNA without any degradation of the array sensitivity.

In Figure 3.29 the block diagrams of the active (*a*), passive (*b*), and hybrid (*c*) array are shown; the term *hybrid* in the present context means that the array is active on the receive mode and passive on the transmit mode. Hybrid arrays that are active on transmit and passive on receive are also possible. The term *hybrid* is likely to introduce some confusion. In the Frank–Coffman proposal an active hybrid array is an array, the aperture of which is divided into two areas: (1) the central area where each antenna element is connected to a T/R module and (2) an annulus that is populated by sets of antenna elements, with each set connected to one T/R module. In the example cited by Frank and Coffman, eight antenna elements were connected to one T/R module. The term *active hybrid* therefore reflects the situation where the array aperture is populated by active elements in the inner part of the aperture and passive in the outer part of the aperture. In this section a *hybrid* array is an array that is active on receive and passive on transmit or vice versa.

Every antenna element of an active array, illustrated in Figure 3.29*a*, is followed by a T/R module; hence the losses incurred between the receiver/transmitter and the antenna elements are minimal.

The fully passive array, illustrated in Figure 3.29*b*, is a low-cost solution, but the designer has to carefully consider the losses between the antenna elements and the LNA/FPA. The passive array option has a low-loss, programmable, phase shifter after each antenna element that is used to implement the scanning–beamforming function of the array.

Similarly, the hybrid array, illustrated in Figure 3.29*c*, has to be examined in some detail to ascertain that the appropriate balance between losses and costs is reached.

If one aims at a low-cost array, then the choices are between a fixed-beam array or a scannable array. The first array, which is a substitute for a

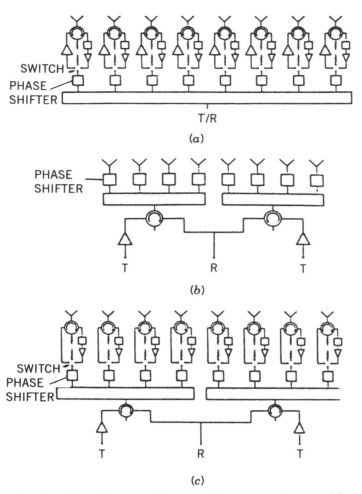

FIGURE 3.29 The block diagram of three archetype phased arrays: (*a*) active; (*b*) passive; and (*c*) hybrid. (*Source*: Tang and Burns [3.62a], © 1992 IEEE.)

monolithic antenna required to have one antenna beam pointed toward the direction of a communication node, such as a geostationary satellite, cannot be classified as a phased array but is included here for completeness.

In many references the advantages of active phased arrays have been emphasized. In what follows we shall support the view that although active phased arrays are very attractive, passive and hybrid phased arrays are worth considering for some applications.

204 PHASED ARRAYS: CANONICAL AND WIDEBAND

3.9.1 The Radar Equation Revisited

There is a need to recast the monostatic radar equations derived in Chapter 1 for the case where a phased array is used instead of a filled aperture.

Let us assume that:

1. The total number of antenna elements is N and the gain of each element G_e.
2. P_A is the average power for each T/R module for an active phased array.
3. P_P is the average power for a passive phased array.
4. L_A and L_P are the losses incurred in transmission including propagation for an active and a passive phased array.
5. T_{AC} and T_P are the system noise temperatures in degrees Kelvin for an active and a passive phased array.

If SNR_A and SNR_P are the SNRs for the active and passive array respectively, the range for the active array is given by the equation

$$R_A^4 = \frac{P_A L_A G_e^2 \lambda^2 N^3 \eta^2 \sigma}{(4\pi)^3 k T_{AC} B \, SNR_A}$$

or

$$R_A^4 \propto \frac{P_A L_A G_e^2 N^3}{T_{AC}} \quad (3.117)$$

if we assume that the aperture efficiency is the same on transmit/receive mode and that B, SNR_A, η, λ, and σ are constants.

For the passive array the R_P is given by the equation

$$R_P^4 = \frac{P_P L_P G_e^2 \lambda^2 N^2 \eta^2 \sigma}{(4\pi)^3 k T_P B \, SNR_P}$$

or

$$R_P^4 \propto \frac{P_P L_P G_e^2 N^2}{T_P} \quad (3.118)$$

Often passive arrays utilize tubes as transmitters while active arrays utilize solid-state transmitters. In these derivations we have assumed that one powerful tube transmitter is used in the transmit mode, so the passive array has no graceful degradation properties. Similar considerations hold for the array when it operates in the receive mode.

Some comparisons between Equations (3.117) and (3.118) are worth considering here; if we adopt the assumption that passive arrays utilize tubes and active arrays utilize solid-state devices, $P_P \gg P_A$ but P_P can be equal to $P_A N$ at cm wavelengths. At mm wavelengths, however, it is often the case that $P_P \gg P_A N$—see Chapter 4. Unless an unforseen breakthrough in high-power solid-state devices occurs, this situation will persist for some time.

For an active phased array the maximization of the range can be attained by the following approaches:

1. An increase in N the number of antenna elements or T/R modules. A small increment in N will result in a substantial increment in the range. Given that costs ultimately limit the number of T/R modules a designer can use, there is a strong incentive to decrease the cost per module. Other considerations, including the available area for the phased array and the resulting array beamwidth, have to be taken into account when N increases.

2. The terms related to power and antenna element gain can be rearranged in terms of the product of the effective isotropic radiated power (EIRP) and the G/T of the array or $\{(P_A N) \times (G_e N)\} \times \{G_e N / T_{AC}\}$. From this rearrangement one can easily deduce the importance of increasing the module's average power (or at least the efficiency of modules) and the antenna element gain. Considerable effort has been expended in increasing the module's average power and efficiency—see Chapter 4. It is possible to increase the gain of the antenna elements by decreasing the array's FOV; however, operational requirements define the FOV of an array, not technological issues.

3. Considerable effort has been expended toward the minimization of the losses incurred between the antenna elements and the LNA—see Chapter 5.

4. Given that the module noise temperature defines the system's noise temperature, considerable effort has been expended toward lowering the module's noise temperature; this is an area where solid-state LNAs reign supreme at cm and mm wavelengths, when compared to their vacuum-tube counterparts.

3.9.2 The EIRP and G/T of Passive and Active Phased Arrays

While the losses L_A for an active phased array are negligible, the losses for passive phased arrays L_P will effectively modify the aperture gain and the effective system noise figure of the passive phased array. Our considerations will be focused on the EIRP and G/T ratio of both active and passive phased arrays.

The SNR of an active phased array is proportional to

$$[\text{EIRP}_A] \times \frac{G_A}{T_A}, \quad \text{where} \quad \text{EIRP}_A = G_A(P_A \times N) \quad \text{and} \quad G_A = G_e \times N$$

Similarly, the SNR of a passive phased array is proportional to

$$[\text{EIRP}_\text{P}] \times \frac{G_\text{P}}{T_\text{P}}, \quad \text{where} \quad \text{EIRP}_\text{P} = G_\text{P} P_\text{P} \quad \text{and} \quad G_\text{P} = G_\text{A} \times (\text{losses})$$

We need to calculate the G/T of the passive and active arrays. For an active array populated by half-wave dipoles, the gain is given by the equation

$$G_\text{A} = \frac{4\pi A_\text{eff}}{\lambda^2} N = \frac{4\pi(\lambda/2)(\lambda/2)}{\lambda^2} N = \pi N \quad (3.119)$$

The effective area of the antenna element is here assumed to be $\lambda/2$ by $\lambda/2$. A passive array will have the same gain but for the losses of the interconnecting transmission lines and those attributed to power dividers.

The line lengths for the square array of patch antennas can be approximated by $(N^{1/2} - 1)d/\lambda$ where d is the interelement spacing in both dimensions [3.63]. Here we have ignored the losses due to the power dividers because these losses are not as high as the line losses; similarly we shall ignore the losses due to the phase shifters because one can add the latter losses to the former losses.

At broadside the gain G_p, for a square aperture having N antenna elements and $\lambda/2$ spacing is given by the equation

$$G_\text{P} = N\pi 10^{-\gamma} \quad (3.120\text{a})$$

where

$$\gamma = (d/\lambda)(N^{1/2} - 1)\frac{\alpha_{\text{dB}/\lambda}}{10} \quad (3.120\text{b})$$

and $\alpha_{\text{dB}/\lambda}$ is the line loss per unit wavelength. Typical losses incurred at millimeter-range wavelengths are 0.065 and 0.25 dB/λ for copper transmission lines on quartz and on GaAs, respectively [3.63].

In Figure 3.30 we illustrate the gain of an array when N takes the values $N = 10$ to $1{,}000{,}000$ for the cases where $\alpha = 0$, 0.065, 0.25, 1, 2, and 3 dB/λ. As N increases, the gain of the passive array reaches a maximum before it decreases significantly.

The case for active arrays, where α is equal to zero is overwhelming; as N takes high values, the gain of the passive array decreases. If the losses due to the phase shifters are taken into account, the gain of the phased array, decreases further.

If T_A is the equivalent antenna temperature of a phased array pointed towards a direction, θ, ϕ, and T_a is the effective antenna temperature seen at

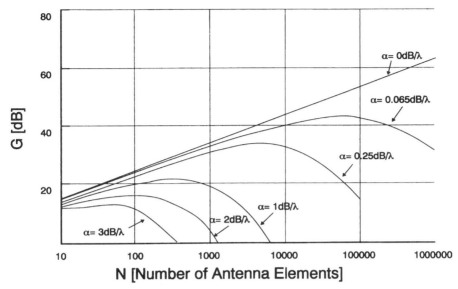

FIGURE 3.30 The gain of active and passive phased arrays as N takes the values from 10 to 1,000,000. For the active phased array the loss α in dB/λ is zero, while for the passive phased arrays $\alpha = 0.065$ and 0.25, 1, 2, and 3 dB/λ.

the receiver input terminals, then the system temperature T_s is given by the equations

$$T_s|_P = \varepsilon_L T_A + T_0(1 - \varepsilon_L) + (F_R - 1)T_0 \qquad (3.121)$$

$$(T_a = \varepsilon_L T_A + T_0(1 - \varepsilon_L))$$

for a passive array and

$$T_s|_A = T_A + (F - 1)T_0 \qquad (3.122)$$

$$(T_a = T_A)$$

for an active array. F_R and F are the noise figures for the receivers in the passive and active arrays, respectively, and ε_L is the feed loss for the passive array. As it can be seen both $T_s|_P$ and $T_s|_A$ are independent of N.

3.9.3 Design Options and Applications

The active array design option can be seen as an extreme design option where the losses are minimum and the cost is highest; this is because the T/R modules are at present expensive and the array cost is proportional to N^3. Let us explore other design options.

If we revisit Figure 3.30, we observe that there is clearly no point in increasing N beyond the point where the array gain for a given value of α is maximum. This condition can be deduced by setting $dG/dN = 0$, which yields

$$N_{\max} = \left[\frac{40}{\alpha \ln 10}\right]^2 \qquad (3.123)$$

Thus when $\alpha = 0.25$, the maximum array gain is attained when $N_{\max} = 4828$; hence, there is no point in increasing the number of antenna elements beyond $N_{\max}$, because the gain will decrease. For a given α value, passive arrays having $N_{\max}$ therefore represent an economical solution that has reasonable losses.

To explore other design options for passive array, we require minimum line losses but high N values. This design objective is met if we subdivide the array into P subarrays of Q antenna elements where $P \times Q = N$.

The COBRA DANE phased array-based radar has 96 subarrays, and each subarray has 160 antenna elements. Each traveling-wave tube (TWT) feeds the 160 antenna elements via a 1:160 power divider [3.43]; with this arrangement the losses are kept to a reasonable level.

Recently the design options have been augmented by the arrival of compact mini-TWTs and microwave power modules that utilize solid-state medium-power amplifiers and miniature high-power vacuum-tube amplifiers—see Chapter 4. Smaller-size subarrays can therefore be fed by either mini-TWTs or microwave power modules, and the losses are further minimized; similar arguments hold for the receive case. Furthermore, this design option offers the designer an improved graceful degradation.

To explore other design options, we have to turn to technological issues; another design option is based on the following issues:

1. The cost of the T/R modules constitutes approximately 50% of the array cost—see Section 3.11.1.
2. The cost of the FPA in a module constitutes a substantial portion of the module cost.
3. The power output of solid-state transmitters at mm wavelengths is minuscule when compared to that attainable from tube transmitters (see Chapter 4); this situation is likely to persist unless an unexpected technological development in solid-state devices takes place.

From the foregoing considerations we can conclude that hybrid architectures, illustrated in Figure 3.29c, which are active-on-receive and passive-on-transmit are worth serious consideration, at least when operation at mm wavelengths is contemplated. This design option is economical, because it has the minimum number of expensive tube transmitters, low losses, and graceful degradation. More importantly, this option results in phased arrays that emanate substantial powers at mm wavelengths. These options, taken in conjunction with the options outlined in Section 3.6, constitute a formidable set.

Finally the losses associated with passive arrays have been considered in some detail in Levine et al. [3.64] because of their importance. The arrays considered were receive-only passive arrays operating at 10 GHz and had an interelement spacing of 0.8λ. The number of array elements was 16, 64, 1024, or 4096, and the resulting gains are tabulated in Table 3.13. The phase array consists of sets of patch antennas, and Figure 3.31 illustrates the cases where $N = 16$ and 64.

All the possible losses, to be explored in Chapter 5, are itemized and tabulated together with the gain associated with a dish. As can be seen, the gain of the passive phased array equals that of the dish when $N = 256$. When $N > 256$ the dish gain is higher. These results are not to be considered as definitive, but illustrative of what gains passive arrays utilizing patch antenna elements interconnected in the way illustrated in Figure 3.31 can offer. Another way of looking at these results is that they form a useful baseline for future improvements.

Given that the array described above is not steerable, phase-shifter losses are not considered. Typical losses for operation at cm wavelengths are in the vicinity of 1 dB [3.65]. At mm wavelengths the same losses are in the neighborhood of 2–3 dB for PIN (positive–intrinsic–negative) diode-based phase shifters and in the vicinity of 8–10 dB for GaAs FET (field-effect transistor) based phased shifters [3.63].

Losses therefore deserve detailed studies; furthermore, considerable efforts have been expended towards their minimization.

TABLE 3.13 Calculated Directivities, Gains, and Losses of Passive Arrays as a Function of the Number of Antenna Elements [3.64][a]

Number of Elements	16	64	256	1024	4096
Directivity without network	20.9	27	33	39.2	45.1
Radiation loss	0.8	1	1.3	1.9	2.6
Surface wave loss	0.3	0.3	0.2	0.2	0.1
Dielectric loss	0.1	0.3	0.5	1	2.1
Ohmic loss	0.1	0.3	0.6	1.2	2.4
Connector loss	0.2	0.2	0.2	0.2	0.2
Calculated gain	19.5	25	30	34.5	37.5
Gain of a dish	18	24	30	36	42

[a] All gains and losses are in decibels.

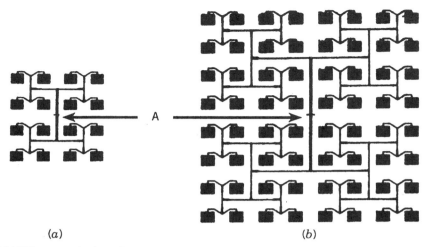

FIGURE 3.31 A phased array utilizing N patch antennas connected to a point A by equal lengths on transmission lines. (*Source*: Levine et al [3.64], © 1989 IEEE.) (*a*) $N = 16$; and (*b*) $N = 64$.

From the foregoing outline of array design options the following conclusions can be drawn:

1. A thorough knowledge of the capabilities of solid-state-based and tube-based T/R modules allows the designers to multiply their options; from this process the solution that meets their array requirements and funds can be selected.
2. Solid-state receivers are relatively inexpensive and remain unchallenged by their tube counterparts at all frequencies.
3. The options for transmitters operating at different frequency bands are not as clear-cut and the designer needs guidance as to when to select solid-state-based transmitters or their tube counterparts; we shall explore this and related issues in Chapter 4.
4. The transmission line and component losses are of critical importance for passive arrays. While the designer can derive some general guidelines from theoretical considerations we have undertaken in this section, a thorough knowledge of the nature of these losses and approaches to minimize them is essential; we shall explore these issues in Chapter 5.

3.10 ARRAY ARCHITECTURES

The array requirements often dictate the array architectures; here we shall consider only the generic architectures from which one can derive variants. Initially we shall explore the front-end architectures before we move on to the beamforming architectures. In all cases we shall outline the regime of applicability for each architecture.

A modern phased array has between 10^3 and 10^5 antenna elements that occupy a certain area; the antenna elements are followed by electronic subsystems, including phase shifters, low-noise amplifiers, transmitters, beamformers, distributed control circuits, and power supplies.

With reference to the radar equation, the array area is determined by the array power aperture product and beamwidth required; the transmitted power, array architecture, sensitivity, scanning requirements, and most importantly, the form of the transmission lines take to distribute signals in an array —determine the volume of the array. There are many ways of filling the required volume; here we shall adopt a hierarchical approach to the options available to the designer and outline the advantages and shortcomings for each option.

The availability of the required real estate is a paramount requirement. The real estate requirements include space for the radiating elements, such as dipoles occupying an area of about $N \times \lambda/2 \times \lambda/2$, feed networks, and the active/passive circuitry of T/R modules, and beamformers.

3.10.1 The Two Basic Array Architectures

With reference to Figure 3.32, there are two basic architectures for phased arrays. We shall briefly describe the two architectures before we consider the options the designer has in interconnecting the array antenna elements and the issues associated with the interconnect options. We shall refer to these two array architectures as brick and tile architectures illustrated in Figure 3.32a and 3.32b, respectively [3.63]. In Section 3.11.3 we shall explore the regimes of application for both these architectures.

3.10.1.1 Brick Architecture The brick architecture, is also known as LITA (Longitudinal Integration and Transverse Assembly); in this architecture the antenna elements and the other subsystems have an orientation that is parallel to the longitudinal direction of the array and are stacked along a transverse direction.

With this architecture the power received by the antenna elements is either coupled electromagnetically or by electric connections to the remaining stages of signal processing blocks assigned to each antenna element.

3.10.1.2 Tile Architecture The tile architecture, is also known as the TILA (Transverse Integration and Longitudinal Assembly); in this architecture the antenna elements and other subsystems, for example, phase shifters, LNAs and FPAs have an orientation that is parallel to the transverse array aperture and are stacked along a longitudinal direction.

With this architecture the power received by the antenna elements located at the top of the tile structure is usually electromagnetically coupled to the

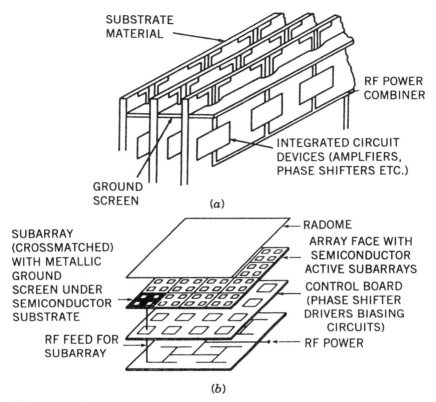

FIGURE 3.32 Phased array architectures: (*a*) tile and (*b*) brick architectures. (*Source*: Mailloux [3.63], © 1992 IEEE.)

second layer, which can consist of phase shifters or active elements depending on whether the array is active or passive. Similarly the third and other layers follow in order to complete the remaining signal processing functions.

3.10.2 Interconnect Approaches

In what follows we shall consider the many ways the phased array antenna elements are interconnected after the designer decides whether:

1. The array is to be active, passive or hybrid.
2. The array is to have a tile or brick architecture.

The interconnect approaches are necessarily of generic nature and many variants of the approaches are possible.

With reference to Figure 3.33 the designer has four basic options:

1. The planar corporate feed, illustrated in Figure 3.33a, is suitable for the tile architecture. Although the array antenna elements shown are dipoles, equivalent interconnect approaches exist for arrays utilizing conventional patch antenna elements and electromagnetically coupled patch antennas. The lengths of lines between the antenna elements and

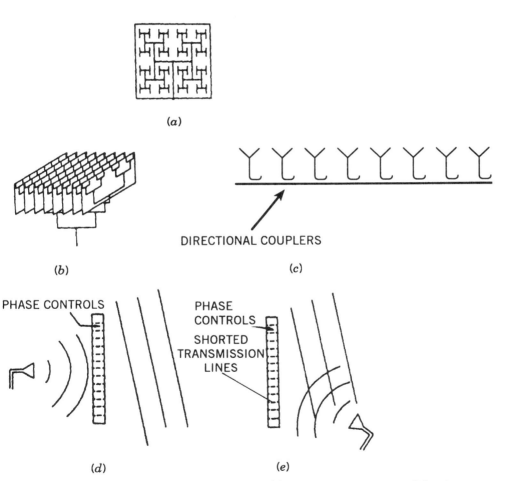

FIGURE 3.33 Five interconnect approaches: (a) planar corporate feed; (b) volume corporate feed; (c) series-fed line array; (d) space-fed array—lens type; and (e) space-fed array—reflect-array type. (*Source*: Mailloux [3.63], © 1992 IEEE.)

the summing point are equal and the incurred losses, have to be considered. If the resulting beam is to be steerable, a programmable phase shifter or a T/R module is inserted between the antenna element and the summing point.

2. The volume corporate feed, illustrated in Figure 3.33*b*, is suitable for the brick architecture. For a passive array, programmable phase shifters are connected in series with each antenna element. This is by far the most popular interconnect approach because a planar phased array can consist of several columns, each having a different length so that the resulting aperture has the required geometric shape.

3. The series or source feed, illustrated in Figure 3.33*c*, appeared in many waveguide and microstrip realizations forming rows or columns of a planar phased array.

4. The space-fed architecture has two realizations: the lens-fed and the reflect array illustrated in Figures 3.33*d* and 3.33*e*, respectively. In the lens-fed version both faces of the array are populated by pairs of antenna elements, and each pair, located on either side of the rectangle shown, is connected to a programmable phase shifter (or a T/R module). In the transmit mode the radiation emanated from the feed horn is received by the antenna elements closest to the feed horn and the appropriate settings of the phase shifters steer the beam to the required direction. Similarly, the returned signals are collimated on to the feed horn, which is usually followed by a T/R module; with this arrangement, the incoming (or outgoing radiation) is not obstructed. Given that the programmable phase shifters take into account the differences in path lengths between the feed shown and the pairs of antenna elements along the length of the rectangle, the rectangle acts as a lens. The resulting array is simpler than its more conventional counterparts but volumetrically unattractive for applications where space is at a premium, as in airborne applications. The reflect array shown in Figure 3.33*e* operates on the same principle as the lens-fed version, except that the programmable phase shifters are terminated in short-circuited transmission lines. The offset geometry of the feed horn is chosen here to minimize the obstruction of the incoming/outgoing radiation; the arrangement shown resembles the geometry of an offset paraboloid.

The Patriot radar, shown in Figure 3.34, is an example of a space-fed radar array [3.66]. The radar is a C-band multifunction, mobile radar designed for tactical air-defense missions. The radar was effective during the Gulf War and enjoyed wide publicity in the press.

FIGURE 3.34 The patriot C-band phased array radar: an example of a space-fed radar array. (*Courtesy*: Raytheon.)

3.11 ARRAY DESIGN CONSIDERATIONS

The array designer should address many issues and tradeoffs in a heuristic manner before detailed designs are undertaken; this is justified on the grounds that many factors are interdependent and requirements can often be met in a variety of ways. In this section we shall broadly outline the many considerations the array designer will explore before arriving at designs worth exploring in detail.

3.11.1 Array Costs

Cohen [3.67] provides us with some interesting statistics: "... More than 80% of existing deployed radar systems are of the mechanically steered variety and of the phase steered systems, all but a very few are of the passive class (i.e., tube-based transmitters)." Furthermore, he points out that the acquisition cost of the active electronically steered array, AESA, is its single most significant problem. Let us conveniently divide the costs into acquisition and life-cycle LCC and explore the former cost. In the following considerations it is clear that future developments, often unforeseen, will necessarily modify the cost estimates that follow. Nonetheless we should outline in broad terms the typical costs for guidance to the designer.

As a baseline comparison, the cost of a forward air control system, the mechanically scanned AN/TPS-43, sold for U.S. $7 M in 1989 [3.59].

The overall cost factors have been divided into their constituent categories for a broad range of ASESAs either currently in production or under development. As a proportion of the array cost, the following conclusions have been reached [3.67]:

1. T/R modules make up 50% of the overall array cost.
2. Subarray manufacturing constitute 40%.
3. Array integration absorbs 10%.

Eighty percent of the module cost is apportioned to the MMIC production and module assembly/test. The subarray cost includes mainly the manufacturing of RF/DC manifolds and thermal management structures.

From the foregoing considerations it is clear that the cost of T/R modules is a fundamental quantity to focus on if one is interested in the realization of affordable active phased arrays. The cost of one T/R module varies:

1. With requirements, and whether the requirements are similar to those of previously realized modules.
2. The manufacturing procedures and yield.
3. The number of modules required.

What follows are again typical 1994 costs for T/R modules and production runs, which will serve the purposes of establishing a baseline from which cost lowering is expected. Raytheon and Texas Instruments are at the time of writing, manufacturing the T/R modules for the ground-based radar (GBR). The system calls for 68,500 T/R modules, to be produced between 1993 and 1995 and the projected cost is to be less than U.S. $1000 per module at the end of the production run. However, full production automation is currently too expensive to justify its use even for this large number of units [3.68]. Caution should be exercised in dealing with module cost because in reality there are considerable deviations from the typical modules.

At present the acquisition cost of an active array radar (AAR) at $1000 per radiating element is comparable to that of a tube system. However, the LCC of the AAR is half that of the tube type [3.68]. Again, caution should be exercised in using the preceding estimates, which reflect current costs; it is fully expected that new developments will decrease the costs cited for the solid-state T/R modules. In Chapter 4 we shall explore detailed approaches for the cost minimization of T/R modules at the system and subsystem levels.

3.11.2 The Array Area/Volume and Thermal Problems

The array area is usually determined by the requirements for a certain power aperture product, or beamwidth or costs. It is usually a tradeoff between many parameters. Similarly, the array volume is determined by the array architecture used, the thermal problems, and cost. For airborne applications the volume is limited and non-negotiable.

While electronics engineers focus attention predominantly on the RF and digital aspects of the array, the thermal problems are often deemphasized. Here we shall demonstrate the importance of thermal problems of arrays operating at centimeter- and millimeter-range wavelengths with the aid of an example.

Let us assume that we utilize T/R modules that are capable of emanating 10 W mean power per element and that a half-wavelength spacing between elements is adopted [3.69]. In Chapter 4 we shall introduce the concept of power-added efficiency (PAE) or η; here we shall just use Equation (3.124) to define the power dissipated P_{dis} in a transmitter as

$$P_{dis} = P_{out}\left[\frac{1}{\eta} - 1\right] \quad (3.124)$$

In Table 3.14 we tabulated the dissipated power when η takes the values of 0.25–0.5 and the frequency of operation varies from 3 to 100 GHz. As can be seen, the dissipated power per square meter becomes significant as the frequency of operation increases and as the PAE of the T/R module decreases.

If that is not enough, the cooling systems and the power supplies required to power the T/R modules also dissipate considerable power.

3.11.2.1 Array Cooling Air cooling is appropriate for low-power phased arrays; the design issues for this option are:

1. The number and construction of fins per module.
2. The number of modules per fan.
3. The heat dissipated by the fans.

TABLE 3.14 Power Dissipated/m² When a Range of PAEs is Assumed for Arrays Operating at cm and mm Wavelengths

η Frequency	W per Element (GHz)	kW/m²				MW/m²	
		3	5.5	10	20	50	100
0.25	30	12	40.3	133.3	533.3	3.3	13.3
0.3	23.3	9.3	31.3	103.5	413.3	2.6	10.3
0.35	18.6	7.4	25	82.7	328.9	2.06	8.3
0.4	15	6	20.2	66.7	266.7	1.7	6.7
0.45	12.2	4.9	16.4	54.2	217.8	1.4	5.4
0.5	10	4	13.4	44.4	177.8	1.1	4.4

218 PHASED ARRAYS: CANONICAL AND WIDEBAND

A 1200-ft^3/min fan produces 1 kW of heat; with one fan per 50 modules, a 3000-element array would require 60 fans, which, in turn, produce 60 kW of waste heat [3.69]. Liquid cooling is more appropriate for high-power arrays as water is 1000 times more dense than air and its heat capacity is 4 times greater. The issues for this option are [3.69]:

1. The flow rates, constrained by erosion.
2. The thermal interface between the module and the cooled channel.
3. Safeguards against a catastrophic failure of the water cooling system and the electric protection of the modules.

Studies linking the mean time between failures (MTBF) of the receiver/STALO or R/S section of the ultrareliable radar (URR) when air or liquid cooling is used, have been undertaken, and the results are presented in Figure 3.35. Typical air and liquid flows are shown and a dramatic increase in the MTBF figures of the receiver/STALO can be seen when liquid cooling is used. Typically with 10°C inlet liquid, the junction temperatures of the majority of the receiver/STALO integrated circuits are less than 55°C with the average about 60°C [3.70].

3.11.2.2 The Humble Power Supplies In the previous section we considered and estimated the power dissipated as heat per solid-state module and

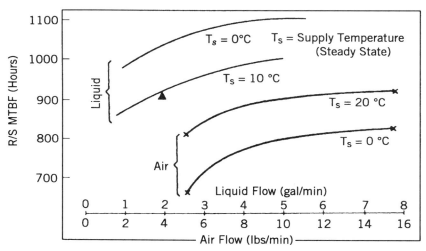

FIGURE 3.35 The MTBF of the receiver/STALO (R/S) of the ultrareliable radar, (URR), as a function of air or liquid flow. The effectiveness of liquid cooling in extending the MTBFs of electronic subsystems is clearly demonstrated. (*Source*: Lingle et al. [3.70], © 1989 IEEE.)

found it to be considerable; here we shall consider the power dissipated in generating the required RF power.

The power supplies we shall consider are the linear and switched-mode power supplies, (SMPSs) (hybrid and high density). Although there is no need to outline how linear power supplies operate, a brief description of SMPS is in order. SMPSs have been used for applications where size and heat dissipation are important considerations; now they are used in lower-power applications, such as computers, as well.

In the SMPS, the input AC, which is usually 50/60 Hz, is rectified and filtered before it is converted into an AC of 20 kHz or higher, for further rectification.

With reference to Table 3.15 [3.71], it is clear that the switched power supplies have the following advantages over the linear power supplies:

1. A significantly higher efficiency. It is worth illustrating this advantage in terms of an example. Suppose we require 200 W of power to be utilized in a phased array. The 80% efficient SMPS requires an input power of 250 W, while the linear power supply, operating at 40% efficiency, requires 500 W. Power radiated as heat is 50 W for the SMPS and 300 W for the linear power supply, a ratio of 6:1. This is by far the most significant advantage SMPSs have in phased array applications where considerable heat has to be dissipated. The increased efficiency is attained by the use of fast-recovery-time Schottky diodes for rectifying or controlling the directions of the output current. The Schottky diodes are required to prevent significant power losses due to the stored charge on the junctions of conventional rectifier diodes. Another factor that further boosts the efficiency is the forward voltage drop of Schottky diodes, which is much lower than that of conventional diodes.

2. The power density is higher for SMPSs, when compared to their linear counterparts, by a factor of 30; this is due to the reduction in size of various components operating at 20 kHz (and above) compared with the linear power supply operating at 50/60 Hz. The large power transformer, for instance, operating at 50/60 Hz used in linear power supplies is substituted by a miniature ferrite core transformer, and the filter capacitors are much smaller.

TABLE 3.15 Comparisons of Power Supply Technologies [3.71]

Parameters	Linear, Discrete	Switching, Discrete	Switching, High Density
Efficiency (%)	20–50	60–90	60–90
Power density (W/in)	0.5–1	1–3	10–30
Weight density (W/lb)	10–20	20–40	100–400
Reliability MTBF	15,000–30,000	10,000–20,000	15,000–30,000

The only large capacitor used by the SMPSs is the input filter capacitor, which should filter a full-wave rectified sine wave at 50/60 Hz.

3. The weight density of the SMPSs is again considerably higher than that corresponding to the linear power supplies, while the reliability of both kinds of power supplies is comparable.

3.11.2.3 Typical Phased Array Switched-Mode Power Supplies Most radio-astronomy phased arrays use SMPSs; here we shall consider two representative examples where SMPSs have been used in conjunction with phased array-based radars.

In phased arrays where power supplies supply the power to subarrays, the failure of one power supply can result in an unacceptable (catastrophic) array performance. While the MTBF for power supplies is important, the mean time between catastrophic failures (MTBCF) is of equal importance for phased array systems. By utilizing a number of redundant power supplies in a variety of configurations, the airborne active element array has a MTBCF of 15,209 h [3.71].

The microwave power module (MPM) (see Section 4.4.1) utilizes a switched power supply or electronic power conditioner (EPC). The MPM consists of a solid-state medium-power amplifier followed by a TWT (traveling-wave tube) that generates 100 W of power in the 6–18-GHz band. Great advances in the miniaturization of the EPC have been reported, including a reduction in volume of 10–20 times compared to currently available supplies [3.72]. The inverter frequencies were 150 and 300 kHz for the two approaches taken for its realization, and efficiencies of about 90% have been reached [3.73]. Interestingly enough, however, the EPC occupies about 81% of the MPM volume.

More reductions in the volume of the EPC will come if the frequency of the EPC is raised further, or the PAE of the power modules is increased. For the MPM, the development of low-profile planar high-voltage transformers will also contribute to the decrease of the volume of the MPM. For the URR, the switching frequency of its power supplies was about 50 kHz [3.70].

From the foregoing considerations we can easily conclude that the thermal problems of phased arrays and their power supplies represent important areas of research, especially if operation at mm wavelength is envisaged.

3.11.3 Applications: Brick or Tile Architectures?

In Figure 3.32 we have illustrated the brick and tile architectures; here we shall compare them and define their respective regimes of application.

3.11.3.1 Brick Architecture—Applications Most of the active phased arrays presently (1994) under development or planned for near-term production use the brick architecture [3.67]. These architectures result in relatively long

T/R modules, which have thin housings containing several MMICs as well as some of the digital control and active device biasing circuitry. The comparisons here are made with respect to the tile architecture—see the next section. A relatively deep active array can, if the space is available, contribute toward easing the thermal problems. The GBR [3.67] can be considered as an archetype for these architectures.

The brick architecture allows the designer to use printed wideband dipoles that have 40% instantaneous fractional bandwidth [3.74] or tapered slotline antennas (also known as Vivaldi antennas) that can operate over 1–3 octaves [3.75, 3.76]—see Chapter 5.

With this architecture full polarization information is not impossible to attain but is not an easy task to undertake; an eggcrate structure, for instance, has been proposed to accommodate printed-circuit-board (PCB) antennas [3.77]; with this scheme half of the antennas receive one polarization while the other half receive the orthogonal polarization. For the brick architecture, replaceability of components or modules is practical.

3.11.3.2 Tile Architecture—Applications The tile architecture is a newcomer to the field of phased arrays; the top face of the array is populated by the antenna elements, such as patch or crossed dipole antennas. Usually the radiation received by the antenna elements is coupled from the top layer to the next level down electromagnetically. Different array functions or sets of functions, including amplification, phase shifting, and switching, are performed in the subsequent layers.

In Figure 3.36 we have illustrated examples of brick and tile architectures. The brick architecture depicted in Figure 3.36a consists of several "sticks" connected to a backplate and electrically connected to the RF, logic, and DC manifold. The ends of the sticks, which have different lengths, are connected to a cooling manifold. The cost of interconnecting the assembly of 2000 T/R modules, shown in Figure 3.36a, is high, the array typically weighs several hundred pounds, and is approximately 12 in. long [3.78, 3.79]. Newberg and Wooldridge [3.78] stated:

> This assembly (of sticks) is not low profile. It cannot be integrated easily or conformed to the skin of an aircraft; placed high on the mast of a ship; put in mobile ground-based systems or the very limited space of a missile; or conformed to the weight and space sensitive design of spacecraft.

One version of a proposed tile architecture, illustrated in Figure 3.36b, utilizes multichip module (MCM) packaging, digital technology, photonic signal manifolding, and true time delay for the realization of affordable, compact, lightweight, and wideband phased arrays.

The array consists of any number of subarrays, four of which are illustrated in Figure 3.36b, interconnected to populate a given aperture. For comparison the weight of a 2000-element array using this new technology is

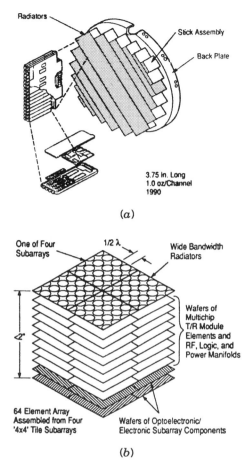

FIGURE 3.36 Practical examples of typical phased arrays using the (*a*) brick and (*b*) tile architectures. (*Source*: Newberg and Wooldridge [3.78], © 1993 IEEE.)

estimated to be less than 75 lb while its depth is less than 2 in. Scalsi et al. [3.80] estimate the ratio of the depths for the brick and tile architectures as 6:1, which concurs with the estimates of Newberg and Wooldridge [3.78, 3.79].

The resulting array performs the functions the assembly of sticks (brick architecture) could not perform and is relatively inexpensive.

Other advantages of the phased arrays having a tile architecture are:

1. It is relatively easy to derive dual-polarization phased arrays [3.81].
2. For narrow-band applications patch and cross dipole antennas can be used as the radiating elements.

3. For applications requiring wide instantaneous bandwidths the following options are available to the designer:
   i. The spiral mode microstrip antennas that can yield a 6 : 1 bandwidth [3.82].
   ii. The microstrip log periodic antennas that can yield a 4 : 1 bandwidth [3.83].
   iii. The planar broad band flared microstrip slot antennas can provide dual polarization over a multioctave band [3.84].

All the above types of antennas are further considered in Chapter 5.

Given that the resulting array is compact, the thermal problems have to be better managed, especially if the transmission of high powers is envisaged.

Although the array can be affordable, the replaceability of components may be difficult [3.81]. However subarrays are easily replaceable.

3.11.4 The Integration of Modern Antenna Elements to the MMICs

Some of the early phased arrays utilized horns or conventional dipoles as array elements; at a later stage arrays utilizing slotted waveguides as array rows or columns were realized. Modern phased arrays use printed-circuit antennas (PCAs) (some are microstrip) in the form of patches, dipoles, or slots. Microstrip antennas are easily and accurately replicated by the use of photolithographic processes; the cost associated with their realization is therefore low; the same antennas are lightweight and can be naturally integrated with the MMIC T/R modules.

It is convenient to use a low relative permittivity (e.g., $\varepsilon_r < 4$), substrate on which to print these antenna elements; such an approach increases the efficiency of the antennas to radiate in free space because space waves are strongly coupled with the radiating elements.

GaAs has been selected by DARPA (Defense Advanced Research Projects Agency, now known as ARPA) the funding body of the MIMIC (MIcrowave and Millimeter Integrated Circuits) Program as the substrate for microwave T/R modules and digital ICs because of its excellent electrical characteristics. However, as the relative permittivity of GaAs is 12.4, it is not particularly suited as a substrate for PCA elements. Given that the T/R modules usually follow the antenna elements, the etching of the antenna element on GaAs although convenient presents many problems, which we shall explore in Chapter 4.

3.11.4.1 Antenna Element Tunability Antennas that satisfy narrowband and wideband applications have been cited. For applications where one requires a narrow bandwidth, the center frequency of which can be moved over a wide bandwidth (e.g., 1–3 octaves), a narrowband antenna element, such as a conventional dipole or monopole, can be used.

As an example, a blade antenna originally designed to operate at 180 MHz was tuned from 30 to 90 MHz with the aid of high-Q passive components, a microprocessor and low-loss switches [3.85]; the bandwidth attained was defined as the bandwidth over which a voltage standing-wave ratio (VSWR) of 2 or less was measured. This realization was possible because the radiation efficiency η_r defined as

$$\eta_r = \frac{P_r}{P_r + P_{loss}} \qquad (3.125)$$

decays slowly at frequencies off the resonance frequency; here P_r is the radiated power and P_{loss} is the power loss, respectively. By contrast, the mismatch efficiency η_m defined as

$$\eta_m = 1 - |\Gamma|^2 \qquad (3.126)$$

decays rapidly at frequencies off the resonant frequency; here Γ is the voltage reflection coefficient. Considerable theoretical and experimental work is required to attain similar results by utilizing patch and printed dipole antennas with minimum losses at centimeter-range wavelengths.

The tuning of a patch antenna over a modest frequency range has been reported [3.86]; the 2-VSWR bandwidth ranged from 1.55 to 1.93 GHz, and the tuning was implemented with a varactor diode attached to the patch.

3.12 WIDEBAND PHASED ARRAYS

The requirement for wideband phased arrays has been formulated in Chapter 1. In this section we shall outline the many prerequisites designers have to satisfy before wideband arrays can be realized and the options design engineers have at their disposal. The realization of wideband arrays depends on the availability of wideband subsystems, such as T/R modules and beamformers, as well as a theoretical framework from which the designer can derive useful guidelines.

Although it is almost impossible to consider wideband phased arrays in a piecemeal fashion, by considering its many subsystems separately, we shall adopt this approach, with the proviso that the many interactions and interrelationships are adequately considered.

3.12.1 General Considerations

Let us suppose that a wideband phased array is required to operate between f_1 and f_2 and where $f_2 = nf_1$, and $1 < n < 10$. The following issues have to be considered.

The selection of appropriate antenna elements that operate satisfactorily over the required frequency range is essential; the considerations here are well-behaved radiation patterns and acceptable VSWRs over the required frequency range and scan angles. Additionally, the radiation patterns should match the required field of view; if dual-polarization operation is required, a high-cross polarization isolation specification is to be met over the entire frequency range and scan angles. Finally, mutual coupling problems, encountered in phased arrays utilizing a variety of antenna elements, have to be addressed so that the array does not suffer any scan angle blindness.

The absence of grating lobes within the required frequency and scan range is mandatory if spatial ambiguities are to be avoided. The problem in hand can be simply defined by the following considerations. If one accepts a $\lambda_1/2$ spacing and follows design procedures to render the mutual coupling problems negligible, grating lobes at f_2 will appear. Conversely, if one accepts a $\lambda_2/2$ spacing, problems related to mutual coupling at λ_1 will emerge; the additional consideration in the latter case is the size of the antenna elements. Meeting these two criteria simultaneously is not a trivial matter, but several approaches have been reported; detailed studies that take into account the particular elements used and the effects of mutual coupling between them are required before experimental models are realized.

The following qualification has to be made here. When the array operates in the radar mode, grating lobes can be tolerable in the transmit and/or receive modes, but no grating lobes are tolerable after the multiplication of the receive and transmit radiation patterns. The implicit assumption here is that the receive and transmit arrays have different radiation patterns.

In Chapter 4 we shall focus attention on T/R modules; here we shall briefly mention that wideband modules have been available for some time now. The T/R module described by Priolo et al. [3.87] has dual channels to accommodate the two polarizations and covers the frequency range from 6 to 18 GHz.

Similarly we have already considered wide-band antenna elements suitable for the brick [3.74 to 3.76] and tile architectures [3.82 to 3.84]; indeed one can easily conclude that there is a plethora of suitable antenna elements for phased array applications.

We have seen that by connecting a set of switchable phase shifters to each antenna element, one can form a beam that in turn is scanned anywhere within the array FOV, if the appropriate phases of the phase shifters are varied accordingly. We have also qualified that this arrangement is ideal for arrays designed to operate over narrow fractional bandwidths.

We will describe several approaches to the realization of narrow-band beamformers in Section 3.14 and the issue of whether one performs the beamforming at some IF or at RF will naturally be explored.

For wideband operation each antenna element is connected to a set of switchable true time delays and scanning is performed by varying the appropriate time delays for each element by predetermined amounts.

Wideband, compact, and lightweight photonics-based beamformers suitable for small arrays have been realized and reported. The RF band of interest, in a photonics-based beamformer, is converted to optical wavelengths and optical fibers provide the necessary time delays for the wideband phased array. The system's wavelength is then converted to a cm band for further signal processing.

This solution provides extremely wide instantaneous bandwidths and the only penalty one pays is high losses, in the vicinity of -40 dB [3.88], due to the several conversions of the system's wavelength; these losses however can be rendered invisible to the system; thus the system's SNR is not significantly degraded. Considerable effort has been directed towards minimizing these losses.

While the above solution is again relatively straightforward, other approaches suitable for large phased arrays, described in the same section, are more appropriate.

3.12.2 Array Considerations

In this section we shall assume that all components for wide-band phased arrays such as antenna elements, T/R modules, and beamformers exist and explore the problems related to the antenna array.

For wideband operation regular spacings that give rise to grating lobes are not appropriate. An aperiodic distribution of antenna elements can be deterministically based on a variety of algorithms or by random placement, a process that has been proposed by Lo [3.89] and adequately treated by Steinberg [3.90]. With respect to the worse sidelobes it has been demonstrated that the random placement is no worse on the average than the best deterministic aperiodic algorithms reported in the literature [3.91].

One-dimensional random arrays result when the N array elements are placed on a straight line randomly at distances x_n from the array origin and the element excitation I_n can be either uniform or tapered. The excitation of the array is (3.90)

$$i(x) = \frac{1}{N} \sum_{n=1}^{N} I_n \delta(x - x_n) \tag{3.127}$$

where $\delta(\ldots)$ is the Dirac function. The far-field pattern of the array $f(u_3)$ can be deduced from the Fourier transform of $i(x)$, which is given by the equation

$$f(u_3) = \frac{1}{N} \sum_{n=1}^{N} I_n \exp(jkx_n u_3) \tag{3.128}$$

where $u_3 = \sin\theta$, θ is measured with respect to the boresight axis of the array and $k = 2\pi/\lambda$. $f(u_3)$ at u_0, is given by the equation

$$f(u_3) = \frac{1}{N}\sum_{n=1}^{N} I_n \exp[jkx_n(u_3 - u_0)] \quad (3.129)$$

which reduces to [3.90]

$$f(u_0) = \frac{1}{N}\sum_{n=1}^{N} I_n = I_{ave} \quad (3.130)$$

at u_0 and where I_{ave}, is the average value of the excitations.

The ratio of the average sidelobe power level (ASL) to the error-free mainlobe power (ML) is given by

$$\frac{ASL}{ML} = \frac{1}{N} \quad (3.131)$$

for the case of uniform excitation.

The peak sidelobe level of a linear random array typically exceeds the ASL by some 10 dB and is about 3 dB higher for two-dimensional arrays [3.92]. The required sidelobe levels therefore determine the number of elements a random array has.

Apart from their wideband capability these arrays have beamwidths that are approximately determined by the size of the aperture. Given that the interelement spacings are usually large, the mutual coupling problems are negligible, cooling of the array becomes more manageable, but the power aperture product decreases.

Skolnik [3.53] provides a good summary of work related to nonuniform arrays and many approaches related to the techniques used to populate linear and planar arrays have been reported.

Let us place the antenna elements of a linear phased array at a canonical spacing c, which renders the mutual coupling at f_1 negligible and monitor the geometric array factor at f_2. Next we move each antenna element to positions $c_1, c_2, c_3, \ldots, c_N$ where $c_n \geq c$ with the aim of decreasing the array sidelobes; the positions of c_n for the array elements are chosen by using multivariable optimal search methods [3.93 and 3.94]. What is observed is the gradual decrease of the first sidelobes and grating lobes to the same level. After several iterations, the array will have the same low sidelobe level as a function of scan angle. Furthermore the array will have the same sidelobe level at all frequencies lower than f_2. If the number of antenna elements is not too high, this approach is appropriate. As the number of antenna elements is increased, however this approach is computationally unattractive.

A computationally simpler approach has been proposed for planar arrays [3.95]; the antenna elements are allowed to occupy positions out of a given set. With this approach only one geometric array factor is calculated at a time, thus large arrays can be accommodated by the computer program.

In Section 3.7.3 we concluded that an economical solution to the problem of realizing a large array is by subarraying a number of antenna elements; each subarray is then connected to a variable delay line and each antenna element to a variable phase shifter. If all subarrays have the same size and are contiguous, grating lobes appear due to the inherent periodicity; furthermore, the angular distance of these lobes will depend on the size of the subarrays—see also Section 3.6.2.3. Often the grating lobes enter the array FOV as the array is being steered and create spatial ambiguities.

If the size of the subarrays is not constant but random and the positions of the subarray centers are also random, the amplitude of the resulting grating lobes is significantly diminished. As an example of the effectiveness of this method the average array factor corresponding to a linear array of 128 element is illustrated in Figure 3.37 [3.96]. In Figure 3.37a the array is divided into 32 subarrays of 4 antenna elements and it is clear that the amplitude of the resulting grating lobes is unacceptable. The same array was subdivided into subarrays that have centers chosen at random and elements that varied from 2 to 6 randomly. The resulting radiation pattern is shown in Figure

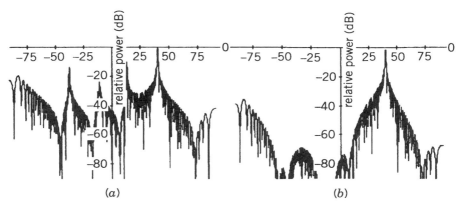

FIGURE 3.37 Grating lobes generated as a result of quantization effects due to subarraying: (a) the array radiation pattern when all the subarrays consist of four antenna elements; (b) the array radiation pattern when the subarray centers are chosen randomly and the number of subarray antenna elements varied from 2 to 6 randomly (from [3.96]).

3.37*b*, which shows that the grating lobes are significantly diminished [3.96]. This approach shows considerable promise for it is economical and effective; it can be seen as the natural progression to placing the elements of the array randomly.

The approaches described are theoretical in nature and detailed and exhaustive calculations are required when specific antenna elements are used. The issues requiring attention are the resulting radiation patterns and input impedance of the antenna elements over the frequency and scan range. An illustrative example, given on Section 3.12.5, will aptly demonstrate the difficulties involved in the design and realization of wideband phased arrays.

3.12.3 Examples of Wideband Arrays

Several approaches to wideband array configurations have been reported; here we shall outline some of the more representative ones.

An experimental 96-element phased array operating from 3.5 to 6.5 GHz with scanning capability from 60° to −60° has been reported [3.97]. The radiating elements were open-ended waveguides placed as close as possible to a triangular configuration. Wideband phase shifters were used and the methods of impedance-matching the array aperture over the required scan angles and frequency range have been described.

A phased array antenna consisting of three arrays of interlaced radiating elements operating at L-, S-, and C-bands has been realized [3.98]. The fractional bandwidths at the three bands were 10, 20, and 20%, respectively. Similarly, an experimental phased array that operated from 2.5 to 10 GHz has been realized by interlacing two sets of Archimedean spiral antennas, one set to cover the 2.5–5-GHz range and the other to cover the 5–10-GHz range [3.99].

Stutzman realized a shared-aperture phased array by stacking three different-size Archimedean spirals in layers, thus creating a three-dimensional array of spirals [3.100]. The array operated over a 2-octave bandwidth, and the blockage due to the partial overlap of elements from different layers of the array was found to be minimal.

Wideband operation has been critically important for ESM and ECM systems for some time; radar on the hand operated over narrow bands and it is only recently that experimental wideband radars are in the process of being realized. For this reason it is appropriate to consider wideband ESM and ECM systems before we return to recent developments related to radars.

3.12.4 Phased Arrays for the ESM and ECM Functions

Curtis [3.101] overviewed the Navy's ESM and ECM developments since World War II and provided glimpses into the future developments in the same areas. To this date it is a remarkably useful contribution.

It is recalled here that the major requirements for the ESM function are high POI (in the spatial and frequency domains), sensitivity and appropriate strategies and algorithms for the de-interleaving of the incoming pulses while the requirements for the ECM function are the timely, intelligent use of high EIRPs and directivity.

Although it is convenient to notionally separate the ESM and ECM functions, in reality the two functions, accommodated in two subsystems, are often integrated. In an ideal case the ESM subsystem surveys the environment and defines as many parameters as possible of a few threat radars and/or missiles, while the ECM subsystem renders these threat radars inoperative by directing as much jamming power (or deception ECM) toward them as is possible.

The threat of antiship cruise missile (ASCMs), which can be launched from many a platform, including submarines, shortened the time the victim ship has to take defensive action to less than 30 s. The reaction times of the ESM/ECM defense, therefore, had to be shortened to cope with this short warning span [3.101].

Often the jamming function can be seen as protecting an attack aircraft, bomber, or ship. Jamming is only one effective way to counteract the opponent's radars; flares and decoys are also effective.

It is noted here that the airborne ESM and ECM functions are challenging because the available raw power is bounded and space is limited. Additionally, the survivability of both aircraft and ships depends on decisions made in quasi-real time during a tactical engagement.

Jamming over an extremely wide surveillance volume is required in two cases: (1) when an aircraft flies at speeds exceeding Mach 1 among too many threats (other aircraft flying at comparable speeds) scattered over its surveillance volume and (2) when an aircraft requires to jam a bistatic or multistatic radar and the locations of the receivers are not known. For these two cases, "fail-safe" [3.102] jamming over the entire surveillance volume is required.

3.12.4.1 Phased Arrays for the ESM Function The efficient deinterleaving of the incoming pulses intercepted by an ESM subsystem depends heavily on specific measurable parameters such as the DOA and emitter frequency, which can be used as discriminants for the sorting strategies and algorithms.

The scanning phased array can be used to provide the ESM subsystem with information related to the DOA for the deinterleaving facility. More explicitly, a phased array provides spatial filtering, which, in turn, yields the following benefits for the ESM subsystem [3.102]:

1. Discrimination of radars that use frequency agility, pulse jitter, and/or spread-spectrum techniques.

2. DOA accuracy.
3. Discrimination of time-coincident pulses.
4. The array gain provides an enhancement of the system's SNR when compared to omnidirectional techniques of deriving DOA information.
5. Jammer excision, so that other weak signals are not masked.

Despite all these benefits, the POI of a scanning phased array is low unless the array can be scanned in the microsecond timeframe [3.103].

Some airborne ESM subsystems use the following omnidirectional techniques to derive DOA information [3.102]:

1. Amplitude comparison DF techniques.
2. Phase interferometers.
3. Correlation interferometers.

Comparing techniques 1 and 2, the phase interferometers offer an order of magnitude of higher accuracy in the derivation of the DOA information [3.102]; that precision, however, is not maintained when the unit is installed on the aircraft. By contrast, the correlation interferometers provide the means of calibrating out any errors caused by the installation of the unit on the aircraft. Given that these systems have omnidirectional antennas (thus providing high POI), the deinterleaving function is aided by frequency filtering. A bank of contiguous filters provides frequency discrimination for the incoming pulses and the possibility for jammer excision.

Another discriminant that can be used for the deinterleaving operation is polarization; the cost and complexity related to the implementation of dual-polarization systems, however, has to be weighted against the benefits provided by such a system.

A multiple-beam array (MBA), illustrated in Figure 3.38a, can provide, in principle, high POI and array gain over the entire surveillance volume. Additionally, it provides the benefits (1–4, at the beginning of this section) already cited. The MBA yields independent, simultaneous, and contiguous beams in the spatial domain and is realized by using a Butler matrix arrangement or the techniques discussed in Section 3.14. The MBA can perform fail-safe jamming or jamming toward two or more directions. The conventional scanning beam phased array is shown in Figure 3.38b.

3.12.4.2 Phased Arrays for the ECM Function The maximization of the EIRP is important, especially for ECM arrays. Given the choice of either combining the powers of several transmitters conventionally, or in space, one

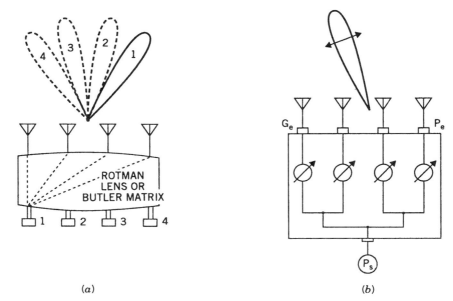

FIGURE 3.38 Two archetypes of phased arrays: (*a*) the multiple beam, or staring array; and (*b*) the scanning beam array.

would normally prefer the latter option because it offers the advantage of aperture gain. In Figure 3.39*a* the power generated by one high power amplifier is divided into the N antenna elements of a phased array. The high-power amplifier is usually a TWT, and this option does not offer graceful degradation. In Figure 3.39*c* the power generated by N low-power amplifiers is combined in space (or on target); in this option, which offers graceful degradation, the amplifiers used are either solid-state, mini-TWTs, or TWTs. In Figure 3.39*b*, M ($< N$) medium-power amplifiers are used to feed the N antenna elements of a phased array; it is an in-between option of the first two.

If one knows the location of a monostatic radar accurately, the use of a planar phased array is ideal to jam the radar's receiver, for all available power can be directed toward the target radar—surgical jamming. The additional advantage this technique provides is stealth, ensuring that the jamming operation is not received by other ESM systems.

In many cases the position of the target radar is not known accurately because the DF system on board planes is not sophisticated (omnidirectional systems already considered), so a linear array generating a fan beam is preferable. The decreased complexity of a linear array, when compared to that of the planar array, is another reason for their attraction.

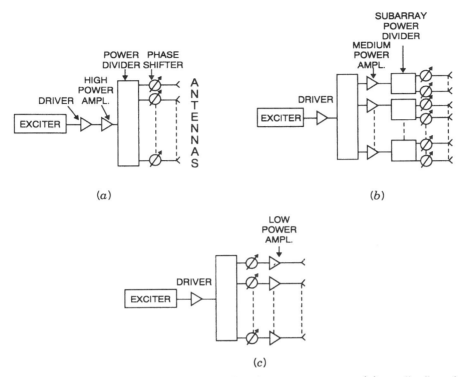

FIGURE 3.39 Three methods of maximizing the on-target power. (*a*) nondistributed (passive); (*b*) semidistributed; and (*c*) fully distributed (active) transmitter power.

At mm wavelengths, where the transmitted powers are low, power addition in space is almost mandatory. The need to use DOA information for the purposes of deinterleaving is, however, deemphasized at mm wave bands because the spectrum is not widely used now.

3.12.4.3 Applications of ESM and ECM Arrays For airborne operations, typical EIRPs levels in the range of 1–10 kW are required for self-protection; for standoff jamming the required levels are between 30 and 300 kW, and in some cases even higher levels are required [3.103]. The generation of jamming powers approaching 1 MW in the I and J bands has been reported [3.103]. This was accomplished by a combination of multiple TWT amplifiers and high-gain phased array antennas. MBAs have been used in conjunction with ESM facilities [3.101–3.103].

Using 16 antenna elements only, an electronically steerable ECM array yielded 15 dB of gain over 40% bandwidth at X-band. The derived radiation was linearly polarized and the reported bandwidth can be extended to 1 octave [3.104].

An active phased array architecture has been shown to be a valid alternative to the traditional ECM power transmitter based on TWTs in the H/J band. The criteria used were power consumption and dissipation. The solid-state solution was by far the most preferable [3.105].

Broad descriptions of the RAPPORT III [rapid alert and programmed power (management) of radar targets] and the ASPJ/ALQ-165 ESM/ECM systems for the F-16, F-18, F-14, AV-8B, and EA-6B have been reported [3.106, 3.107]. The jammers were self-protection jammers, and the accent was on speed of operation and power management. The systems provide programmable power management, which permits optimum ECM technique selection matched to the threat.

3.12.5 The U.S. Navy's Airborne Early Warning (AEW) Radar

The Navy's current AEW capability is based on the APS-139 radar installed aboard the E-2C aircraft. The radar operates at the UHF band in conjunction with a mechanically scanned antenna implemented within a rotodome configuration [3.108]. Current and future threats, threat scenarios, and technological advancements support the need for a wideband AEW radar system with an electronically agile beam. Operation at the UHF band offers detection advantages over L-band against reduced RCS targets and device technology, which offers higher powers. The new system, which is at the proof-of-concept stage, utilizes an active phased array that has the following characteristics [3.108, 3.109]:

- Frequency range: 0.4–1.4 GHz, a ratio of 3.5 : 1
- Number of elements: 153 (9 rows and 17 columns)
- FOV: $\pm 60°$
- Antenna element: suspended circular disk
- Root-mean square sidelobe level: 45 dB or better
- Provision for two orthogonal polarizations

Theoretical work was centered on the derivation of the array pattern and elements' input impedance over the entire frequency range of operation and scan angle.

The calculations involved the generation and subsequent solution of matrix equations with the number of unknowns of the order of 6000 at the high end of the frequency band. By using the reflection and translation symmetries of the array as well as computational shortcuts, the CPU (central processing unit) time for individual runs per frequency was reduced to 2–4 h! Preliminary results are in excellent agreement with measurements [3.109]. The active array area is 1.81 × 0.96 m, and the physical area of the array is 2.67 × 1.39 m. The periphery around the active array is populated by antenna elements terminated by matched loads in order to damp any surface waves.

The T/R module has a peak transmit power of 500 W and an instantaneous IF bandwidth of 2 MHz on receive. In order to achieve maximum power capability, the T/R module is of a dual-band design, 400–850 and 850–1400 MHz. Other system and subsystem parameters are outlined in Teti et al. [3.108].

3.12.6 Multifunction, Wideband, or Shared-Aperture Systems

The case for multifunction, wideband phased arrays has been made in Chapter 1. The same arrays are often referred to as *shared-aperture arrays*, where one aperture can support a phased array operating over several octaves and performing a diverse range of interrelated functions.

Steps toward the realization of truly multifunction wideband arrays include the following activities and parallel other activities related to wideband modules and beamformers already outlined:

1. Programs such as ICNIA and PAVE PILLAR focusing on shared aperture systems for tactical aircraft.
2. The realization of a small-scale model [3.110].
3. Studies related to wideband antenna elements [3.75–3.77, 3.111].
4. System approaches [3.110].

The small-scale model operated in the 6–18-GHz band and utilized flared slot antennas [3.110]; the array yields the two orthogonal polarizations by collocating, at 90°, a pair of two-element subarrays, each fed with a Wilkinson power divider. A simulation of active VSWR with azimuth scan on a line array was performed using a central row of an 8×8 arrangement of orthogonal subarray element pairs and the average active VSWR was below 2:1 over a $\pm 60°$ in X-band. Full-polarization diversity has been accomplished, and the T/R circuit can be packaged in densities consistent with 18-GHz performance. The occurrence of high active VSWR with scan at lower frequencies needs to be investigated and understood.

A set of shared apertures around a tactical aircraft providing 360° coverage can offer significant operational benefits such as radar target track and semiactive missile support during aircraft maneuvers and high-gain ECM and ESM through large frontal arrays [3.113]. The marriage of the shared aperture concept with conformal active arrays and the devices offered by the MIMIC program will enable the realization of aircraft having truly smart skins, which will have overwhelming advantages over their more conventional counterparts. The same reference offers glimpses of the choices and system options one should consider before embarking toward the realization of any prototypes. The issues covered are [3.113] as follows:

1. The support of a diverse set of functions, some of which interfere with others.

2. Wideband circuits increase susceptibility to RFI from on-board sources and off-board threats or sources.
3. High isolation is therefore required between subsystems performing ESM/RWR (radar warning receiver) and ECM functions.
4. Wideband circuits have demonstrated poor performance; for instance, the efficiency and power output of wideband high-power amplifiers are lower than those corresponding to their narrowband counterparts. Similar arguments apply to low-noise circuits. Finally, higher costs are usually associated with wideband circuits.
5. There is a need to maintain spectral purity and parallel operations.
6. Multiplexed narrowband modular subsystems spanning a wide bandwidth seem to have considerable advantages over their wideband counterparts; furthermore, the system configuration allows the designer to update modules and/or expand the system with nominal costs.
7. There is a need for hardware commonality and functional parallelism.

These are only preliminary studies on a topic that is destined to occupy the minds of many array designers for a considerable length of time.

3.12.7 The Cottony and M. N. Cohen Approaches for Linear Wideband Phased Arrays

Many over-the-horizon radars (OTHRs) are linear phased arrays and operate in the 3–30-MHz band. Typically the arrays are not monostatic, so the designer is free to arrange the grating lobes of the receive array at different positions from the grating lobes of the transmit array. For radar operation the two radiation patterns are multiplied, and in the resulting radiation pattern the sidelobe levels at the positions of the grating lobes are tolerable [3.114].

In Figures 3.40a–3.40c typical receive, transmit, and resulting radar radiation patterns are shown at 25 MHz. At lower frequencies of operation, the high sidelobes aligned with positions of the grating lobes of the transmit radiation patterns move away from the main beam. A design procedure for a HF OTHR has been outlined [3.114]; here we shall briefly state the key concepts.

Let us assume that the total number of antenna elements of the receive and transmit arrays is limited and equal to p, and m and n are the numbers of elements in the receive and transmit arrays, respectively. The angular distances between lobes will be λ/S_r and λ/S_t (and multiples thereof) for the receive and transmit arrays, espectively; here S_r and S_t are the interelement spacings for the receive and transmit arrays, respectively.

For the condition that the angular space between the axis of the first grating lobe of the receiving array and that of the transmitting array $\lambda/S_r - \lambda/S_t$ be just wide enough for the combined half-beamwidths $\lambda/mS_r - \lambda/nS_t$

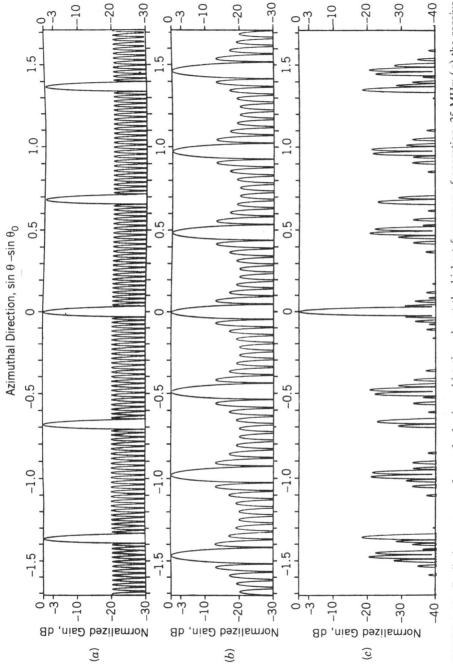

FIGURE 3.40 Radiation patterns of over-the-horizon, bistatic radar at the highest frequency of operation, 25 MHz: (*a*) the receive radiation pattern; (*b*) the transmit radiation pattern; (*c*) the resultant radar radiation pattern. (*Source*: Cottony [3.114], © 1970 IEEE.)

to fit into that space, the following equations have been deduced for the case where $m = n - 1$

$$\frac{S_t}{S_r} = \frac{n^2}{(n-1)^2} = \frac{(m+1)^2}{m^2} \qquad (3.132)$$

Design tables for a range of m/n from $\frac{4}{5}$ to $\frac{49}{50}$ are given in Reference [3.114].

Cohen [3.115] derived a number of interdependent Diophantine equations, which describe the condition such that over a specified range of frequency, scanning angles, and allowable beamwidths, the two synthesized arrays will not generate grating lobes that overlap beyond a prespecified angular distant. Unfortunately, the paper [3.115] is in a summary form and contains only guidelines for the design of wideband arrays.

There is a real opportunity to extend this method to planar monostatic and bistatic arrays. In the former case the receive and transmit arrays, although separate, nevertheless occupy the same real estate.

3.13 SPECIAL-PURPOSE PHASED ARRAYS

In this section we shall consider some special-purpose phased arrays. Although the arrays we shall consider do not enjoy the popularity of canonical or wideband arrays, they are nevertheless complementary to the arrays we have already considered. The phased arrays we shall consider here are the minimum or null-redundancy arrays. While the applications for the former arrays are obvious, the applications for the latter arrays need some elaboration.

3.13.1 Minimum and Null-Redundancy Arrays

Arsac first proposed the minimum- and null-redundancy arrays (M/NRAs) for radio telescopes in 1955 [3.116], and theoretical work continues to this date [3.117–3.121]. Additionally, applications for these arrays continue to appear [3.122–3.131].

Arsac observed that canonical phased arrays have too many redundant spacings. In Figures 3.41a and 3.41b the spacings of an 8-element canonical array is shown together with the corresponding spacings for null-redundancy arrays having 1, 2, 3, and 4 elements shown in Figures 3.41c and 3.41d. It can be seen that the spatial sensitivity of the canonical array contains too many short spacings. More explicitly, a canonical array with N antenna elements separated by a unit spacing u_0 will have $(N-1)$, u_0 spacings, $(N-2)$, $2u_0$

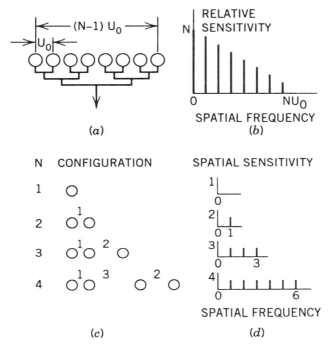

FIGURE 3.41 The occurrence of spacings between array elements of canonical and MRAs: (a) the topology of a canonical array having eight elements and (b) its spatial frequencies; (c) the topology of MRAs having 1, 2, 3, and 4 antenna elements; and (d) their spatial frequencies.

spacings and so on up to the longest spacing $(N-1)u_0$, which occurs only once. By contrast, all the spacings of the NRAs occur only once.

Simply put, each interferometer acts as a spatial filter; many interferometers having different separations and orientations, available from a phased array, will therefore act as different spatial filters. The image of a scene can be reconstructed from the available information by Fourier inversion.

In a more technical language each interferometer pair yields a complex number, and a phased array yields many complex numbers that are values of the Fourier transform of the observed brightness distribution function, also known as the *complex visibility function* [3.121], in the different points of the (u,v) plane corresponding to all the possible projected baselines. The u,v plane is the plane normal to the source direction. For a point source, the complex visibility function becomes the array sampling function or the spatial [3.117] sensitivity of the array.

Canonical arrays are relatively easy to realize; the powers received by the antenna elements are added, their spectral sensitivity contains a predominance of short spacings, and their resulting radiation pattern has the form of $[\sin(Nx)/N\sin(x)]^2$.

By contrast, the powers received by M/NRAs are correlated, their sensitivity is uniform, and their resulting radiation pattern has the form $\{[\sin(2Nx)/2N\sin(x)] + \text{constant}\}$ [3.117]. M/NRAs therefore yield the highest spatial resolution for a given aperture length. These arrays are therefore eminently suitable for airborne platforms, for example, airplanes, where space is at a premium.

While it is relatively easy to define the spacings between antenna elements of NRAs when the number of antenna elements is low (e.g., 4 or 5), considerable research has been channeled toward the definition of the spacings between elements of M/NRAs when N is considerably higher [3.119–3.121]. Element spacings when N is equal up to 30 have been deduced [3.121].

3.13.2 Applications of M/NRAs

The first M/NRAs were used for radio-astronomical observations [3.116, 3.117, and 3.122]; for a given number of antenna elements, the maximum number of nonredundant spacings are obtained and the arrays have the highest resolution possible. Furthermore the Earth's rotation is used to obtain additional spacings [3.123].

For some applications an adequate sampling of the spatial frequency spectrum is required almost instantly (when the time to form an image is short). For these applications the imaging is also referred to as "snapshot" imaging. Given that M/NRAs provide the maximum number of spacings instantaneously and possess the highest possible spatial resolution, their selection for these applications is not surprising. M/NRAs have been used for imaging the Earth's surface with maximum spatial resolution for remote sensing applications [3.124–3.127].

Recently a prototype array operating at 1.4 GHz has been flown [3.127] to radiometrically sense the Earth's surface. We have briefly considered this system, which is referred to as the ESTAR, in Chapter 1.

With reference to Figure 3.42, ESTAR's array consisted of eight "sticks" and each stick consisted of a linear array of eight crossed dipoles. By combining the dipoles on a stick into rows, an array of eight parallel stick antennas resulted. For each stick the resulting FOV extends approximately $\pm 45°$ wide in the cross-track dimension and about $\pm 6°$ in the along the track dimension. The fan beam is swept along the Earth's surface as the plane moves and spatial resolution in the across-track dimension is achieved by cross-correlating the outputs of pairs of the stick antennas; as can be expected all correlations are measured simultaneously [3.127]. The measured minimum detectable signal Δt for the above system is 0.53 K and compares favorably with the the theoretical value of 0.32 K [1.126].

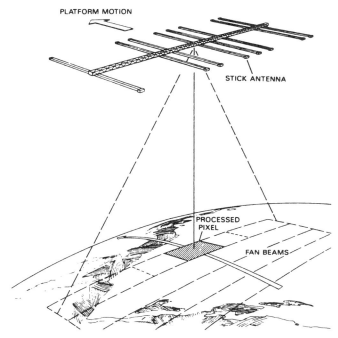

FIGURE 3.42 Airborne radiometric mapping of the Earth's surface using a minimum-redundancy phased array. The array FOV perpendicular to the flight path is shown. The array beams formed within the array's FOV are then swept by the aircraft's movement to form an image. (*Source*: Le Vine et al. [3.127], © 1990 IEEE.)

M/NRAs have been considered for DF/DOA applications [3.126 and 3.128] and interference cancellation applications [3.128 and 3.129].

3.14 BEAMFORMERS

The term beamformer applies to the steering of the antenna beam in many directions and the adjustment of the beam shape (sidelobe suppression and selective nulling) through phase shifters and gain controllers.

One of the many advantages of active phased arrays is that the beamformers though lossy, can be invisible to the phased array system. More explicitly the designer can use a lossy beamformer provided enough amplification exists in the system prior to the beamformer. With this arrangement a plethora of beamforming techniques can be used to meet the many diverse requirements. Naturally the designer aims to minimize the beamformer losses.

From an RF system point of view, the beamformer is the subsystem that delivers the end product to the array designer for further processing. In order to cope with so much diversity, there is a need to establish an appropriate nomenclature which will serve the purposes of classifying the many types of beamformers and recall the many system requirements that are placed on them.

Let us consider a phased array operating at a radio frequency f_0; the designer can form beams either at f_0 or at an IF, f_{IF}. Furthermore f_{IF} can be lower or higher than f_0. If $f_{IF} \gg f_0$ beams are formed at optical wavelengths. Similarly if $f_{IF} \ll f_0$, beams are formed by the use of digital techniques.

Another useful criterion that applies to beamformers is whether one requires one or more antenna beams to be agile or whether the derived beams are staring, a shorthand expression for independent, simultaneous, and contiguous beams. The term independent needs some elaboration for while one can generate a multitude of antenna beams, usually there is no point in generating antenna beams that have redundant information. In the radio-astronomy context the separation between the beams ought to be equal to the half Rayleigh limit of resolution ($\lambda/2D$, where D is the array diameter) [3.132]. In the final analysis, however it is up to the designer to have some redundancy.

For the case where one beam is required one would like to know whether the array HPBW is required to be constant, or whether its width is required to change to perform different functions such as target tracking (narrowest possible beam) or terrain following (a broader beam). One of course cannot neglect the array foreshortening due to scanning.

Stationary beams fill some part or the entire surveillance volume and do not move in space. An array can for instance yield eight or more staring beams (in elevation), which are scanned (mechanically) in azimuth to cover the required surveillance volume.

The case where a canonical phased array is required to yield one agile beam that scans the surveillance volume is to be considered in detail in Chapter 4. The beamformer takes the form of programmable phase-shifters, located within each T/R module and a central computer implements the scanning process.

Two important specifications for beamformers are the instantaneous and total bandwidths the beamformer is required to operate.

While we shall consider beamforming techniques here, the generation of nulls toward jammers is an equivalent operation. However we shall not cover in this book the topic of adaptive nulling toward jammers, which is adequately covered in other textbooks.

We shall proceed from some general considerations to the simplest beam-formers until we consider progressively more complicated ones and introduce new concepts as required. Applications ranging from radio astronomy to ESM/ECM and radar systems utilizing the different beamforming techniques will also be outlined.

3.14.1 General Considerations

Hall and Vetterlein [1.133] rightly asserted that "IF, digital and optical beamforming are major topics in their own right" before they expound on a review of beamforming techniques operating at radio frequencies; in passing it is worth noting that their review paper contains 115 references.

This being the case, our aim here is to provide a roadmap of the many diverse beamforming techniques available to the array designer, outline their regime of applicability for the different techniques, and provide suitable references (review and specific papers) that will gently lead the reader into greater depths. Naturally, novel approaches to beamforming will be treated at some length.

Phased array losses due to aperture tapering, resistive losses, and manufacturing tolerances have been considered elsewhere; an additional loss related to the beamspacing that reduces to zero for orthogonal beam sets [3.134–3.138] has to be considered for beamformers. It is noted here that mutually orthogonal beams satisfy the condition that the average value over all angles of the product of one beam response with the conjugate of the other must be zero. Most practical phased arrays will not have exactly orthogonal beams and will therefore incur this additional loss.

When contiguous antenna beams are required, the level of the crossover points is of importance. A crossover level of -3 dB from the beam maximum is typical if no information is to be lost. Avoiding significant coupling losses and consequent reduction in radar sensitivity requires that the beams be relatively far apart, resulting in lower crossover levels. Beams generated by a uniform amplitude distribution call for a -4-dB crossover point. In practice, beams with distributions yielding sidelobe levels lower than -13 dB call for much lower crossover levels [3.137].

The maintenance of the system's dynamic range, which is intrinsically related to the array SNR, is an issue worthy of the designer's consideration; this is especially true when beamformers operating at optical wavelengths are used.

For a linear array beam shaping can be implemented by a suitable phase distribution across the array. At the array center the phase is zero while the maximum phase is reached at either ends of the array. The phase progression can be linear or can follow a quadratic (or higher order) function.

Many valuable contributions related to beamforming methods are included in References [3.138, 3.139]. The dates of the references cited will aid the readers in assessing the maturity and/or originality of the methods used.

3.14.2 The Formation of Stationary (Staring) Beams at RF and IF and Applications

We have already considered the case where a line of stationary beams perpendicular to the line of the aircraft's flight are formed and an image of the scene is formed as the aircraft moves [3.127].

Similarly, for some radio-telescopes image formation is implemented as a two-step process. First a row of stationary beams, say along the NS direction, is formed and directed toward the source of interest. The row of beams is then electronically scanned along an EW direction.

3.14.2.1 Realizations Using Cables Let as consider the case where we need to form M antenna beams from a phased array having N antenna elements. With reference to Figure 3.43 the M beams are formed if the N antennas are connected to the M beams via $N \times M$ cables. Every beam port

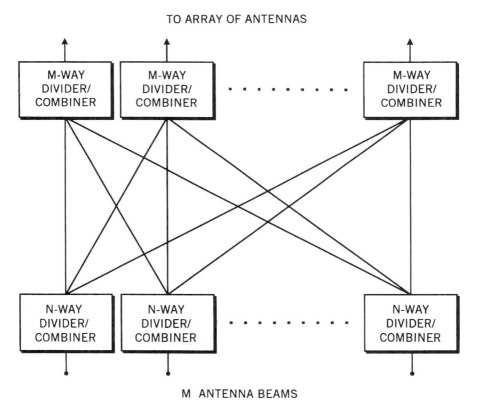

FIGURE 3.43 The block diagram of a beamformer yielding M staring beams from an array having N antenna elements by using $N \times M$ lengths of cables.

is connected to the antenna elements via the appropriate set of coaxial cables, the length of which depend on the beam direction and the array geometry. The beamformer is therefore a real-time analog Fourier transform in the spatial domain [3.140].

Blum [3.141] formed 15 stationary beams for a radio-astronomical application using this method. Fourikis [3.140] formed 48 stationary pencil beams derived from the Culgoora Radioheliograph, along a N–S line by using the same method—see Chapter 1 also. The derived beams were in turn swept along an E–W direction with aid of phase shifters (changers) introduced in the local oscillator (73 MHz) used to down-convert the RF (80 MHz) into the IF. With this approach two radio pictures of the Sun, each consisting of 48×64 picture points, corresponding to the two opposite circular polarizations, were produced every second. The beamformer consisted of 48×48 (half the number of antenna elements) = 2304 accurately cut coaxial cables that occupied a volume of $1 \times 3.2 \times 2$ m.

The bulk and weight of these beamformers prevented designers from realizing many more antenna beams from phased arrays.

3.14.2.2 Photonics-Based Beamformers
The maturing of the photonics technology allowed researchers to consider the realization of stationary beams by using optical fibers instead of coaxial cables. Using these architectures, Cardone [3.142] realized three beams when the number of antenna elements was eight. Using this approach, studies related to the realization of 1024 staring antenna beams resulting from an array having as many antenna elements have been reported [3.143–3.146].

The beamformer was part of a proposal [3.147] for a phased array-based radar that generated 1024 staring beams filling the entire surveillance volume; furthermore, the radar was capable of imaging targets of interest by using the ISAR technique—see Chapter 2 also. The specifications called for the beamformer to be:

1. As compact and lightweight as possible.
2. Invisible to the array system; the insertion losses had to be offset before the beamformer so that the system SNR was unaltered.
3. The spurious free dynamic range to be as high as possible when the instantaneous bandwidth was typically 1, 0.1, and 0.01 GHz.

Requirements (2) and (3) are met by using an external modulation scheme and erbium doped fiber amplifiers (EDFAs) as illustrated in Figure 3.44. The Mach–Zehnder electro-optic modulators (EOMs) suggested by the studies have a truly wide bandwidth (40 GHz) and lower relative intensity noise (RIN) in the vicinity of less than -170 dB/Hz compared to direct modulation methods.

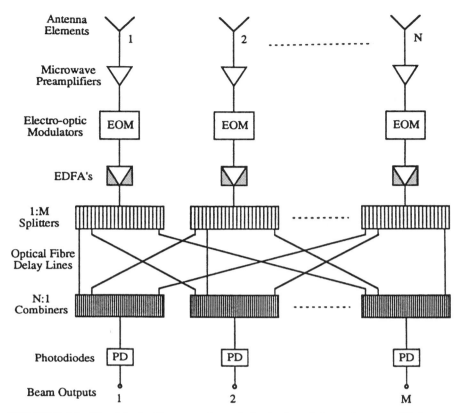

FIGURE 3.44 The block diagram of a photonics-based beamformer utilizing an external modulation scheme and EDFAs. With this arrangements the beamformer is invisible to the system.

With this arrangement a system noise figure of 3.5 dB resulted when the front-end noise figure was 3 dB; furthermore a spurious free dynamic range (SFDR) [3.148] of 134 dBHz$^{4/5}$ or 62, 70 and 78 dB results when the bandwidth is 1, 0.1, 0.01 GHz, respectively, [3.144 and 3.145]. Other researchers either proposed or used EDFAs in conjunction with beamformers (e.g., [3.149]).

Requirement (1) is met by combining two important approaches for reducing the hardware required for beamformers: (i) the partitioning of the beamformer [3.150]; and (ii) the use of a wavelength division multiplexing (WDM) scheme. By partitioning the beamformer, the minimum number of delays required, $K_{min}|_P$ is given by the equation

$$K_{min}|_P = 2M\sqrt{N} \qquad (3.133)$$

The WMD scheme further reduces the number of interconnections which is given by the equation

$$K_{min} = 2M \log_2 N \qquad (3.134)$$

For the WMD scheme there exists an optimum number of wavelengths that minimizes K_{min} [3.144, 3.146]. For the case where $M = N = 1024$, the optimum number of wavelengths required is 256.

When $M = N = 1024$, the minimum number of interconnections is reduced from 1,048,576 to 20,480 [3.144], a massive reduction indeed; this reduction represents a 98% decrease on the number of interconnections.

Multiple simultaneous beams can also be generated by using an acousto-optic system [3.151].

3.14.2.3 Realizations Using Resistive Networks B. Y. Mills and his colleagues proposed and realized several cross-type radio telescopes. In 1963 Mills et al. [3.152] proposed the one-mile-long Sydney University cross-type radio telescope that had 11 independent pencil beams [3.153, 3.154]; the beams were formed by using resistive dividers between the appropriate antenna and beam lines, shown in Figure 3.45. The required phase-shift between an antenna and a beam was selected by connecting the beam lines to

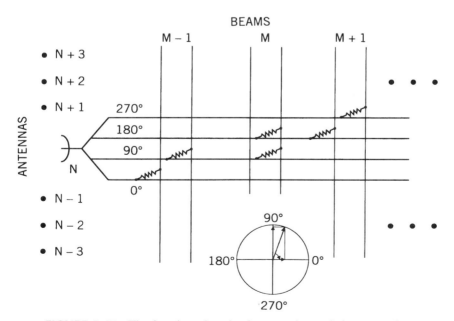

FIGURE 3.45 The forming of staring beams using resistive networks.

the appropriate antenna lines via a resistor-dividing network. In Figure 3.44 the required phase between beam $M - 1$ and antenna Nth is in the first quadrant; therefore, the resistive dividers are between the lines corresponding to that beam and the 0° and 90° lines corresponding to the Nth antenna. The resistive dividers are then selected to match the required 0° and 90° vectors. A similar technique of beamforming and an extension of the method that allows the generation of several agile beams was proposed to the radar community in 1989 [3.154a].

The simplicity, versatility, and elegance of the architecture described above was recognized by the researchers of the COMSAT laboratories [3.155] as well.

3.14.2.4 Realizations Using the Blass and Butler Matrices—Applications
The Blass [3.156–3.161] and Butler [3.162–3.168] matrices, shown in Figures 3.46a and 3.46b, respectively, have been used for many applications to generate a number of staring beams.

The Blass matrix [3.156] consists of a set of traveling-wave feed lines connected to a linear array crossing another set of lines and directional couplers interconnecting the two sets of lines. A signal applied at a beam port progresses along the feed line to the end termination. At each crossover point a small signal will be coupled into each element line, which excites the corresponding radiating element. The topology of the matrix is such that beams are formed in different directions and the aperture illumination is defined by the coupling coefficients of the couplers.

In designing Blass matrices one attempts to minimize the powers dissipated in the terminations; however, in practice the range of coupling values available to the designer is limited [3.157]. A synthesis procedure which estimates the efficiency of arbitrary beam crossover levels has been reported [3.158]. Shaped beams have been synthesized using Blass matrices [3.159]. The Blass matrix concept has been extended to form a planar, two-dimensional multiple-beam microstrip patch array [3.160, 3.161].

The Butler matrix [3.162] consists of fixed phase shifts interconnected to hybrids and yields orthogonal beams having −3.9 dB-beam crossovers. The introduction of 180° hybrids reduces the number of phase shifters significantly; in addition, the minimum phase shift is twice that for a 90° hybrid matrix, and this property facilitates the maximum matrix size constraint [3.163]. A further reduction in the hybrid count can be achieved using reflective matrices [3.164, 3.166]. The crossovers shown in Figure 3.45b can be eliminated by reconfiguring the matrix into a checkerboard network [3.166]. Wide-bandwidth matrices with up to an octave frequency range have also been reported [3.167, 3.168]. Finally, Butler matrices have been realized in different media, including waveguides for high-power use [3.169], microstrip [3.170], and integrated optic form [3.171].

A 64 × 64 Butler matrix is perhaps the largest possible, using low relative permittivity microstrip technology [3.133].

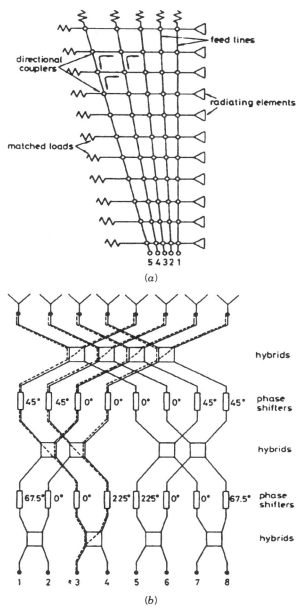

FIGURE 3.46 Conventional beamforming approaches yielding staring beams: (*a*) the Blass beamformer; and (*b*) the Butler beamformer.

3.14.3 The Formation of One Agile Beam—Applications

For radar applications one agile and inertialess beam is required to scan the surveillance volume; for active phased arrays, the beam is formed by the following arrangements:

1. Each T/R module contains a programmable phase shifter that is set to a specific value so that the array points to the required direction θ, ϕ; in some cases amplitude control signals and corrective amplitude or phase adjustments to offset intermodule variations are required.

2. The steering and correction commands for each T/R module are generated in a central computer and then distributed to each T/R module, or a central computer transmits minimal information to each module. For example, the required θ and ϕ values are transmitted and the element controllers in each T/R module perform the remaining calculations and functions; the transmission of the essential information from the central computer to the T/R modules is implemented conventionally [3.172] or via optical lines [3.173]. As the arrays become larger and the requirements for sidelobe-level control become more stringent, the latter approaches become more attractive. The fiber-optic link envisaged to perform the required functions [3.173] is small and lightweight, provides excellent isolation, has inherently wide bandwidth, and is by and large immune to EMI (electromagnetic interference) and EMP (electromagnetic pulse). A compact electrooptic controller for microwave phased array antennas using nematic liquid-crystal display (LCD) technology has been experimentally demonstrated [3.174].

3. All outputs of the T/R modules are suitably processed so that sum and difference beams are generated.

The issue of whether a beamformer should operate at RF or at IF is an interesting one. A couple of recent developments have a significant impact on beamformers: (1) the use of optical fibers to distribute RF and/or digital signals to and from the T/R modules located all over the array aperture [3.78, 3.79] and (2) the emergence of affordable MMICs that are manufactured to close tolerances. These two developments render beamforming at some IF not only possible but affordable and attractive. Before these developments took place, it was difficult to distribute a stable LO signal to the T/R modules for the required frequency down conversion (from RF to the IF), and the introduction of a down converter and an IF amplifier with enough gain and phase and amplitude stability, increased the system's complexity.

Now it is relatively easy to implement such a scheme that offers the designer the following advantages. The RF stage needs to have minimal gain for the bulk of the required gain before the beamformer is attained at the IF. Similarly, phase shifters are more accurate, have minimal insertion loss and are less expensive if they are realized at the IF rather than at the RF. As the

RF of the array is increased, the validity of this statement is strengthened. The IF is not set for all radars but can be at a frequency that can accommodate the array's instantaneous bandwidth, B requirement, so that the ratio B/IF is about 10%. The only phased array radar employing IF beamforming is the ground based ELRA (electronic steerable radar) operating at FGAN-FFM [3.175].

3.14.4 The Formation of Several Agile Beams—Applications

With reference to Figure 3.45, it is recalled that a number of staring beams can be realized by using the arrangement of crossed lines connecting the antennas and beams, and resistors. If the resistors are replaced by programmable vector modulators, which we shall consider in Chapter 4, we will have the arrangement shown in Figure 3.47 [3.154a, 3.155]. The programmable vector modulators can produce any phase shift between the antenna and beam lines by varying the X and Y vector weights.

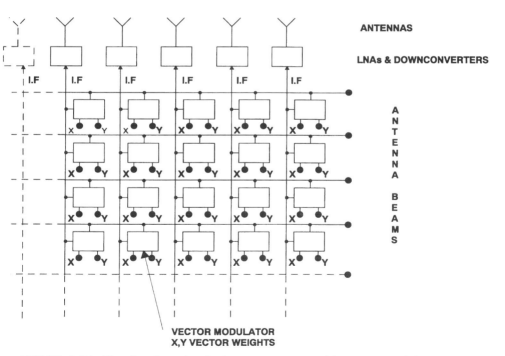

FIGURE 3.47 The forming of agile beams; programmable vector modulators are placed at the cross-over points of the lines connecting the antenna elements and the antenna beams.

When inputs X and Y take the sets of values X_1 and Y_1 or X_2 and Y_2, the resulting beams point toward the direction θ_1, ϕ_1 or θ_2, ϕ_2, respectively. Thus a cluster of otherwise staring beams is scanned anywhere within the surveillance volume. Alternatively each beam can be directed to any direction within the surveillance volume.

As the beamforming operation is undertaken at the IF, the same beamformer can be used at RF_1, RF_2, and so on, provided a suitable LO down converts the RFs to the same IF. With this arrangement the same beamformer can operate at several spot RFs on a time-sharing mode and can accommodate an instantaneous fractional bandwidth that is dependent on the vector modulators used but is usually 10% or more.

The resulting beams can be shaped [3.155] and a slight variation of the same basic architecture allows the generation of hopping beams [3.155].

From the foregoing considerations it is clear that the beamforming architectures we have explored have obvious applications in radar and communication [3.155] systems; applications for the former systems have not been fully explored.

3.14.5 Photonics-Based Wideband Beamformers

Early work on photonics-based wideband beamformers [3.176] demonstrated the feasibility of photonics beamforming at L and X band by using the same beamformer. Later work [3.177] demonstrated the significant advantages of photonics-based beamformers, which are:

1. Instantaneous wideband operation at 3–6 GHz.
2. The beamformer was 75% smaller and lighter than its electronic counterpart.
3. The loss incurred by the beamformer was 14 ± 0.4 dB ($\lambda = 1.3$ μm), which is tolerable.
4. EMI immunity.

The beamformer utilized optical fibers, and other realizations use acoustooptics devices to attain true time delays [3.178].

The weight advantage of photonics-based beamformers makes them attractive for airborne and satelliteborne applications.

Considerable work toward closing the knowledge gap between electronic engineers and photonics experts has been reported [3.179–3.182]. Additionally, work that explores new photonics techniques by academics and communication engineers is to be found in the open literature [3.183–3.184]. The application of photonics to phased array problems, for example, beamforming and signal distribution across the array face, represents a fertile field for continuing R & D.

It is true however that photonics researchers have to compete against other established beamforming techniques [3.185]. There are enough traditional beamforming techniques to satisfy most requirements; indeed some of the techniques we have explored have unprecedented versatility [3.154a–3.155]. However it is useful to reiterate that optical techniques have a weight advantage compared to conventional approaches and offer immunity to EMI and EMP.

3.14.6 Digital Beamforming

Once the array SNR is established by the front-end subsystems, the formation of the antenna beams can be implemented by many means. Given the popularity of digital signal processing, it is not surprising that digital beamforming is widespread in many fields.

The essential advantages of digital beamforming techniques are [3.186, 3.187]:

1. Improved adaptive nulling.
2. The generation of multiple beams.
3. Array element pattern correction.
4. Antenna self-calibration and ultralow sidelobes.
5. Flexible radar power and time management.

Their limitations are centered mainly around costs and instantaneous bandwidth. So long as the cost of A/D (analog-to-digital) converters is high, digital beamformers will be used at the subarray level. In the early realizations the bandwidth of digital beamformers was adequate for narrowband applications [3.188]. However, with the passage of time, their bandwidth increased and costs decreased.

The advantages of digital beamforming for radar and communications have been recognized for many years. The technology required to implement a digital beamformer, however, has not been available until recently [3.189]. By using hybrid wafer-scale integrated-circuit (HWSIC) technology, a practical 8-channel, 4-beam, 10-MHz bandwidth beamformer can be implemented in a package only 2 in. on the side [3.190]. Array sidelobes can also be suppressed because compensations due to amplitude and phase errors can be implemented [3.189, 3.190]. Sidelobe levels of −45 dB have been attained from a 64-element phased array [3.189].

The requirements for a phased array antenna designated to operate in conjunction with a bistatic receiver at X-band have been outlined [3.191].

3.14.7 Nonlinear Beamformers

Most if not all, beamformers are based on linear models that, terms of statistics, make use of the first- and second-order moment information of the data only, including the mean and the variance. Natural phenomena underlying the generation of many processes are often nonlinear; the quality of linear modeling is therefore inadequate.

In recent years, a nonlinear model based on radial basis function (RBFs) has drawn considerable attention from the signal processing community. The use of RBF was introduced by Powell [3.192] for multivariable interpolation. The RBF method has been used for time-series prediction, and the method can be implemented using a neural network structure [3.193].

The RBF technique can also be used in the applications of array processing and beamforming. The technique is developed on the basis of nonlinear functions and is model independent. Therefore it possesses the following desirable characteristics:

1. It can estimate signals in different kinds of environments of Gaussian, non-Gaussian or colored noise.
2. It is robust in terms of resolving multiple coherent signals.
3. It is capable of resolving signals separated by less than an antenna beamwidth; in fact, some preliminary results have been reported [3.194].

3.15 ARRAY PERFORMANCE MONITORING, FAULT ISOLATION, AND CORRECTION APPROACHES

Phased arrays have graceful degradation properties and relatively long times before catastrophic failures occur. The following operational requirements however are worth considering:

1. To monitor the array performance regularly.
2. To isolate any faulty units (usually due to phase-shifter settings and faults due to other causes).
3. To implement a correction procedure that restores the array performance to a state that is as close as possible to the free-fault array performance, in an operational environment that does not allow the immediate replacement of faulty units.

The systems proposed in Liu et al. and Lee et al. [3.195, 3.196] are considered here. Using transmission-line signal injection, the T/R modules are fed with a train of pulses with a selected PRF (pulse repetition frequency). In order to isolate the element under test from the other elements, the phase shifter of the module under test is toggled by changing the phase shifter phase by 180° from pulse to pulse. The output of the phase toggled element can be picked up using a (PRF/2) filter while the other stationary element outputs are filtered out; a similar technique is used to pick up moving targets from stationary clutter.

If a faulty element is discovered, then the fault correction method is used to compensate for the failed element by readjusting the complex weights of the remaining operating elements. These techniques are often referred to as "self-healing." Shore's sidelobe sector nulling method is used [3.197].

3.16 AFFORDABLE PHASED ARRAYS: SYSTEMS APPROACHES

The cost of phased arrays is destined to decrease with the passage of time, due to many innovative approaches for the realization of many subsystems. Given that T/R modules constitute 50% of the array cost, considerable effort has been devoted to the reduction of their cost [3.198, 3.199]. More details related to the many different approaches used to reduce module costs will be explored in Chapter 4.

Array cost can also be reduced by considering different phased array configurations and by the evolution of phased arrays performing a multitude of required functions. In the latter case economies result by the substitution of many radars performing a set of functions with one radar performing the same set of interrelated functions.

While many designers consider what can be done to reduce the cost of one phased array and then multiply the derived cost by 4 (to calculate the cost of a four-faced phased array), P. M. N. Keizer, in a remarkable paper [3.200], proposed an approach for reducing the cost of a four-faced phased array. The approach he proposed was to utilize the same transmitter for the four faces of the array on a time-sharing basis and the power-sharing arrangement is illustrated in Figure 3.48; to minimize leakages between arrays, each array has to operate at a different frequency. The CW transmitter therefore generates four different frequencies—F_1, F_2, F_3, and F_4—which are, in turn, interleaved in the arrangement shown in Figure 3.49. Given that the transmitter cost constitutes a significant portion of the T/R module, the savings gained by the adoption of Keiser's proposal are considerable.

At present a variety of radar-related functions, such as surveillance, fire control, and tracking, are undertaken by the same phased array; future wideband phased arrays will be capable of performing a much larger set of functions. The following examples will serve the purposes of illustrating the trends:

1. The phased array included in the shared-aperture proposal will perform a variety of functions including radar, ESM, ECM, and communication functions [3.108–3.113]; furthermore, the radar functions will be performed at the optimum frequencies for surveillance and tracking, and the radar will have have ECCM capabilities [3.111, 3.112].
2. At present phased arrays perform a set of functions, while the identification function of certain targets is performed by a conventional

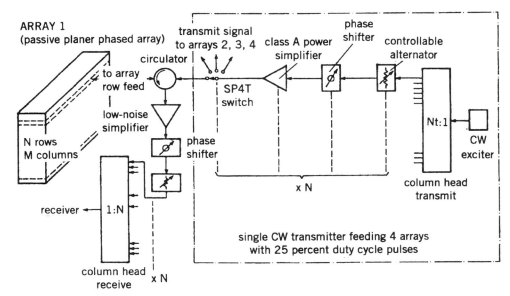

FIGURE 3.48 Block diagram of a four-faced array configuration driven by a semi-distributed solid-state CW transmitter as proposed by Keizer [3.200].

ISAR; the possibility of performing the ISAR function with the same phased array has been proposed [1.147]. The proposal called for the array to operate in the staring mode so that the ISAR operation can be performed under LPI conditions. The array generates enough staring beams (about 1000) to fill the required surveillance volume $\pm 45°$ in azimuth and $45°$ in elevation when the HPBW is $2°$, and the scene is either flood-illuminated or illuminated with the aid of a bidirectional beamformer. For cases where the scene is flood-illuminated, a compact and lightweight photonics-based beamformer capable of yielding the required number of beam has been proposed [3.143–3.146].

3. Active arrays to perform a set of functions related to air traffic control and safety [3.201, 3.202]—see Chapter 1 also.

4. Ultrawideband active radar arrays on board unmanned air vehicles to perform the long acquisition range surveillance and foliage penetrating SAR functions [3.203]—see Chapter 1 also.

This trend is destined to continue because the cost apportioned to many instruments performing a set of functions is higher than or equal to the cost of one truly multifunction array. What is more important however is that the set of interrelated functions performed by truly multifunction systems will be performed more efficiently.

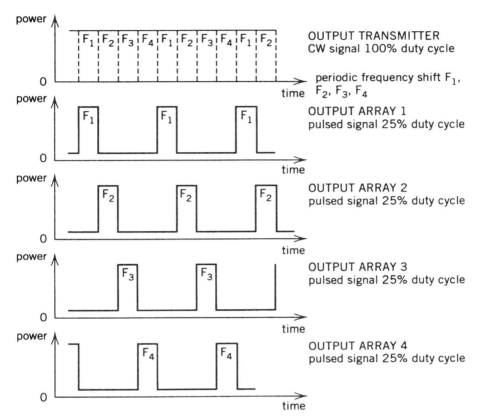

FIGURE 3.49 The CW signal generated by the transmitter of a four-faced phased array, is split-up in four trains of interleaved pulses, as proposed by Keizer [3.200].

3.17 CONCLUDING REMARKS

This chapter is dedicated to various narrow-band or canonical and wideband phased array-based systems. The chapter begins with an outline of the essential characteristics of linear and planar phased arrays and the derivation of modern synthesis procedures yielding the sum and difference radiation patterns of linear arrays; these procedures are founded on the premise that if a suitable amplitude taper is established across the array, the resulting far-field radiation pattern will, at least theoretically, have a specified sidelobe level.

Given that current T/R modules having arbitrary power levels in the transmit mode cannot be economically produced, the designer is unable to implement the required amplitude taper; several researchers however came up with ingenious solutions most of which involve spatial tapers across the arrays. The concerns with these approaches have been the derivation of the conditions under which the equivalence between amplitude and spatial tapers holds and the resulting sidelobe and grating lobe levels.

In a more general way, the consequences of having quantization of amplitude, phase or delay errors in a phased array, have been assessed and defined; these errors give rise to grating lobes and can be likened to the systematic errors encountered in filled apertures only because of their severity; methods of managing the resulting grating lobes of an array are also outlined. Random errors on the other hand give rise to increased sidelobe levels, broaden the main beam and decrease the array gain. If the amplitude and phase variations in a set of T/R modules are tightened, the resulting array sidelobe level decreases; we have outlined the tradeoffs at the disposal of the designer.

The regimes of applications for active and passive arrays are defined before array architectures and interconnects are examined. The heating resulting from the multitude of active circuits in phased arrays is a significant problem, especially for arrays designated to operate at millimeter-range wavelengths. The designer has to manage not only the heat emanated from active circuits and their power supplies but also minimize the volumes of the T/R modules and their power supplies. Current solutions to these problems have been reviewed.

In Chapter 1 we made a case for truly multifunctional wideband phased arrays; while the current interest is toward shared aperture systems, which are considered in some detail, several examples of wideband arrays drawn from radar, HF, OTHRs, ESM, and ECM systems are considered. Substantial R & D effort is required before the realization of shared aperture systems takes places; this is mainly because the challenges imposed by shared aperture phased arrays are different from previous challenges. There is no doubt however that this area of research activity is destined to be the next frontier for array designers.

A theoretical framework for minimum or null redundancy radiometric phased arrays is given before the systems utilizing these arrays are reviewed. While current interest is in applied science applications and the arrays have relatively low spatial resolution, arrays operating at mm wavelengths will have higher resolutions. Inter alia the applications envisaged for the latter systems are aerial terrain following under cloud conditions and reconnaissance under LPI conditions.

The chapter ends with a section on beamformers and encompasses most of the current techniques and approaches. Photonics-based beamformers have a weight advantage over their conventional counterparts and can be used in conjunction with wideband phased arrays. More conventional beamformers, including digital beamformers, on the other hand offer a suite of attractive features.

CHAPTER FOUR

Transmit/Receive Modules

> Nothing in the world can take the place of persistence. Talent will not; nothing is more common than unsuccessful men with talent. Genius will not; unrewarded genius is almost a proverb. Education will not; the world is full of educated derelicts. Persistence and determination alone are omnipotent.
>
> Calvin Coolidge (1872–1933)

It is very nice for system designers to consider phased arrays that are affordable, can operate over wide bandwidths and are compact. In the remaining two chapters of the book we shall explore in some detail how realizable some of these systems are. We shall deal with the complex issues related to T/R modules in this chapter, while Chapter 5 is devoted to phased array antenna elements.

From an economic and technological point of view it is not an exaggeration to state that the T/R modules are the cores of phased arrays. While many scientists and engineers focus attention on the technological issues, which we shall explore in this chapter, the economic issues are just as important. In the civilian arena economic issues have always been important; lately the same issues are important for the military because the defense budgets of many countries have shrunk and continue to shrink.

The wider acceptance of phased arrays into systems assigned for military and civilian applications is therefore based wholly on the thesis that the cost of T/R modules is bound to decrease in the not-too-distant future. In this chapter we shall consider and examine the underlying reasons on which this thesis is based. More explicitly, the infrastructure required for the production of affordable solid-state MMICs is in place in many countries, and array designers are about to reap the benefits resulting from the significant efforts that started as early as 1987 when the U.S. Defense Department's Advanced Research Project Agency (ARPA) initiated the MIMIC program. Within this infrastructure, manufacturing processes that increased the yield, availability, and uniformity of MMICs were developed. Cost reductions will therefore result from the breakthroughs that came about from a long period of continuous learning.

From a systems point of view, the T/R modules define, inter alia, the equivalent-noise temperature, the power output and dynamic range of the array. The radar range is therefore significantly dependent on the T/R modules. Other important array parameters, such as the average sidelobe level, heat dissipation, and bandwidth (instantaneous and total), also depend on the T/R modules.

It is impossible to study T/R modules without coming face-to-face with the dilemma: Should the array designer choose solid-state MMIC or vacuum-tube-based modules? There is widespread agreement that solid-state devices (SSDs) are eminently suitable for most subassemblies related to phased arrays, such as LNAs, phase shifters, logic circuits and microprocessors, medium-power amplification, phase-locking subassemblies, and fixed- and variable-frequency oscillators. It is even true that solid-state power amplifiers have replaced some high-power tube amplifiers at frequencies ranging from a few megahertz to 1.5 GHz of the spectrum [4.1]. Recently 22 kW of peak power was generated by solid-state devices in the 2.7–2.9-GHz range by using conventional power addition and state-of-the-art power combiners [4.2].

From these premises, however, one cannot draw the conclusion that high-power solid-state-based amplifiers and oscillators will soon substitute vacuum tubes at all wavelengths. There is, therefore, a need to define the regimes of applicability for these competing technologies; furthermore, we shall define the regimes of applicability for the many kinds of active solid-state devices not only for medium- to high-power generation but also for low-noise amplification.

Array designers faced the either (solid-state) or (vacuum-tube) dilemma for too long; recent developments, however, urge the designer to consider not only the solid-state but also the vacuum-tube option. More specifically, the microwave power module (MPM) utilizes solid-state and vacuum-tube technology for the purposes of offering the designer high power over relatively large bandwidths. These developments and modern developments related to vacuum technology hold the promise for new approaches to the problem of generating high powers, especially at mm wavelengths.

While the transmission of high powers is important to maximize the radar's range, we are naturally interested in transmitters that can be phased-locked so that a further increase in the radar's range can be attained by the integration of several pulses. If this is not possible, for some devices, we would like to explore the techniques developed to track the phase and frequency of each transmitted and received pulse.

It is important to stress that comparisons between devices, necessary as they might be, are time-dependent in the sense that new technologies will necessarily modify our conclusions. While it is true that the technology will change with the passage of time, it is also true that the new devices have to meet the criteria we have defined in this chapter and compete against the established devices considered here.

It is beyond the scope of this book to describe in detail the structures and the principle of operation of the many SSDs and tubes that we will consider here; we shall, however, succinctly outline the principle of operation of the most popular vacuum tubes and SSDs and refer the reader to published work where more information can be found.

4.1 VACUUM-TUBE AMPLIFIERS AND OSCILLATORS

Vacuum tube amplifiers and oscillators are usually classified into two categories, slow- and fast-wave devices; slow-wave devices are further subdivided into linear-beam and cross-field categories. Klystrons, TWTs, and backward-wave oscillators (BWOs) are slow-wave, linear-beam devices, while magnetrons and cross-field amplifiers (CFAs) are examples of slow-wave, cross-field devices; finally, gyrotrons are examples of fast-wave devices.

For some time slow-wave, linear beam devices enjoyed considerable popularity mainly because both the intrapulse and interpulse phases were controllable with the aid of appropriate phase-locked loops. These transmitters were therefore useful for a variety of radars, where a number of transmitted pulses are integrated for the purposes of increasing the radar's SNR.

Magnetrons and gyrotrons did not enjoy as much popularity because the phase and frequency of pulses emanating from these devices were time-variant within one pulse. Recently these devices have been used in a variety of applications because modern signal processing techniques allow the designer to overcame the above shortcomings—see Section 4.1.4.1.

What follows are descriptions of the most popular tubes used for radar applications; brief descriptions follow of novel and promising devices that are likely to gain prominence in the not-too-distant future.

4.1.1 Slow-Wave, Linear-Beam Tubes (LBTs)

A linear electron beam is established between the cathode and the collector of the devices in this category; additionally, the electric and magnetic fields are parallel to the electron beam. A microwave slow-wave circuit supports a traveling wave of electromagnetic energy which interacts with the electron beam.

The most popular tubes used for radar applications are klystrons, extended-interaction klystron oscillators and amplifiers (EIKOs, EIKAs), and TWT amplifiers (TWTAs).

4.1.1.1 The Klystron Family Figure 4.1*a* illustrates the operation of a modern klystron that was invented by the Varian brothers in the late 1930s.

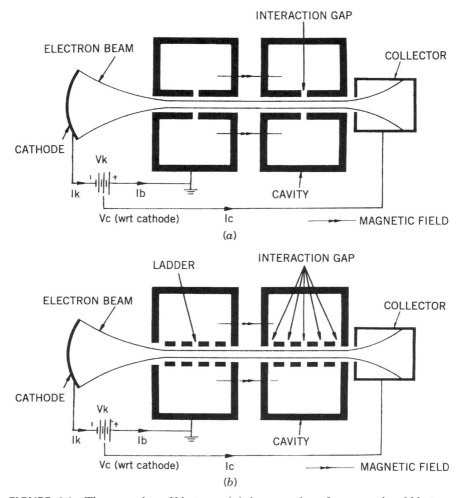

FIGURE 4.1 The operation of klystrons: (*a*) the operation of a conventional klystron; (*b*) the operation of an extended interaction klystron.

The cathode emits a stream of electrons when it is heated to about 1100°C. The electrons are focused electrostatically into a narrow-waist beam, the diameter of which is maintained through the application of a collinear magnetic field. The electrons can have a terminal velocity of about 15–30% of the velocity of light as they enter the cavities. The cavities are used to either set the frequency of operation for the klystron oscillator or as input/output ports for the klystron amplifier. The interaction between the RF field and the electron beam causes velocity modulation and bunching of

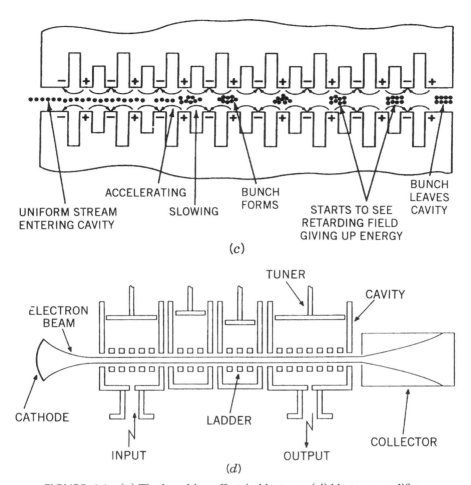

FIGURE 4.1 (c) The bunching effect in klystrons; (d) klystron amplifier.

the electrons. The bunched beam causes induced currents to flow in the output circuit of the klystron, and energy is thus extracted from the device.

While a conventional klystron has one interaction gap per cavity, the EIK illustrated in Figure 4.1b has several interaction gaps per cavity. The distributed or extended interaction provides more efficient bunching of the electrons, which leads to higher interaction efficiency or gain bandwidth products. The bunching action of electrons is illustrated in Figure 4.1c, while klystron-based oscillators and amplifiers are illustrated in Figures 4.1d and 4.1e, respectively.

Figure 4.2 illustrates the highest peak powers available from EIK as a function of collector voltage and frequency in solid lines; highest peak output

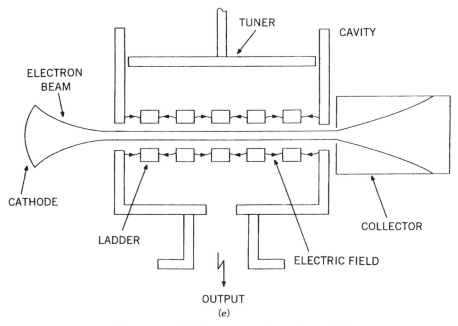

FIGURE 4.1 (*e*) Klystron oscillator (from [4.3]).

powers achieved are indicated by crosses. Klystrons are essentially high-power devices operating over narrow bandwidths [4.3].

4.1.1.2 The TWT Family The traveling-wave tube was invented by Rudolph Kompfner toward the end of World War II. Figure 4.3a illustrates a helix TWT amplifier that consists of a cathode, a collector, and a helix that allows the RF to interact with the electron beam, which is again confined within the helix by an axial magnetic field; the electron beam is again velocity-modulated by the injected RF signal. The modulation or electron bunching induces higher-amplitude RF currents on the slow-wave structure.

The attenuator shown in Figure 4.3a suppresses oscillations caused by reflections within the slow-wave circuit but does not affect the forward-traveling electron bunches, which begin the amplification process anew as they emerge from the attenuator.

The primary function of the collector is to collect the spent electron beam after it has interacted with the signal on the slow-wave circuit. If the collector is at the same potential as the body of the TWT, the thermal dissipation in the collector will be extremely high and the overall tube efficiency low. A successful way to overcome this problem is to decelerate the electrons prior to collecting them. Figure 4.3b illustrates the method adopted by operating the collector at a negative voltage with respect to the slow-wave circuit; this

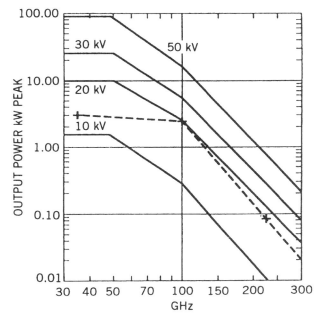

FIGURE 4.2 An illustration of the highest peak powers available from extended interaction klystrons as a function of collector voltage and frequency (solid lines); the highest peak output powers achieved are indicated by crosses (from [4.3]).

mode of operation is termed *depressed collector operation*. Space–charge forces between the electrons cause the beam to spread after its exit from the helix. The depressed collector has one, two, or three stages as illustrated in Figure 4.3b. The overall efficiency of the TWT improves as a function of its collector stages.

The great attraction of the TWTs is that the designer can have either moderate power outputs over a relatively wide bandwidth by using TWTs or maximum power outputs over a narrow bandwidth, using coupled-cavity TWTs (CCTWTs).

4.1.1.2.1 The Coupled Cavity TWTs The coupled cavity TWT, illustrated in Figure 4.3c, is ideal for narrowband, high-power applications. The power output of the CCTWT is typically one order of magnitude higher than that of a helix TWT. The bandwidth of the CCTWT is, however, typically in the range of 2–10% and can be extended to 40% by the introduction of lossy material in strategic locations to suppress out-of-band spurious propagation modes [4.4].

Figure 4.4 illustrates the output power versus frequency for TWTs designated for (*a*) space communications, (*b*) terrestrial communications, and (*c*) broadband communications [4.4], respectively.

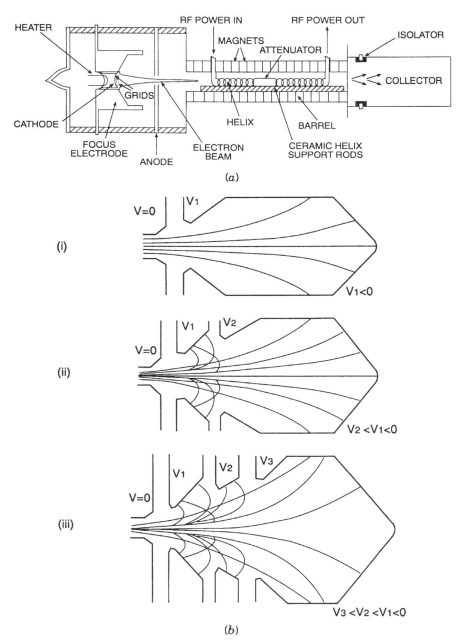

FIGURE 4.3 Traveling-wave tubes: (*a*) the helix TWT; (*b*) depressed collector of TWTs as a means to attaining higher efficiencies [(i) single-stage depressed collector; (ii) two-stage depressed collector; (iii) three-stage depressed collector].

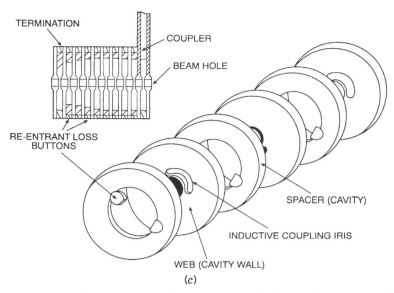

FIGURE 4.3 (c) The CCTWT. (*Source*: Hansen [4.4], © 1989 *Microwave J.*)

4.1.2 Slow-Wave, Cross-Field Tubes (CFTs)

The electric and magnetic fields in CFTs are perpendicular to each other, and the electrons can take a linear or circular path. All practical CFTs use a circular path. Cross-field devices include cross-field amplifiers (CFAs), magnetrons, and carcinotrons. Given that the magnetron is a cross-field device, the electrons emitted by the cathode take a cycloidal path to the anode. The anode can be a copper block and contains a resonant RF structure, such as slot and hole cavities. The magnetic field can be generated by a permanent magnet [4.5].

Magnetrons operating at the low-GHz region are used in microwave ovens because of their simplicity, efficiency, and low cost.

The CFA operates in a fashion similar to that of the magnetron but as an amplifier rather than an oscillator. The use of nonresonant slow-wave structure enables the CFAs to have bandwidths in excess of 15%. Their gain, however, is limited to between 10 and 20 dB. For high-power applications a CFA usually follows a TWT amplifier [4.6].

4.1.3 Fast-Wave Devices

Fast-wave devices include gyrotrons, laddertrons, and peniotrons [4.5] and use very high-velocity electron beams; only recently have gyrotrons been used for radar applications for the reasons outlined in Section 4.1.4.1.

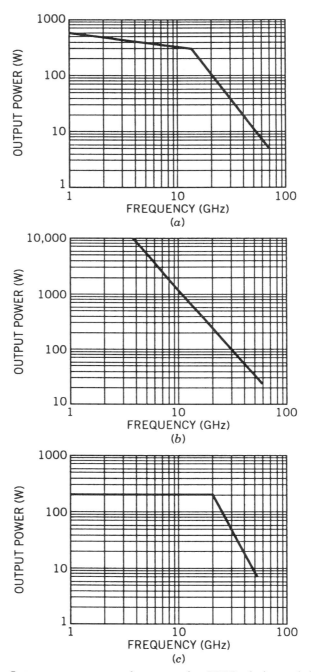

FIGURE 4.4 Output power versus frequency for TWTs designated for (*a*) space communications; (*b*) terrestial communications; (*c*) broadband instrumentation. (*Source*: Hansen [4.4], © 1989 *Microwave J.*)

4.1.3.1 Gyrotrons

For gyrotrons energy is transferred from the electron beam to the RF field by angular electron bunching, in contradistinction to longitudinal bunching, which occurs in the slow-wave, linear tubes. The RF output is available through a window.

Although gyrotrons can yield higher RF powers than other tubes, they are relatively bulky. Only some ground applications utilize gyrotrons. We have already delineated some of the pertinent characteristics of EIK and TWT operating at millimeter-range wavelengths.

In Table 4.1 the essential characteristics of klystrons, CCTWTs, magnetrons, and CFAs operating at centimeter-range wavelengths are tabulated [4.6].

4.1.4 R & D toward High-Power Vacuum-Tube Devices and the Microwave Power Module (MPM)

While improvements in performance for devices first conceived between 1930 to 1960 will continue, novel and promising devices are being developed as a result of many government initiatives and programs in the United States and research efforts undertaken outside the United States.

TABLE 4.1 Essential Characteristics of Klystrons, CCTWTs, Magnetrons, and CFAs Operating at cm Wavelengths (after Sivan [4.6])

Characteristics	Klystron—Amplifier	CCTWTA	Magnetron—Oscillator	CFA
Average power at L-band	1 MW	12 kW	1.2 kW	1.3 kW
Average power at X-band	> 10 kW[a]	10 kW	100 W	2 kW
Percent bandwidth	1–10%	5–15%[b]	1%[c]	5–15%
Gain (dB)	30–65	30–65	10[d]	10–20
Efficiency (%)	≤ 65	≤ 60	≤ 70	≤ 80
Cathode voltage for peak power	≤ 125 kV 5 MW, L-band	≤ 42 kV 0.2 MW, L-band	≤ 60 kV 1 MW, L-band	≤ 105 kV 5 MW, L-band
Thermal noise	Typically −90 dBc in a 1-MHz bandwidth	Typically −90 dBc in a 1-MHz bandwidth	20 dB higher than a comparable LBT	20 dB higher than a comparable LBT

[a] 250-kW klystron amplifiers operating at X-band have been reported by Freiley et al. [4.7].
[b] Up to 100% for helix TWTs.
[c] With injection locking; 15% with mechanical tuning.
[d] Approximately 10 dB for injection locking.

270 TRANSMIT/RECEIVE MODULES

The Tri-Service/DARPA (Defence Advanced Research Program Agency) initiative resulted after several efforts by the Advisory Group on Electron Devices (AGED), which commissioned the Special Technology Area Review (STAR) [4.8]. The initiative (1990/1991), which came after some 20 years of "benign neglect" [4.9], was directed toward the following high-risk payoff areas:

1. Wideband RF amplifiers based on vacuum microelectronics;
2. Second-generation fast-wave amplifiers for mm-wave radar.

The requirement for a microwave power module (MPM) surfaced in October 1988 in the STAR and was consequently supported by the Naval Research Laboratory (NRL). The MPM employs solid-state preamplifiers, while the final power amplifier, (FPA) is a miniature TWT especially developed for this application. Thus the MPMs combine the advantages of both solid-state devices (SSDs) and vacuum tubes.

Some R & D activity on gyroklystrons has been funded by the U.S. Department of Energy [4.10]. Devices first conceived in the former Soviet Union, such as the gyrocon and magnicon, are now the foci of research activities in the West [4.9]. Notable efforts undertaken at the Institute of Applied Physics, Gorky include a 500-kW, 167-GHz, 29-ms pulsed gyrotron and a 500-MW, 10-GHz, 20-nS pulsed relativistic BWO (backward-wave oscillator) [4.9].

Other modern developments are outlined in the following section.

4.1.4.1 High-Power Gyrotrons and Magnetrons Recent developments, spurred by meteorological applications, have been undertaken mainly along two directions. The derivation of ever-increasing output powers from magnetrons and gyrotrons and the application of modern signal processing techniques render the above-mentioned high-power sources useful for low-cost, single-aperture radar applications. We are reviewing this work here because of its potential relevance to passive phased arrays.

High-power gyrotrons have been developed mostly for heating of fusion plasmas. Recently the possibility of using gyrotrons in conjunction with modern signal processing techniques to obtain Doppler radar returns from clouds and other particulates has been investigated [4.11]; Doppler images of clouds have already been reported by using low-power radars [4.12]. The higher power available from gyrotrons operating at millimeter-range wavelengths would allow much greater range and range resolution; thus a greater variety of clouds can be studied with the aid of high-power Doppler radars operating at millimeter-range wavelengths. Other possible applications include the remote sensing of clear air turbulence and of humidity profiles of clouds [4.11].

The use of modern signal processing techniques in conjunction with gyrotrons was necessary for the following reasons:

1. For a gyrotron oscillator, not only is the phase random from pulse to pulse; the waveforms may also vary.
2. The desired waveform may not be a perfect sine wave with a random phase, but may have a frequency chirp to obtain finer range resolution.

The signal processing technique consists of recording the amplitude and frequency modulation of the transmitted and received signals; the recorded signals are then compared with a reference signal [4.11]. This done, normal signal processing techniques appropriate to Doppler radar are used.

Given the cost of klystrons and TWTs, researchers are turning to magnetrons for low-cost Doppler radars operating at centimeter-range wavelengths for meteorological applications. The magnetrons, like the gyrotrons, have a random starting phase, and their frequency chirps during the transmit pulse. Modern signal processing techniques, similar to those described for the gyrotron, have to be implemented for magnetrons [4.13]; additionally advances in solid-state magnetron modulator technology have been reported [4.14].

We have noted these developments here to illustrate that vacuum-tube technology enjoys a resurgence of interest and sustained R & D efforts will lead to powerful and reliable RF sources; while most of these sources will be used by the high-energy physics research accelerators, some spinoffs of generic value will ultimately benefit the radar community. We can trace two important synergies that will bring the above-mentioned high-power devices to the radar, EW (electronic warfare), communications, and the remote-sensing communities.

1. The synergy between modern, fast, and inexpensive signal processing capabilities and high-power devices already reported at cm and mm wavelengths.
2. The synergy between capabilities availed by solid-state devices and high-power tubes to solve important and pressing problems in the above-mentioned disciplines.

4.2 SOLID-STATE TRANSMITTERS AND LOW-NOISE AMPLIFIERS

There is a plethora of SSDs used as oscillators or high-power amplifiers, which are adequately described in References [4.15]–[4.17]. SSDs can be conveniently divided into two- and three-terminal devices. The former category includes IMPATT (impact ionization avalanche and transit time) and Gunn devices; initially the latter category included bipolar junction transistors and metal–semiconductor field-effect transistors (MESFETs).

Recently other three-terminal devices, including GaAs heterojunction bipolar transistors (HBTs) and high-electron-mobility transistors (HEMTs), joined the latter category. Additionally, InP (indium phosphate)-based HEMTs and HBTs have evolved and complement their predecessor SSDs in the low-noise [4.16] and high-power output [4.17] areas, respectively. The term high power is here related to SSDs only and will be quantified in Section 4.3.1.

The HEMT, an evolutionary improvement in the GaAs FET, is variously known as a modulation-doped FET (MODFET), two-dimensional electron gas FET (TEG-FET), selectively doped heterostructure FET (SDHFET), or heterostructure FET (HFET).

The HEMT and FET differ in where the source–drain current flows. With reference to Figure 4.5, in the FET, the electrons flow in the doped channel layer, through the donor ions, and therefore undergo considerable scattering. In the HEMT, however, a potential well is created on the GaAs side of the AlGaAs/GaAs interface, or heterojunction, due to the different conduction band energies for the two materials. With this arrangement ionized impurity scattering is greatly reduced and the electrons move with higher mobility and velocity. It is this high-electron mobility and velocity that makes the RF performance of HEMTs superior to their FET counterparts [4.18].

HEMTs exhibit very high maximum frequency of oscillation f_{max}—the frequency at which power gain drops off to unity or 0 dB. Typically HEMTs

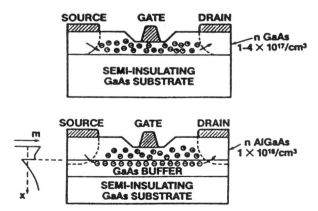

FIGURE 4.5 Comparison of the conventional GaAs FET (top) with the HEMT (bottom) [4.18].

can yield maximum power gains of 11 dB at 94 GHz, and the f_{max} is extrapolated to be 350–380 GHz [4.18].

Today there are three types of HEMTs: the conventional HEMT (based on the AlGaAs/GaAs heterojunction), the pseudomorphic HEMT (based on the InGaAs/AlGaAs heterojunction), and the InP-based HEMT, which contains an InAlAs/InGaAs heterojunction. The three basic channel structures of HEMT types are illustrated in Figure 4.6. The HEMT will undoubtedly have a profound effect on future microwave and mm-wave systems.

Although current mm-wave HEMTs are small, generating only modest levels of power, multifinger power devices capable of much higher power are under development [4.18]. Output powers of 0.5 W at 35 GHz were reported in 1989 and the same authors predicted output powers of 0.5 W at 60 GHz and 0.2 W at 94 GHz [4.18].

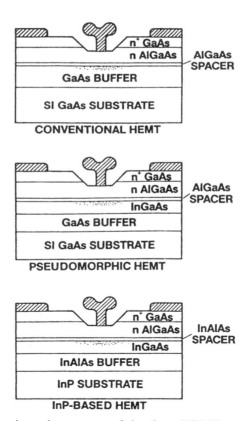

FIGURE 4.6 Basic channel structures of the three HEMT types, the conventional, pseudomorphic and InP-based HEMT (from [4.18]).

4.3 IMPORTANT COMPARISONS BETWEEN SSDs AND VACUUM TUBES

We wish to compare the capabilities of SSDs with those of vacuum tubes because these comparisons will go some way toward the definition of the regimes of applicability for vacuum tubes and SSDs. The definition of these regimes is often straightforward and needs no elaboration; we can cite the following immediate examples. There is widespread agreement that SSDs are eminently suitable for most subassemblies related to phased arrays, such as LNAs, phase shifters, logic circuits and microprocessors, medium-power amplification, phase-locking subassemblies, and fixed- and variable-frequency oscillators.

However, the generation of high powers by solid-state and vacuum-tube FPAs is an area where comparisons are most useful. We shall use the following criteria in our comparisons.

1. Power output as a function of frequency
2. Noise figure, thermal power, and noise density
3. The MTBF issue
4. Relative initial and life-cycle costs
5. PAE of devices
6. The moving-target indicator (MTI) stability
7. Graceful-degradation considerations
8. Duty-factor considerations
9. Power improvement per decade
10. Specific-weight factor
11. Miscellaneous considerations.

In undertaking these comparisons we are fully cognizant that technological changes will modify our conclusions. While it is true that the technology will change with the passage of time, it is also true that the new devices have to meet the criteria we have defined in this chapter and compete against the established devices considered here.

While it is convenient to notionally discuss each criterion separately, it is often the case that other considerations have to be taken into account in our discussions.

4.3.1 Power Output as a Function of Frequency

The output power of devices considered in this chapter is related to narrowband applications. Figure 4.7 illustrates the power output of solid-state devices and vacuum tubes as a function of frequency [4.19]. The regimes of

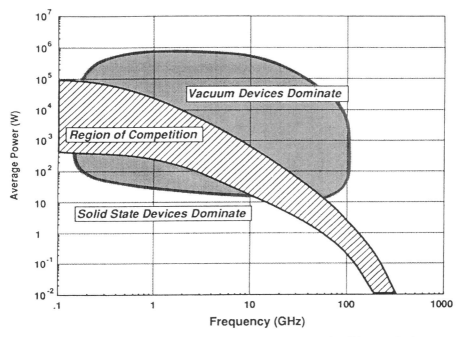

FIGURE 4.7 The average power output of vacuum tubes and solid-state devices as a function frequency [4.19].

application are clearly drawn and the region of complementarity (or competition) defined.

Figure 4.8 illustrates a more detailed outline of the capabilities of a variety of tubes and of SSDs; the power output of a single device (SS or tube) is used here [4.19]. Conventional or quasi-optical power addition techniques can boost the output levels shown in the graph for both the SSDs [4.1] and for the tubes, e.g., [4.20]. Such an approach can introduce graceful degradation for transmitters based on either SSDs or tubes.

Conventional power addition of 448 power transistors that produce a peak power output of 28 kW at L-band has been reported [4.1]. Similarly, power addition was used to generate 22-kW peak power in the 2.7–2.9-GHz frequency band [4.2].

Other examples where conventional power addition of several SSDs was used to produce considerable powers for many a phased array are tabulated in Table 4.2 [4.1].

Considerable effort has been expended to increase the power of SSDs at mm wavelengths by using quasi-optical power addition techniques [4.21]; we shall consider a variant of these techniques in Section 4.7.1.1.

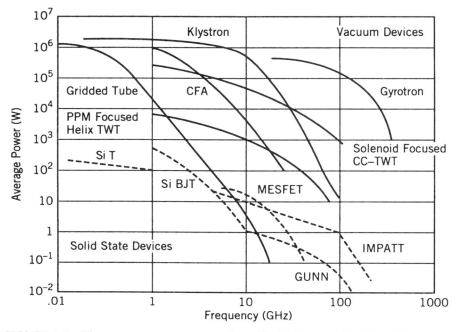

FIGURE 4.8 The average power output of specific devices as a function of frequency (from [4.19]).

From Figure 4.7 it is not hard to deduce that the average power output advantage of vacuum tubes over the SSDs is significant and that it becomes more significant as the frequency increases. A more detailed examination of the output power–frequency characteristics of SSDs is therefore warranted.

Let us examine the output power–frequency characteristics of SSDs in the 1–100-GHz and 100–1000-GHz regions separately. Figure 4.9 illustrates the power output of several SSDs as a function of frequency. Below 5 GHz the bipolar transistors yield maximum powers, while the power output of MESFET-based MMICs has significantly increased during the past few years in the 5–30-GHz frequency range.

In 1994, a 5–10-GHz, 1-W HBT amplifier having a peak PAE of 58% was realized [4.22]. Turning to narrowband operation, a 7-W monolithic HEMT amplifier was reported having a PAE of 30.8% and a 1-dB bandwidth of 1 GHz centered at 16 GHz. The power density per unit chip area for this monolithic is a respectable 0.68 W/mm² [4.23]. A 100-W CW power source operating at X-band has also been reported; it consists of a spatial field power combiner that combines the power output of 10 GaAs FET amplifiers having a 27% efficiency, to yield the available power over 10% bandwidth

TABLE 4.2 Fielded Solid-State Transmitters below 2 GHz [4.1]

System	Frequency (GHz)	Peak/Average Power (kW)	Number of Modules	Peak Power per Module (kW)	Year Finished
ROTHR	0.005–0.03	210/210	84	3	1986
NAVSPASUR[a]	0.218	850/850	2666	0.32	1986
SPS-40[a]	0.4–0.450	250/4	112	2.5	1983
PAVE PAWS[b]	0.42–0.450	600/150	1792	0.34	1978
BMEWS[a]	0.42–0.450	850/255	2500	0.34	1986
TPS-59	1.2–1.4	54/9.7	1080	0.05	1975
TPS-59[c]	1.2–1.4	54/9.7	540	0.1	1982
SEEK IGLOO	1.2–1.4	29/5.2	292	0.1	1980
MARTELLO[a]	1.25–1.35	132/5	40	3.3	1985
RAMP	1.25–1.35	28/1.9	14	2	1986
SOWRBALL	1.25–1.35	30/1.2	72	0.7	1987

[a] Solid-state replacements of prior tube-type transmitters.
[b] Parameters per array face.
[c] Upgraded with 100-W-peak power modules.

[4.24]. In the long term HBTs and HEMPs are destined to dominate the 10–100-GHz frequency range [4.15]—see Section 4.7.1.1 for more detailed considerations.

Figure 4.10 illustrates the power output of IMPATT oscillators as a function of frequency from 1 to 1000 GHz; while the power output is proportional to f^{-1}, at 5–100 GHz, the power output is proportional to f^{-3} (instead of the theoretical dependence of f^{-2}) above 100 GHz. The fast power rolloff above 100 GHz is due to technological difficulties and the basic electronic device limits [4.15].

It is clear that maximum powers at different frequency bands are attained by the utilization of different semiconductors and processes [4.15]. Barring an unexpected technological breakthrough, vacuum tubes will be used for some time as transmitters in many mm-wave phased array-based radars designated to have long ranges.

4.3.2 Noise Figure and Thermal Power Noise Density of Devices

While it is desirable to have as much power output as possible, it is worth investigating the noise figure and thermal power noise density of the devices we have been considering (tubes and SSDs).

In Table 4.1 we have tabulated the thermal noise of several tubes, and in this section we shall focus on the noise figure and thermal power noise density of TWTs and SSDs.

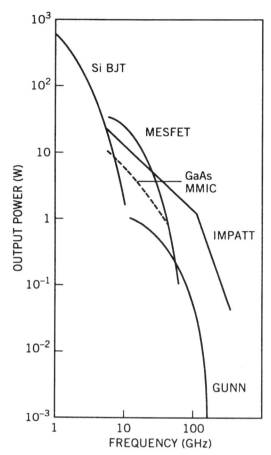

FIGURE 4.9 The output power of several SSDs as a function of frequency. (*Source*: Shih and Kuno [4.15], © 1989 *Microwave J.*)

The noise figure of a TWT is defined as the ratio of the signal-to-noise level at the input of the TWT and the signal-to-noise level at the output of the TWT. Given the linearity of the tube, the input RF thermal noise is amplified linearly but a noise term due to electron velocity fluctuation phenomena is usually added, so that the noise figure will be greater than one and will increase as the tube gain increases.

Practical multitude anode low-noise TWTs can have noise figures in the range of 8–25 dB for low-power (low-current-density) and low-temperature cathodes. Higher-power, higher-temperature cathode TWTs with a single anode can have noise figures in the range of 30–45 dB [4.6]. By comparison, typical noise figures for solid-state power amplifiers are about 8 dB.

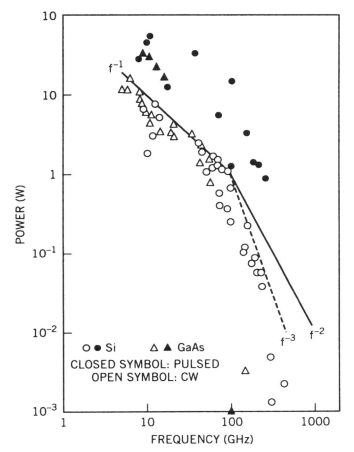

FIGURE 4.10 The output power of IMPATT oscillators as a function of frequency. (*Source*: Shih and Kuno [4.15], © 1989 *Microwave J.*)

Thermal noise power density (NPD) is defined as the noise power per unit frequency bandwidth and is given by the equation [4.6]

$$\text{NPD} = kTB(\text{NF})G \qquad (4.1)$$

where NF and G are the noise figure of the tube and its gain. In a typical TWT where NF = 40 dB and G = 55 dB,

$$\text{NPD} = -19 \text{ dBm/MHz} \qquad (4.2)$$

An effective way to decrease the NPD of TWTs is to precede it by a solid-state preamplifier; with this approach the total noise figure is defined by the solid-state preamplifier and the gain of the TWT can be decreased because of the gain contribution of the preamplifier. A figure of -45 dBm/MHz has been reported with the MPM, which is considered in some detail in Section 4.4—see also Table 4.7.

4.3.3 The MTBF Issue

The mean time between (or before) failures applies to repairable items. For a stated period of time of an item, the MTBF is defined as the mean value of the length of time between consecutive failures, computed as the ratio of the cumulative observed time to the number of failures under stated conditions [4.25]. Here cumulative time is the sum of the times during which each individual item has been performing its required function under the stated conditions.

A general impression among microwave engineers is that SSDs have longer MTBFs than tubes; this impression is based mainly on horror stories about unreliable tubes [4.26]. And yet tubes can have long MTBFs; operational life of megawatt-range EIK, of 30,000 h has been reported [4.26, 4.27], and Woods [4.26] reports that most of TWTs used in NASA deep-space missions operated for more than 7 years (60,000 h) and some lasted for as long as 14 years (120,000 h). Woods states that the reliability of microwave tubes depends strongly on two factors:

1. The attention paid to finding and eliminating failure mechanisms, that is, the learning curve for a given tube
2. The "stress" level introduced by the working environment, which can exceed the design margins

The following comments shed some light on the long lives of spaceborne TWTs [4.10]. When failures occur in the highly publicized area of space exploration and communications, the costs are high and visibility is worldwide. Thus considerable emphasis is placed on the reliability of spaceborne equipment, including TWTs.

The manufacturing documentation for the TWTs described above totals close to 1500 pages, weighing 28 lb or 14 times the weight of the tube itself [4.10]; this explanation nicely illustrates factor 1 made by Woods [4.26]. Factor 2 is clarified by observing that the environment for spaceborne TWTs (performing the communication function) is benign in the sense that the tube operates uninterrupted for long periods of time.

The MTBF of modern RF and microwave transistors is on the order of 1–10 million hours (when calculated using MIL-HDBK-217E); turning to figures related to systems that have had long operational lives, we have the following reports [4.1]:

1. 7200 T/R modules that use some 50,000 transistors have been running at the PAVE PAWS sites 1 and 2 for 10 years with an operational MTBF in excess of 77,000 h; the arrays operate at 420–450 MHz.
2. The transmitter portion of the module (PAVE PAWS system) has an MTBF well in excess of 100,000 h.
3. MTBFs of 100,000 h have been reached for the transmit module used in the Naval Space Surveillance (NAVSPASUR) arrays; the main array is 2 mi long and has approximately 2600 elements with a transmitter module at each element. The figure cited was derived after one year of operation and is expected to reach the 200,000-h mark within 3 years.

Typical calculated failure rates are eight per million hours for the transmit function of X-band T/R modules and four per million hours for the receive function [4.28]. While LNAs utilizing GaAs-based and InP-based HEMTs have comparable mean time to failure (MTTF) figures—1.5×10^7 h at 125°C with an activation energy of 1.78 eV [4.29], the MTTF of three-stage, low-noise pseudomorphic (PM) HEMT MMICs operating at 26 GHz was 2×10^6 h [4.30]. These figures for the reliability of modern MMICs are impressive.

4.3.4 Relative Initial and Life-Cycle Costs

For some time there was some agreement that the acquisition costs for solid-state-based phased arrays were higher than those corresponding to phased arrays utilizing tubes; the high costs are due mainly to the high parts count and greater assembly labor required for the former arrays [4.1, 4.26].

Experience with several arrays, including PAVE PAWS and BMEWS, supports the thesis that the initial costs were offset by the less attentive maintenance and therefore a reduction in operating personnel and a significant decrease in life-cycle support [4.1].

The NAVSPASUR systems provide useful insights to the issues under consideration because the systems utilized tubes initially before they were converted to all-solid-state systems. The mission of the NAVSPASUR is to detect and track satellite and other space objects in orbit, as they pass over the continental United States. There are three transmitter sites, located at Lake Kickapoo (main site), Jordan Lake, and Gila River, and five receiver sites positioned on a great circle in the southern part of the United States. The radar operates in a bistatic mode and locates targets by triangulation at 217 MHz. The sensitivity of the system allows it to detect targets out to 15,000-mi (24,140-km) orbits and beyond (depending on the target's RCS) [4.31]. The arrays are linear, and the main array is 2 mi (3.22 km) long [4.1].

The system provides an unaltered detection capability and has had a high traffic capacity over the past 30 years; the objectives of the NAVSPASUR transmitter modernization program at the three transmitter sites were, inter alia, to [4.31]:

1. Configure the new system for minimum life-cycle cost consistent with reliable operation at the required performance levels
2. Utilize the existing antenna and corporate RF distribution system to the greatest extent possible
3. Improve signal quality to state of the art

The approach selected to achieve the fundamental modernization objectives was to replace the high-power tube transmitter with solid-state modules distributed along the array. In Table 4.3 we have summarized [4.31] the key performance criteria of the NAVSPASUR system before and after the modernization took effect.

The operational and maintenance costs of the updated system are significantly lower than those attributed to the original system; similarly, other performance improvements include greater efficiency in transmitting power, greater system availability, and improved signal purity.

Solid-state transmitters are cost-competitive from the HF (5 MHz) to Q band (45 GHz) and offer many advantages over conventional tube transmitters, in particular, a significant decrease in life-cycle costs [4.1].

Recently the claim that acquisition costs of solid-state transmitters are higher than those corresponding to tubes has been challenged. The acquisition cost of the solid-state transmitter is lower, it is claimed, because of the highly integrated design of the low-cost RF modules and because it replaces dual-tube transmitter channels [4.2]. We have already mentioned that the acquisition cost of an active array radar (AAR) at $1000 per radiating element is comparable to that of a tube system. However, the LCC of the AAR is half that of a tube type [3.68, in Chapter 3].

4.3.5 The PAE of Devices

The efficiency of converting DC or RF input power to higher RF power output is of great importance for SSDs and tubes; if the conversion efficiency is too low, excessive heat is generated, which adversely affects the operation of the device generating the high RF power. More power therefore has to be expended to remove the generated heat.

The PAE is the accepted measure of efficiency and is given by the equation

$$\mathrm{PAE} = \frac{P_{\mathrm{out}} - P_{\mathrm{in}}}{P_{\mathrm{DC}}} \qquad (4.3)$$

TABLE 4.3 NAVSPASUR System—Performance Comparisons for Tube Solid-State System [4.31]

Performance Measure	Tube System Performance	Solid-State Performance	NAVSPASUR System Improvement
Maintenance (shift technicians)	28	19	Lower operation and maintenance
Reliability	Replace 80 RF, 10-kW final amplifier tubes per year	Replace 250 low-power components per year	Low operation and maintenance costs
Power output[a]	576 kW	767 kW	1.2 dB better performance
Safety	10-kV HVPS; extensive	28 V LVPS; no interlocks	Lower operation and maintenance costs; greater personnel safety
Prime power usage[a]	26.9% of station prime power is radiated	52.6% of station prime power is radiated	Lower operation and maintenance costs
Spectral purity	5 Hz BW; 5×10^{-10} stability	0.2 Hz BW; 5×10^{-12} stability	System detectability improvement; increased target discrimination
Array-mounted RF amplifiers	Nominal 1-dB loss between amplifier and antenna	Loss reduced to zero	20% increase in RF power delivered to dipole per RF W generated
Beam steering[a] and focusing	None	Individual phase control at bay level	Supports high-precision single-penetration orbital detection and increased single-pass target discrimination processing
Availability[a]	0.8	0.9998	Higher reliability

[a] At Lake Kickapoo.

where P_{out} is the RF output power, P_{in} is the RF input power, and P_{DC} is the power supplied to the device.

The dissipated power is, in turn, given by the equation

$$P_{dis} = P_{DC} - P_{out} + P_{in}$$

$$= P_{out} \frac{[(1 - \text{PAE}) - (1 - \text{PAE})/G]}{\text{PAE}} \quad (4.4)$$

where G is the RF gain of the device usually greater than unity; for $G > 10$, Equation (4.4) can be approximated to

$$P_{dis} = P_{out}(PAE^{-1} - 1) \qquad (4.5)$$

Thus, for a PAE of 50%, the dissipated power is equal to the RF output power.

In Table 4.4 we have tabulated the dissipated power P_{dis} as a function of PAE when the output power is 100 W. At 10% PAE one has to dissipate 900 W to generate 100 W of RF power. By contrast, only 25 W of power is dissipated when the PAE is 80%. Some authors use the term *overall efficiency* E as the ratio of the RF power output P_{out} to the DC power P_{DC}. From Equation (4.3) we easily deduce that $E = PAE$ when P_{in} is too low.

Typical efficiencies for klystrons, CCTWTs, magnetrons and CFAs operating at cm wavelengths are tabulated in Table 4.1. Gyrotrons operating at mm-range wavelengths usually have the highest PAE of current vacuum devices e.g., 40% at a CW power level of 200 kW at 60 GHz (4.5) and their efficiency typically drops to 15% at 375 GHz [4.5 and references therein]. Gyrotrons generating up to 200 kW of CW at 140 GHz have been reported [4.10].

The PAE for a modern high power klystron amplifier delivering 250 kW at X-band is 45% [4.7] and an EIA having a peak power output of 2.8 kW and a PAE of 16% at 95 GHz is reported [4.5].

Typical overall efficiencies for TWT amplifiers operating at mm wavelengths are: 41, 47 and 50% when the tube has one, two or three depressed collector stages respectively [4.4].

With reference to Figure 4.11a, which illustrates the PAE of HEMT devices as a function of frequency, PAEs of up to 60% have been reported at 10 GHz [4.15]; as the frequency of operation increases, their PAE drops to about 45 and 18% at 35 and 95 GHz, respectively. In Figure 4.11b the PAE

TABLE 4.4 Dissipated Power P_{dis} as a Function of PAE, When the Output Power is 100 W

PAE (%)	P_{dis} (W)	P_{RF}/P_{dis}
10	900	0.11
20	400	0.25
30	233.3	0.43
40	150	0.67
50	100	1
60	66.7	1.5
70	42.9	2.33
80	25	4

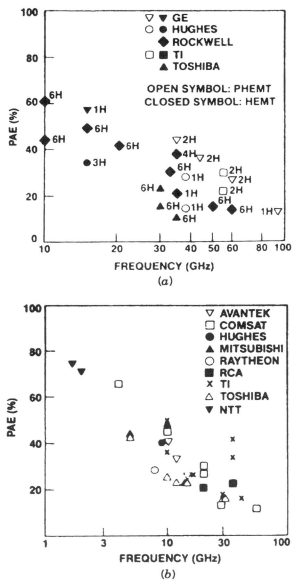

FIGURE 4.11 The PAE of SSDs as a function of frequency for (*a*) HEMTs (the symbol H refers to the number of heterojunctions); (*b*) MESFET amplifiers. (*Source*: Shih and Kuno [4.15], © 1989 *Microwave J.*)

of MESFET amplifiers is illustrated as a function of frequency. A PAE of 70% has been reported in the frequency range 5–6 GHz by a MESFET MMIC [4.32].

4.3.6 The MTI Stability

The MTI stability is calculated from measurements of the transmitter's residual SSB phase noise and AM noise [4.2].

The MTI stability of a solid-state transmitter generating 22 kW at S-band was 73.8 dB, when the same figure for its synthesizer was 75.4 dB [4.2]; the transmitter is used for airport surveillance radars.

This level of stability, it is claimed, exceeds that of any available tube type or solid-state transmitter. Higher radar stability is required for enhanced target detection and to improve detection of low RCS returns from gust fronts and dry microbursts–weather phenomena that have caused airplane accidents on landing and takeoff.

4.3.7 Graceful Degradation Considerations

The power from several sources can be combined to meet the high-power requirements for phased arrays; graceful degradation of the array performance is also attained when the designer adopts this approach.

Power combining has been used to combine the powers of a multitude of SSDs; recently the powers from several mini-TWTs have been combined [4.20] and the trend is destined to continue.

For SSDs and tubes, power combining can be implemented at the module level or on target, also known as quasi-optical power addition/combining. For SSDs however power combining can be implemented at the transistor level also; this characteristic of SSDs increases the options a designer has to meet the system's power output requirements.

The power combiner should have the following features:

1. Low insertion loss or high combining efficiency
2. High isolation between amplifier modules
3. Adequate heat sinking
4. Graceful degradation

Feature 1 implies good impedance matching at the output of all amplifiers and minimal differential phase and/or amplitude variations between the amplifiers used. Feature 4 implies that if F failures occur when N amplifiers are used, the measured output power of the combiner is in accord with theoretical predictions that have been derived. More explicitly, the output power of the combiner P_{out} is related to the input power P_{in} by the equation [4.33]

$$P_{\text{out}} = \eta_c \left[\frac{N-F}{N}\right]^2 \quad \text{when} \quad F = 1, 2, 3, \ldots \quad (4.6)$$

where η_c is the efficiency of the combiner and the squared term is the loss factor when F out of N amplifiers fail. It is assumed here that no phase or amplitude variations exist in all amplifiers used.

In Table 4.5 we have tabulated the loss factors when F takes the values $1, 2, 3, 4, 10, \ldots, 2000$ and N takes the values $4, 8, 16, \ldots, 32{,}768$. Blank entries are attributed to either nonapplicable cases or to loss factors below 0.5 dB. From this tabulation it is clear that as N increases to its limiting value, the loss factors become insignificant even when $F = 2000$. The same table can be used for conventional or quasi-optical power combining. It is clear from Table 4.5 that graceful degradation can be accurately defined in terms of the losses incurred when N sources are combined and F sources fail.

If 1 or 16 high-power tubes are used in a passive array, the losses incurred when one or two tubes fail can be either catastrophic or reasonable (e.g., 0.6 or 1.2 dB). When the same transmitted power is attained by using, say, 512 modules, the failure of one or two modules will incur negligible losses. Therefore, if graceful degradation is a principal design requirement, the option of using MMIC-based T/R modules is an attractive alternative.

TABLE 4.5 A Measure of Graceful Degradation Loss Factors in dB When F Amplifiers Out N Fail[a]

$N \setminus F$	1	2	3	4	10	20	100	200	500	1000	2000
4	2.5	6	12	∞[b]							
8	1.2	2.5	4	6							
16	0.6	1.2	1.8	2.5	8.5						
32		0.6	0.9	1.2	3.2	8.5					
64				0.6	1.5	3.2					
128					0.7	1.5	13.2				
256						0.7	4.3	13.2			
512							1.9	4.3	32.6		
1,024							0.9	1.9	5.8	32.6	
2,048								0.9	2.4	5.8	32.6
4,096									1.1	2.4	5.8
8,192									0.55	1.1	2.4
16,384										0.55	1.1
32,768											0.55

[a] No differential phase and/or amplitude variations exist between the amplifiers used.
[b] Catastrophic failure.

Maximum combining efficiencies of up to 90–95% have been reported when four mini-TWTAs operating in the 2–8-GHz band were combined [4.20]; the same combiner can be used to combine the powers of SSDs.

When the power of many modules is combined with the aid of combiners, a concept of "hot maintenance"—the ability to safely remove and replace any module without shutting down the system—naturally emerges. This capability is invaluable when the radar function, in the context of performing the surveillance function in an airport, has to be performed uninterrupted on a 24-h/day, 365-day/year basis. Other applications include the radar functions performed in periods of military confrontations lasting for an indefinite period of time.

In the next section we shall treat the cases where there are mismatches or where amplitude and/or phase variations between amplifiers exist. Analytic expressions for the efficiency of combiners under realistic conditions, and the roles and importance of the many variables involved are also delineated.

4.3.7.1 A Definition of the Combining Efficiency under Realistic Conditions

Let us explore the situation where N amplifiers are placed between an N-way divider and an N-way combiner. The efficiency of an N-way combiner η_c is the measure of how well the combiner performs its function of arithmetically adding the powers P_{av} from each individual signal source, and it is defined by the equation

$$P_0 = \eta_c \sum_{k=1}^{n} P_{av,k} \tag{4.7}$$

The following conditions apply if η_c is to truly stand for the intrinsic efficiency of the combiner:

1. P_0, the power output, is maximized by impedance matching of each of the N sources and the load connected to the combiner ports.
2. η_c is maximized and is thereafter denoted as η_{max}, when the signals to be combined are identical with each other in both phase and amplitude.

Realistically, the signals to be added do not normally comply with condition 2, so we should accept that there are amplitude and/or phase variations between the signal sources, in which case the powers are added vectorially (when the N-way power combiner is linear—a reasonable assumption).

If $P_{av,k}$ and θ_k, respectively, are the available power and phase angle of the kth input signal with respect to some reference, Equation (4.7) can be written as

$$P_0 = \frac{\eta_{max}}{N}\left[\left(\sum_{k=1}^{N} \sqrt{P_{av,k}}\cos\theta_k\right)^2 + \left(\sum_{k=1}^{N} \sqrt{P_{av,k}}\sin\theta_k\right)^2\right] \tag{4.8}$$

IMPORTANT COMPARISONS BETWEEN SSDs AND VACUUM TUBES 289

A measure of the combiner's efficiency is therefore deduced by combining equations (4.7) and (4.8) and is given by

$$\frac{\eta_c}{\eta_{max}} = \frac{\left[\left(\sum_{k=1}^{N} \sqrt{P_{av,k}} \cos \theta_k\right)^2 + \left(\sum_{k=1}^{N} \sqrt{P_{av,k}} \sin \theta_k\right)^2\right]}{N \sum_{k=1}^{N} P_{av,k}} \quad (4.9)$$

Although Equation (4.9) is accurate, it is not useful because the design engineer does not always know the exact phase and amplitude deviations from a reference for all signal sources. What is usually known is the range within which the amplitudes and phases of all sources lie.

What follows is the approach taken by Gupta [4.34]; if we stipulate that the reflection coefficients of the terminations at the load and source ports defined with respect to a reference R_0 are all equal to zero, η_c is given by the equation

$$\eta_c = \frac{P_0}{\sum P_{av,k}}\bigg|_{\substack{\Gamma_L = \Gamma_{s,k} = 0 \\ k=1,2,\ldots}} = \frac{\left|\sum s_{0,k} b_{g,k}\right|^2}{\sum |b_{g,k}|^2} \quad (4.10)$$

where all summations are carried over the range of $k = 1$ to N and

where $\Gamma_L, \Gamma_{s,k}$ = reflection coefficients of the terminations at the load and source ports, defined with respect to R_0

$b_{g,k}$ = complex amplitude of the power wave, which is launched on a transmission line of characteristic impedance R_0 by the source connected at the kth input port

$s_{0,k}$ = an element of the scattering matrix of the $(N + 1)$ port combiner, referenced to R_0, representing the transmission from the kth port to the output port designated as 0

We intend to derive the minimum value of η_c when the amplitudes and phases of the N sources are allowed to vary within known limits.

If the following substitutions are made:

$$B_k = |b_{g,k}|$$

$$S_k = |s_{0,k}|$$

$$\rho_k = \arg(b_{g,k} s_{0,k})$$

Equation (4.10) is rewritten as

$$\eta_c = \frac{\left|\sum B_k S_k \exp(j\rho_k)\right|^2}{\sum B_k^2} \qquad (4.11)$$

The lowering of η_c is due to the scatter in the values of the three variables B_k, S_k, and ρ_k from port to port; in a typical production situation the component tolerances will be known. More explicitly the amplifiers will have gains and phase angles ranging within the bounds $\pm \Delta G$ and $\pm \Phi$, respectively. If we use the substitutions

$$M_b = 10^{\Delta G/10} \quad \text{and} \quad \delta_{\max} = \Phi_{\max} \qquad (4.12)$$

the following equation has been derived [4.34]

$$\eta_c \geq \frac{4 M_s M_b \cos^2 \delta_{\max}}{(1 + M_s M_b)^2} \sum |s_{0,k}|^2 \qquad (4.13)$$

where M_s is a parameter related to the power combiner. In most cases the asymmetry caused by the power combiner can be safely neglected (i.e., $M_s = 1$); this result can also be expressed in terms of the maximum efficiency $\eta_{\max}$, which can be deduced by making the n input signals $b_{g,k}$ identical to each other in Equation (4.10):

$$\eta_{\max} = \sum |s_{0,k}|^2 \qquad (4.14)$$

Equation (4.13) can therefore be rewritten as

$$\Delta\eta = \frac{\min[\eta_c]}{\eta_{\max}} = \frac{4 M_b \cos^2 \delta_{\max}}{(1 + M_b)^2} \qquad (4.15)$$

where $\Delta\eta$ is the worst-case degradation in combining efficiency. Figure 4.12 illustrates the dependence of $\Delta\eta$ on ΔG and $\Phi_{\max}$ when perfect combiners are used. If the maximum phase and gain deviations are 13° and 2 dB, respectively, $\Delta\eta$ is a respectable 0.95.

4.3.8 Duty-Factor Considerations

In the context of pulsed radar, the duty factor (or cycle) of a transmitter is the ratio of the pulse duration over the pulse period. A short duty factor during which all available power from the source is transmitted is desirable if returns from nearby targets are not to be missed. The peak and average power for a tube increase with an increase of the duty factor until the design

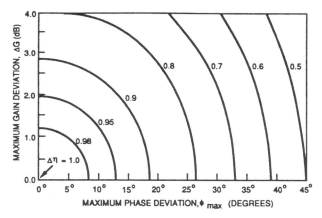

FIGURE 4.12 The maximum gain deviation as a function of the maximum phase deviation when $\Delta\eta$ takes the values of 1, 0.98, 0.95, 0.9, 0.8, 0.7, 0.6, and 0.5. (*Source*: Gupta [4.34], © 1992 IEEE.)

limits of the transmitter are reached. Any increase of the duty factor after the device reaches this limit will result in a decrease of output device power.

The average power limit for TWTs operating at C- to Ku-bands with peak power capability up to 150 kW is reached when the duty factor is in the 2–5% region [4.35]. Duty factors of 30–18% have been reported for some early SSDs [4.1]; similarly, duty factors as low as 1.6% have been reported for SSDs operating in 400–450-MHz band and yielding 4 kW of average power [4.1].

Operation at high duty cycle requires the use of high pulse-compression ratios to provide unambiguous range with fine range resolution. To cover the short ranges, short filler pulses are usually interleaved with the long pulses [4.1]. A similar approach was used for the S-band radar dedicated to airport surveillance [4.2].

Solid-state transmitters tend to favor high-duty-cycle, long-pulse applications.

4.3.9 Power Output Improvement per Decade

Kosmahl [4.36] in 1985 and Woods [4.26] in 1986 observed an improvement in average power of 20 dB per decade for tubes operating in W-band, while the same improvement for SSDs was 3 dB per decade. The timeframe they considered for the SSDs was from the late 1960s to the early 1980s. We have already observed that the power output of SSDs is relatively low at mm wavelengths. The following comments, however, are in order.

The development of a particular semiconductor process, such as the realization of InP-based HEMTs and HBTs took one decade to mature [4.29].

After a process matures, however, developments take place at a fast rate. The unity current gain for HEMT, for instance, increased from 80 GHz in 1987 to 250 GHz in 1991 [4.16]. While the efficiency of SSDs at X-band was 25% in 1985 [4.36], efficiencies of 61% were reported in 1992 [4.32].

The output power of SSDs is continually increasing with the passage of time because:

1. New materials and/or processes, enhance the performance of SSDs.
2. The evolution of efficient power combining techniques within the chip.
3. The evolution of conventional, efficient power combining techniques.
4. The evolution of quasi-optical power combining techniques.

Often the designer faces the dilemma of choosing between what is available now and what will be available in the near future and in distinguishing near future availability from future availability.

4.3.10 Specific-Weight Factor

At low powers, tubes had a built-in weight [grams per watt (g/W)] penalty; at higher powers, however, the tubes overcome their weight disadvantage [4.26]. For pulsed coupled-cavity TWTs, the specific-weight factor W_{tube}, expressed in g/W, decreases by about 25% per 3-dB increase in RF output capability and is approximately given by the equation [4.26, 4.37]

$$W_{\text{tube}} \approx 25.4 P_{\text{av}}^{-0.915} \ (\text{g/W})$$

Below 20 W the SSD-based transmitters had a weight advantage over their tube counterparts in 1985 [4.36].

These specific-weight formulations and claims have to be revisited now in the light of the following:

1. McQuiddy, Jr. et al. [4.28] have shown that an active phased array has a substantial weight advantage over a passive array. The details of these comparisons between passive and active arrays, operating at X-band, when the radar range is equal, are tabulated in Table 4.6. All arrays considered have 1500 array modules that constitute a phased array designated for fighter aircraft. The authors considered two passive arrays, one using a TWT and the other using a solid-state transmitter. The active phased array has a weight advantage mainly because the losses between the array elements and the active circuits are minimized. As a corollary the active array generates lower RF power, which in turn requires a less weighty power supply.

This example demonstrates the value of detailed considerations that encompass the weight of systems that meet many specific requirements such as the radar range and prime power. Conversely the example illustrates the

TABLE 4.6 Comparisons between Passive and Active Phased Arrays Operating at X-Band When the Radar Ranges are Equal [4.28]

Parameter	Passive TWT Tx	Passive SS Tx	Active Array
Peak power [kW]	45	66	15
Transmit duty cycle (%)	33	33	33
System noise figure (dB)	2	2	3^a
Peak two-way sidelobes (dB)	−60	−60	−60
Aircraft prime power Tx + array (kW)	53	105	27
Weight for Tx + array + array power supply (lb)	1300	1500	440

^{a} The noise figure is high because the LNA gain is limited and the losses incurred between the LNA in the front-end module and the radar receiver are considerable.

futility of considering the weight of the transmitter in isolation, or without any reference to systems that are built to meet a range of requirements. It is important to stress that these comparisons are valid for airborne phased arrays operating at X-band.

2. The weight and volume of TWTs has decreased dramatically with the introduction of the miniaturized TWTs and the MPM, which is considered in detail in the next section.

We would like to conclude that we are experiencing a rapidly changing and dynamic development environment; furthermore, weight is closely related to the array architecture. Under these conditions caution has to be applied in conjunction with timely and detailed considerations for every specific application.

4.3.11 Miscellaneous Considerations

It is acknowledged that tubes have higher resistance to EMP radiation than SSDs [4.26]. For military applications, this tube characteristic is valuable. For coherent processing of radar return signals, it is required to have regulation of tube voltages on a pulse to pulse basis.

A phase stability specification can be translated directly into applied voltage variation limits on the various tube electrodes by means of the tube pushing factors, which relate changes in electrode voltages to disturbance of the output tube pushing phase. Derivations and listings of the pushing factors for the different tube types are given in Sivan [4.6] and Ewell [4.38].

4.4 RECENT DEVELOPMENTS TOWARD HIGH-POWER MODULES

To increase the power of an MMIC-based active array, one has to increase the number of its modules. An increase in the number of modules implies an increase in the geometric area and resolution of the array. For airborne applications the array geometric area is limited; furthermore, an increase in resolution is not welcome in ECM applications, where the location of the radar to be jammed is not precisely known. For these applications one would like to increase the array power without an attendant increase in its resolution. The development of high-power modules is the natural corollary of the preceding requirement. The development of high-power modules will also benefit other types of phased arrays and single aperture systems.

Ideally the designer would like to combine the advantages offered by tubes, such as high output power and those offered by SSDs, such as graceful degradation into the next-generation phased arrays. The development of the microwave power module (MPM) goes some way toward meeting these requirements.

4.4.1 The MPM Module: Concepts, Applications, and Goals

We have mentioned some ECM applications that will benefit from high-power modules. Other applications are [4.39]:

1. Missile seekers
2. Long-range ground-based radars
3. Airborne shared-aperture radars (which perform, inter alia, the radar, EW, and communications functions—see Section 1.6.6.1)
4. Airborne standoff or support jammers and intercept radar
5. Civilian communications, ground and space segments
6. Expendable jammers and decoys

Given the totality of these applications, the production costs of the MPM will be significantly reduced. While the bandwidth of canonical arrays is narrow, it is desirable to develop high-power, wide-bandwidth modules for future active phased arrays that can perform a multiplicity of functions. A case has been made, in Chapter 1, for wideband phased array-based systems and a brief description of the 100-W MPM will follow. The power output of a single tube operating in the 6–18-GHz band was about two orders of magnitude higher than the output power of a SSD operating over the same frequency band. This was the fundamental premise on which the MPM development was based [4.39].

In Table 4.7 we have tabulated the essential characteristics of the 100-W MPM and the demonstrated achievements and goals of the coordinated programs to support vacuum technology R & D activities [4.8, 4.40].

The MPM module consists of MMIC-based medium-power amplifiers, a miniature TWT as the FPA, and the electronic power conditioner (EPC), which satisfies the power requirements of the module. Typically the MMIC

TABLE 4.7 100-W MPM Characteristics: Goals and Achievements [4.41]

Goals and Achievements Characteristics	Demonstrated (1992)	Demonstrated (1993)	Goals (1994)
Frequency (GHz)	6–18	6–18	6–18
RF power (W)	50–100	≥ 100	50–100
Gain (dB)	50	> 60	50
Duty factor	0 to CW	CW	0 to CW
Efficiency	15–40% (1 stage)	> 30%[a]	> 33%[b] (3–5 stages)
Noise power density (dBm/Hz)	−45		−45
Noise figure (dB)[c]		7/12	
Volume (cm^3)	123–655.5	125.6	122
Thickness (cm)	0.787–2.16	0.813	0.762

[a] Booster efficiency.
[b] 35–40% efficiency goal.
[c] Noise figure of the SSA/module.

amplifier provides 30 dBm of RF power with 30-dB gain and 10–15-dB gain equalization over the frequency band; the FPA, also known as the *vacuum power booster* (VPB), provides a minimum additional gain of 20–30 dB [4.41]. This combination allows the MPM to reach the power level and efficiency required while its overall noise figure is typically 12 dB, a figure that is more than 20 dB lower than the noise figure of typical TWTs; the noise figure of the MMIC is less than 10 dB [4.41].

4.4.1.1 The Vacuum Power Booster (VPB) Several manufacturers have been working toward the realization of the first MPMs, so their approaches have some divergence [4.41].

The VPB is a miniaturized helix TWT contained in a package of only 19.7 cm^3. The small size is achieved by designing for the lowest possible beam voltage (3.9–4.1 kV), minimizing the gain and using a very small multistage depressed collector to enhance its efficiency. A multistage depressed collector is used to increase the efficiency of the miniature TWT; similarly, the helix pitch is modified near the output to increase the RF to beam interaction for increased power output. For pulsed operation, beam modulation capability is provided by either a grid, or focus electrode. The realization of a high-efficiency VPB was significantly aided by the use of state-of-the-art computer design codes.

The MPM is designed for either liquid or conduction cooling.

4.4.1.2 The MMIC Amplifier The main function of the MMIC amplifier is to provide low-noise amplification to the signal generated by a phased-locked oscillator (radar application) to a level approximately equal to 1 W; the output of the amplifier is fed to the VPB, which provides an additional 20 to 30 dB of high-power amplification. Its secondary function is to equalize the output of the VPB across the band of operation (6–18 GHz).

4.4.1.3 The Electronic Power Conditioner (EPC) The EPC supplies power for the MMIC and VPB; in Table 4.8 we have tabulated the essential characteristics of typical EPCs [4.41]. The EPCs use high-frequency inverters of excellent efficiency and occupy exceedingly small volume; more explicitly, their efficiency represents a 5–10% increase in efficiency and a reduction in volume of 10–20 times compared to currently available supplies at this level.

In Table 4.9 we have tabulated the volumetric characteristics of the MPMs during the FYs (fiscal years) 1991–1994 [4.8]. It is clear that significant progress has been made during that period in maximizing the output power and minimizing the volume of the 100-W MPM.

From existing realizations of the MPM we can deduce that the volume of 122 cm^3 (minimum thickness 0.752 cm) consists of the following components:

1. The EPC occupies 98.3 cm^3 or 80.6% of the MPM volume.
2. The VPB occupies 19.7 cm^3 or 16.1% of the MPM volume.
3. The MMIC driver amplifier occupies 4 cm^3 or 3.3% of the MPM volume.

As can be seen, the EPC occupies the bulk of the MPM volume despite the considerable efforts made to minimize its volume. A further reduction of MPM volume will come about if the VPB efficiency is improved and/or the following developments related to the EPC are undertaken [4.41]:

1. The development of low-profile planar high-voltage transformers
2. Higher-frequency EPC designs, which will require advances in high-speed switching diodes and transistors

TABLE 4.8 Essential Electronic Power Conditioner Characteristics for the 100-W MPM [4.41]

Characteristics	Varian–Westinghouse	Northrop	MPM Goal
Beam voltage (kV)	4.1	3.9	—
Output power	320	300	300
Inverter frequency (kHz)	150	300	—
Efficiency (%)	91	90	90
Duty factor	0 to CW	1%	0 to CW
Volume (cm^3)	229.4	98.3	98.3

TABLE 4.9 Volumetric Characteristics of the 100-W MPMs during FYs 1991–1994 [4.8]

Features: Years	Volume (cm³)	Power/Volume (W/cm³)	Weight (kg)	Power/Weight (W/kg)
1991	1311	0.08	2.95	33.9
1992	655.5	0.15	1.5	66.7
1994	122	0.8	1.1	90.9

As the efficiency of the EPC is already 90%, the scope for considerable reductions in its size is not great; an increase in the VPB efficiency, on the other hand, will ease the burden of cooling and the demands on the EPC.

4.4.1.4 System Considerations Figure 4.13 illustrates the MPM, which can be used in linear high-power, wide-bandwidth phased arrays. If common EPCs can be used, the modified MPMs can be used in planar phased arrays.

The other consideration is that one cannot assume that the power output derived from SSAs will be lower than that derived from the MPMs by one or two orders of magnitude for an indefinite period of time in the frequency range 6–18 GHz. The competition from SSAs operating at mm wavelengths is far less acute.

4.5 SUMMARY, DISCUSSION, AND FUTURE TRENDS

From the foregoing considerations it is clear that both tubes and SSDs offer some unique and attractive features to phased array systems. In this section we shall summarize our considerations related to tubes and SSDs.

If we focus attention on the array power output, the designer has basically two fundamental options:

1. The passive array, which utilizes one or a small number of high-power tubes. The array has some losses but considerable output power, especially at frequencies above 100 GHz.
2. The MMIC-based active phased array has minimum losses at all frequencies and considerable output power below 30 GHz; above 100 GHz their power output is proportional to f^{-2} or even f^{-3}.

Graceful degradation can be a major requirement, in which case the option of using one high-power tube in conjunction with a passive array is not attractive. Passive arrays utilizing a number of tubes are to be preferred. The choice between these two options will depend on how much graceful degradation is required and other requirements, such as reliability, costs, (acquisition and LCC) frequency of operation, and the required radar range.

298 TRANSMIT/RECEIVE MODULES

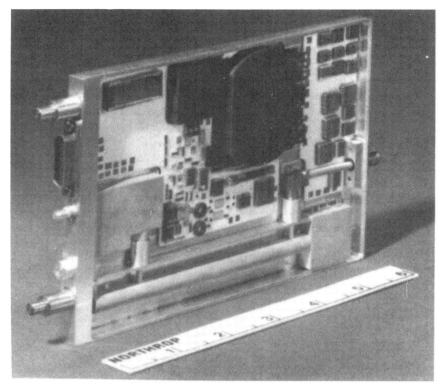

FIGURE 4.13 A photograph of the Microwave Power Module. (*Source*: Christensen [4.41], © 1993 IEEE.)

If the required range is to be maximized and operation at mm wavelengths is envisaged, the passive array utilizing tubes is a clear option. A minimization of the passive array losses can be attained if the designer chooses a passive-on-transmit and active-on-receive architecture.

4.5.1 Advantages Offered by Solid-State-Based Systems

It is widely recognized that MMIC-based modules are the undisputed contenders for all electronic functions except for the generation of high powers. For this function MMICs and tubes share the domains of output power versus frequency we have delineated in Section 4.3.1. Additionally, MMICs modules offer a suite of advantages we shall summarize in this section.

The dominance of MMICs in the area of low-noise amplification extends from a few hundred meters (HF band) to the mm wavelengths. For a long time yield was the barrier to inexpensive MMICs; in Section 4.6.4 we shall outline methods and approaches that increased the yield of MMICs significantly for GaAs MESFETs and work in progress to increase the yield of other semiconductors.

MMICs have long lives, and active phased arrays have excellent graceful-degradation characteristics. Active phased arrays utilizing MMICs necessarily combine all these advantages to attain a cost advantage over phased array systems utilizing tubes. Additionally, MMIC-based radars have exceptional MTI stability factors compared to their tube counterparts.

If we define the 35–60-GHz range as a frequency band of convenience, we could make the following statements:

1. The PAEs of SSDs and those attributed to tubes are comparable at frequencies below the band we have defined.
2. High powers can be conveniently generated by SSDs or by combining the power available from several SSDs operating below the band we have defined. Here convenience is related to low-voltage power supplies, volume and weight.

With the passage of time we expect this band to move upward in frequency.

4.5.2 Advantages Offered by Systems Utilizing Tubes

Conventional systems usually utilize one tube and offer high powers over narrow bands (canonical radar operation) or over one or more octaves (EW operation). Above the band we have stipulated (35–60 GHz) the power advantage of tubes over SSDs becomes more pronounced. Given that the power of several tubes can be combined, this power advantage of tubes over SSDs is exaggerated. Lastly, tubes offer protection from EMP radiation, a characteristic that is significant for military applications.

There is a discernible trend away from phased arrays utilizing one bulky tube that supplies all the array power required and no graceful degradation. Phased arrays having graceful degradation by using a multitude of small tubes are now preferred.

4.5.3 The Future

The resurgence of interest in tubes and the developments related to the MPM are important developments for future phased arrays because these modules combine the advantages of both SSDs and tubes. If a separate power supply is required for each MPM (current approach), there is a spatial limit to the power a module can have in a fully populated planar array. In the long term, wideband RF amplifiers based on vacuum microelectronics and fast-wave amplifiers are likely to yield novel power sources. Quasi-optical power combining is bound to yield higher powers at mm wavelengths. Recent progress in SSDs has been impressive. There is scope for both tubes and SSDs to improve their PAEs at all frequency bands; these improvements are especially needed at mm wavelengths, where the spacing between antenna elements for canonical arrays is narrow and the heat dissipation problem acute.

4.6 THE SOLID-STATE T/R MODULES

A typical phased array has some 500–100,000 antenna elements, and active phased arrays will have as many T/R modules if polarization information is not required. For polarimetric phased arrays, the number of modules is doubled. It is widely acknowledged that the realization of so many subsystems —such as antennas, T/R modules, and matching structures between the antenna and T/R modules—can be economically implemented by using MMICs. The other important aspect of monolithic integration is space. If we were to realize all these subsystems from discrete components, we could not meet the $\lambda/2$ separation between elements constraint, especially at mm wavelengths.

Despite these advantages, MMICs have the following serious drawbacks:

1. High nonrecurring costs generally limit MMICs to applications requiring large quantities. The MIMIC program in the United States, started in 1987, has contributed significantly to the establishment of many GaAs facilities in several semiconductor firms. Thus the required infrastructure that is economically desirable [4.42] for the development of several manufacturing approaches was created. The overall program objectives has been described [4.43] as follows: "Provide the needed microwave and millimetre-wave products at a price that will allow their use in fielded Department of Defense systems, that meet all required electrical, mechanical and environmental parameters and that continue to operate reliably for the time necessary to fulfil their intended application." While there are many criteria to assess the impact of the MIMIC program and the development of MMICs, cost is a useful criterion. The cost per square millimeter of producing an MMIC chip has dropped from $20/mm^2 in 1987 to $3–8/mm^2 in 1989; The $3/mm^2 cost corresponds to a small-signal amplifier, while the $8/mm^2 cost corresponds to a high-output power amplifier. The cost estimate corresponding to a small-signal amplifier has been corroborated by van den Bogaart and Bij de Vaate [4.44].

2. Circuit tunability. It is common practice to optimize the electrical performance of complex hybrid circuits by tuning. Given that there can virtually be no adjustment in MMICs, designs should be tolerant of process variations.

Many design practices have evolved to offset this impediment, and recently on-chip tunability has been reported [4.45]. Airbridges are used to vary the inductance of critical circuits; similarly, airbridges are used to vary the capacitance or resistance of circuits. Several capacitors or resistors are paralleled to attain the required value, which can be altered with the aid of airbridges.

4.6.1 MMIC Options

It is relatively easy to accept that MMIC-based T/R modules offer the most appropriate solution for the realization of affordable active phased arrays. Apart from the economic aspects, the monolithic implementation guarantees repeatability and high reliability. One can go as far as to say that we know of no better way to realize the above-mentioned subsystems in the quantities envisaged. Once this thesis is accepted, however, one is confronted with a few realization options, which we shall consider in this section.

The repeatability is derived from the controlled monolithic processes, while the reliability aspect needs some elaboration. Process yield is a central issue for it directly influences costs and the affordability of phased arrays.

The designer has basically two main options from which other options can be derived:

Option 1 The realization of all the building blocks of the module, including the antenna element and its matching structure, the LNA and post-LNA amplifiers, the power amplifiers and FPA, the programmable phase-shifters and their switches in one chip (on a given substrate).

Option 2 To realize each of these building blocks of the T/R module in separate chips and interconnect them to the antenna and matching structure. The issue here will be how small or large a building block is to be.

Option 1 is preferred by production engineers because production costs and the costs related to the interconnections are minimized; similarly, the unreliability caused by the interconnections usually associated with option 1 is minimized. In terms of yield, however, option 1 is not attractive. Option 2 is preferred by electronics engineers, who are more concerned with the optimization of the particular functions of the module's building blocks. One very important advantage of multichip modules is that the designer can theoretically meet a diversity of array requirements by assembling a finite number of basic chips.

In Section 4.6.4.1, we shall demonstrate that, to a first approximation, process yield is proportional to the chip area. Therefore, option 2 is attractive because each individual building block occupies a relatively small area.

Interconnections between the building blocks are, however, costly to implement and tend to be unreliable [4.46].

It is important to note here that the preceding statements are made for the purposes of initiating discussion and are therefore not set in a tablet form. Naturally there are several in-between options; one can, for instance, arrange to have the LNA, phase shifter, medium-power amplifier, and low-power switches in one chip while another chip can be dedicated to the FPA; the choices are many.

In what follows we shall explore whether the antenna element should be included in the T/R module before we address the issues related to options 1 and 2.

4.6.2 MMIC T/R Module with or without the Antenna?

The great attraction of microstrip antenna elements is that they can be readily integrated with the MMIC module; there is, however, a major drawback with such a proposition.

For efficient operation, microstrip antenna elements are realized on substrates that have a low-dielectric constant (e.g., < 4)—see Chapter 5. With this arrangement the fields associated with the conductors are loosely coupled to the substrate material. The EM waves are therefore easily launched from the substrate to free space.

GaAs (gallium arsenide), which has a dielectric constant of about 12, on the other hand, is the preferred substrate for many MMICs. The fields associated with the conductors etched on GaAs (or other high-dielectric substrates) are tightly coupled to the substrate. This arrangement is ideal for transmission lines but far from ideal for antennas that have to launch electromagnetic radiation into free space.

Teflon and other similar substrates are used for antenna elements operating at cm wavelengths. It is well known that the Z-cut quartz, or similar substrates, are the preferred substrates for antenna elements operating at mm wavelengths, because of their low-loss and low-permittivity characteristics; additionally, the substrates have excellent mechanical characteristics, including considerable temperature stability and can be easily machined.

This is only one dimension of the conflict; let us explore another. Let us accept the tile architecture and place the T/R module next to its antenna element on the radiating surface. This is a convenient approach because the interconnections between the antenna elements and their T/R modules are easy to implement and short in length. The radiating area, however, is not fully utilized and grating lobes will appear in the visible space. These considerations take on a central importance if operation at mm wavelengths is contemplated; this is because the space between modules has to decrease.

From the foregoing considerations designers do not usually realize the antenna elements on the MMIC substrate; we shall therefore treat antenna elements and their matching structures in Chapter 5. This said, we will add that in some experimental arrays the antennas were fully integrated on the GaAs substrate [e.g., 4.47] for demonstration purposes.

In summary, it is appropriate to state that microstrip antennas can be more readily integrated with MMICs than say, waveguide slotted antennas; the proviso here is that different substrates have to be used if maximum efficiency antennas and MMICs are required.

4.6.3 T/R Module Realization Approaches

We have already decided to separate the antenna elements from the T/R module. The many building blocks representing the many functions required to take place after the antenna elements are shown in Figure 4.14. A major problem the designer has, is to decide what to include and what not to include in the T/R modules that are directly behind the antenna elements. While the choices are to some extent application-dependent, we shall endeavor to draw some generic guidelines.

It is recalled that the area directly behind the antenna elements is limited by the $\lambda/2$ spacing requirement, and we aim toward minimizing the system losses. There is widespread agreement that the majority of the building blocks listed in Figure 4.14 ought to be included in the T/R module; there are, however, two exceptions:

1. The phase shifters or vector modulator, hereafter referred to as the *vector modulator subassembly*
2. The polarization diversity subassembly

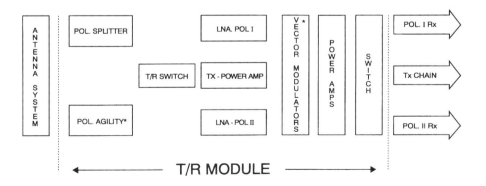

*POLARIZATION AGILITY AND VECTOR MODULATOR FUNCTIONS ARE UNDERTAKEN AT.R.F.

FIGURE 4.14 The building blocks of T/R modules.

It is recalled that the module phase and amplitude errors constitute a large portion of the array errors, which in turn, define the essential characteristics of the array radiation pattern. The module phase and amplitude errors, in turn, are defined by the vector modulator subassembly.

The following advantages can be gained if the vector modulator subassembly is at the IF, and not at the RF, of the transceiver [4.48]:

1. Moving the vector modulator subassembly from the RF to the IF will decrease costs and increase the accuracy of the vector modulator subassembly because there is negligible interaction between amplitude and phase settings [4.37, 4.48]. The amplitude and phase errors of modules where the phase shifting is implemented at RF are typically ± 1 dB and $\pm 5°$, respectively, for current T/R modules and ± 0.25 dB and $\pm 3°$ for future modules [4.1], while the corresponding errors for modules where the phase shifting is implemented at the IF are 0.02 dB and $0.2°$ in phase rms, respectively; these are significant improvements. Here we have assumed that it is relatively easy and economical to bring the required phase-locked LO signals to the modules via optical fibers and that the instantaneous array bandwidth B is related to the IF by the relationship $B/\text{IF} \leq 0.1$.

2. If standardization of vector modulator subassemblies operating at the several IFs can be agreed on, several phased arrays operating at many wavelengths ranging from cm to mm can use the same low-cost, vector modulator subassemblies.

Ultimately detailed considerations must be made before a decision can be reached as to where the vector modulator subassembly is placed. As the frequency of operation increases, placing the vector modulator subassembly at the IF of the system becomes more attractive.

It is recalled here that if the array has a full-polarization diversity capability, the user obtains all the benefits resulting from receiving/transmitting and processing the full-polarization information. For radio-astronomy and satellite communications applications, almost all the complex functions related to polarization processing are performed at the IF to minimize the RF losses. In the phased array context the same considerations apply, although several authors reported experimental models where different polarizations are switched at the RF—see Chapter 5.

Again, as the frequency of operation increases, the attraction of this approach increases. From a systems point of view, while the implementation of full-polarization diversity in conjunction with a single aperture has been reported, the implementation of the same capability for radar phased arrays is nontrivial and costly. We shall continue this discussion in Chapter 5, for the implementation of full-polarization diversity in the phased array context is intrinsically related to the antenna elements.

Let us examine the proposition that all building blocks of the T/R modules are realized in one GaAs chip (called *die* by the semiconductor community). We have already seen in Figure 4.9 that if one is to attain maximum power below 5 GHz, one ought to use silicon bipolar junction transistors, (BJTs), and not GaAs-based transistors. This is not such a fundamental problem because tradeoffs can be made and modern pseudomorphic HEMTs can compete with silicon BJTs.

At present the microprocessor chips, which are part of the distributed-logic approach, are separated from the main T/R module; the future integration of microwave and digital subsystems will further decrease costs and increase reliability.

Next we shall focus attention on the problem of achieving high production yields, which is fundamental to affordable T/R modules and phased arrays.

4.6.4 The Yield of MMICs

The issues related to the yield of MMICs are important for the appreciation of the costs issues associated with MMICs, which, in turn, are central to the affordability of active phased arrays. Typical MMIC yields were about 10% in 1982 [4.49], a dismally low figure that impeded the widespread use of active phased arrays.

The reasons for low yields are complex but are related to the tight tolerances expected of T/R modules, to electromagnetic phenomena that can be neglected when the ICs operate at low frequencies but cannot be neglected at higher frequencies, the large number of components that constitute a T/R module, and the adoption of the same fabrication techniques used for discrete transistors to realize transistors within the MMICs.

In this section we shall focus on the statistical and deterministic approaches, that dramatically increased the yield of MMICs to 60% or even 70%. Indeed, it is impossible to overemphasize the importance of the following approaches, which will significantly lower the costs of phased array-based systems.

Initially we shall investigate approaches which maximize the yield of the individual building blocks of a T/R module; at a later stage we shall investigate the approaches taken to maximize the yield not only of the T/R module but also of several modules forming one subarray.

4.6.4.1 Statistical Approaches to Increase Yield and Decrease Cost In Table 4.10 we have tabulated the number of chip sites per wafer when the wafer diameter is 2, 3, and 4 inches and the chip size varies from 1×1 mm to 10×10 mm.

Chip costs are usually proportional to the chip real estate, so small chips are much cheaper than large ones. The costs we are considering here, illustrated in Figure 4.15, do not include costs due to electrical and packaging yield, test costs, amortization of design costs or overheads.

TABLE 4.10 Number of Chip Sites per Wafer When the Wafer Diameter Varies [4.50]

Chip Size	Wafer Diameter		
(mm × mm)	2 in.	3 in.	4 in.
1 × 1	1800	4000	7000
2 × 2	450	1000	1800
3 × 3	200	450	800
4 × 4	110	250	450
5 × 5	70	160	280
6 × 6	50	110	200
7 × 7	36	80	145
8 × 8	28	60	110
9 × 9	22	50	85
10 × 10	18	40	70

The most compelling reason, however, for using small chips is not related to the real estate area used by the chip but is strongly related to yield. The larger the chip, the greater the probability of having critical parts of the circuit located over a significant defect. Conversely, the smaller the chip, the lower the probability that a defect would be included within the chip area [4.50]. Small chips therefore have higher yields, and this is illustrated in Figure 4.15. Although the abscissa and ordinate of the graph are epoch-dependent, the figure illustrates the point. With the passage of time and as manufacturing processes are controlled better, the yield die cost approaches the die-site cost.

4.6.4.2 Deterministic Approaches to Increase Yield and Reduce Cost In this section we shall explore the deterministic approaches the semiconductor community, aided by design engineers, have taken in order to increase the yield of MMICs.

Ab initio, the approaches are conveniently divided into system tradeoff considerations and design guidelines These important topics define macro and micro issues related to the design of MMICs and in many ways define their costs and performance.

Finally, the multifunction self-aligned gate (MSAG) fabrication process, which is by and large responsible for the phenomenal increase in the yield of MMICs, is outlined.

4.6.4.3 Design Guidelines for Increased Yield To increase the yield and reduce the cost of GaAs ICs, special design guidelines are required. The design guidelines outlined here are based on the premise that there are considerable variations in the many transistor parameters and the sheet

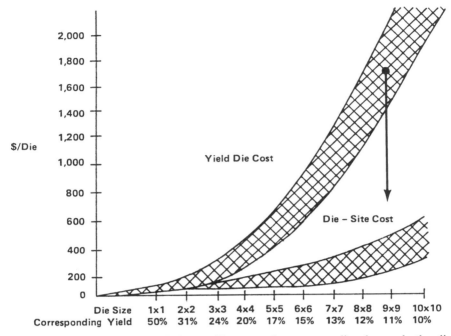

FIGURE 4.15 The lower shaded area illustrates the cost per die when only the die site cost is taken into account. It is assumed here that a 3-in. wafer costs $10,000. The upper shaded area illustrates the cost per die when yield is taken into account. In both graphs the die varies from 1×1 to 10×10 mm. The two graphs illustrate the state of the art in 1986; as the yield increases, the upper shaded area moves closer to the lower shaded area (from [4.50]).

resistance both intrawafer and interwafer; these variations determine the process windows for the derived MMICs.

The following are some of the important guidelines for yield-tolerant designs [4.50]:

1. Design for a reduced gain per amplifying stage by the use of feedback; some of the gain available is applied as feedback. Variations in the transistor's transconductance therefore become less significant.
2. Prefer staggered tuned gain stages in a multistage amplifier. The resulting amplifier will have a flat gain characteristic even though the cutoff frequency of the transistors changes from wafer to wafer.
3. Avoid yield-reducing circuit elements, such as large MIM capacitors; use transformer-coupled circuits or other techniques instead.

4. Use the ratio of resistors as a design parameter instead of absolute resistor values. Sheet resistance will vary across a wafer and from wafer to wafer; resistor values will therefore vary as well. If, however, the resistor ratio is used as a design parameter, the ratio will not change significantly either intra- or interwafer.
5. Do not impose on MMICs low noise figures and/or extremely tight gain flatness requirements over wide bandwidths. Similarly, do not impose hybrid MIC architectures and performance requirements on MMICs.

4.6.4.4 The MSAG Fabrication Process for MESFETS: A High-Yield Process
For high-performance MESFETs some of the most critical factors required are proper thickness of the thin active channel layer under the gate (typically 0.1 µm), minimum source resistance R_{gs}, and low gate resistance R_g.

The recessed-gate method, illustrated in Figure 4.16, is the traditional method to fabricating GaAs MESFETs. A relatively thick active layer ensures high conductivity between source and gate. Proper active layer thickness under the gate is attained by wet-etching a recess into which the gate metal is subsequently placed. This technique, which has been used by the semiconductor industry, ensures good FET performance. The active channel thickness and R_{gs} are, however, hard to control because of the limitations in the control of the wet-etching process and in the maintenance of photolithography tolerances.

For discrete FETs where individual devices meeting particular specifications can be selected from a wide distribution of characteristics, these difficulties do not present problems. The same difficulties, however, create a barrier to high yield across a wafer and from wafer to wafer for circuits with tens, hundreds, or thousands transistors. The self-aligned gate process eliminated the need for a gate recess, the single most important yield and

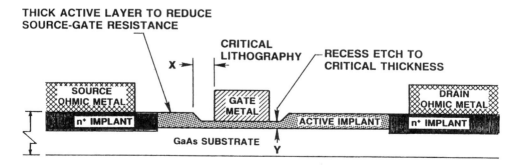

FIGURE 4.16 Illustration of the conventional fabrication of the recessed-gate FET.

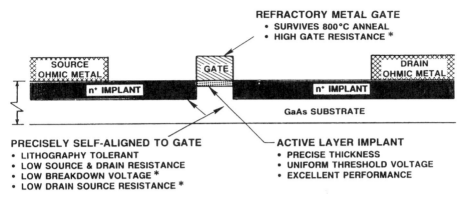

FIGURE 4.17 Illustration of the self-aligned gate (SAG) FET process for digital applications. (*Source*: Bahl et al. [4.52], © 1990 IEEE.)

reproducibility limiting step. Figure 4.17 illustrates the digital SAG FET structure, which eliminates these problems and increases yield dramatically [4.51, 4.52]; a slightly modified SAG process has been used for the realization of MMICs designated to perform analog signal processing functions with as much success; hence the term *multifunction* in the acronym MSAG. Figure 4.18 illustrates the MSAG process.

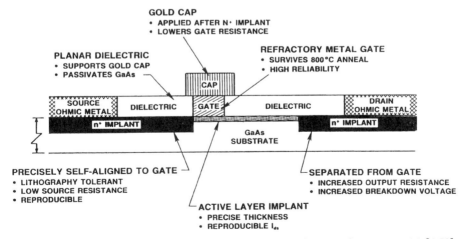

FIGURE 4.18 Multifunction self-aligned gate (MSAG) FET. (*Source*: Bahl [4.52], © 1990 IEEE.)

The self-aligned implant is confined to the source side, while the drain-side implant is spaced away from the gate during processing by a photoresist strip that provides both good voltage breakdown and sufficiently high output resistance. A high conductivity gold cap is applied to reduce gate resistance after annealing has occurred. The lithography tolerance related to this step is relaxed since the dialectic layer that covers the entire GaAs surface can support any gold overhang.

The MSAG fabrication process has been applied to MESFETs used in low-noise and power amplifiers with great success [4.53]. During a 3-month period, using the process described above, 500,000 FET transistors were fabricated and 3000 transistors were sampled and RF probe-tested. The average maximum available gain (MAG) was 15 and 9.5 dB at 10 and 18 GHz, respectively. More importantly, however, the standard deviation in MAG was 4.7% at 10 GHz and confirmed the high degree of reproducibility.

Similarly, 173 out of 80,000 power FETs operating at C-band were randomly sampled, diced, packaged, and tested. The resulting average gain, 1-dB output power, and efficiency were 8 dB, 1.9 W, and 38%, respectively. Again, the standard deviation for the power output was 5.3% only. These results indicate that the fabrication process is reliable and can yield MESFETs that have a high degree of uniformity.

Other researchers [e.g. 4.52] reported excellent results by using the multi-function SAG (self-aligned gate) processes to attain high-yield, low-cost GaAs MMICs and similar procedures for the fabrication of high-yield GaAs MMICs have been reported [4.32].

4.6.4.5 Other Approaches to Maximize Yield Several approaches to maximize the yield of the active and passive components of MMICs have been reported [4.54–4.60]; given the plethora of passive components in MMICs, the work related to passive components [4.54] is significant. The yields attained, tabulated in Table 4.11, range from 80–100% for passive components and 30–90% for MMICs. These improvements in yield came about by the use of design-technological process optimization [4.55], better materials and processes [4.54, 4.55], and comprehensive test procedures at critical stages of the MMICs. The following testing procedure is often used [4.55]. After fabrication and before scribing, the wafers are DC probed using an automatic DC test system. The measurements of 10 critical parameters are undertaken in order to evaluate the process yield of functional circuits or individual elements, for example, transistors, or passive components; from these measurements the dispersion of each parameter is derived. Typical DC yields for a medium power amplifer, having a chip size 2.3×4.1 mm^2, can vary from 49% to 60% [4.44]. The RF yield for DC-good circuits soldered on carriers can be better than 90% for the same medium power amplifier [4.44].

TABLE 4.11 Typical Component and MMIC Yields

Component/MMIC	Area (mm × mm)	Yield DC or RF (%)	Reference
MIM capacitor		80	[4.54]
Interdigital		97	[4.54]
Spiral inductor		100	[4.54]
Thin-film resistors[a]		92	[4.54]
Local oscillator, X-band		60	[4.54]
LNA—X band	1.3 × 2.1	70	[4.44]
Vector modulator	5 × 2.7	30	[4.44]
Medium-power amplifier	2.3 × 4.1	49–60 DC 90 RF	[4.44]
LNA—ion-implanted MESFETs—Ka-band	1.08 × 2.53	Max 70 RF	[4.59]
LNA + IRD[b] HEMT—W-band	5 × 2	46–56 RF	[4.60]

[a] High or low value.
[b] Image-rejection down converter.

The concept of "intrinsic yield," which is defined as

$$\text{Intrinsic Yield} = \frac{\text{Chips within microwave specification}}{\text{Chips which are DC functional}}$$

is often used [4.56]. The intrinsic yield of X-band GaAs MESFET, MMIC amplifiers can be as high as 62.5% [4.57]. The associated DC–RF correlation is usually found to be over 70% for LNAs operating at X-band and when the chip area is less than 2 mm^2 [4.55].

Many developments related to the manufacturing technology and fabrication processes of several transistor types have taken place; we shall mention only a few here. The realization of high-yield pseudomorphic HEMTs has been aided by developments related to the associated manufacturing technology [4.58]. Similarly, high-yield fabrication processes for low-noise MMICs using ion-implanted MESFETs operating in K-band have been reported [4.59] and the maximum RF yield attained was 70%. High-yield fabrication processes for HEMT-based low-noise amplifiers operating in W-band have been outlined [4.60]; depending on the criteria used, the RF yield varied from 46 to 56%. All the yield measurements cited here are tabulated in Table 4.11. It is evident that the RF yield is strongly dependent on the chip area it occupies and the complexity of its functions; the yield of a low-noise amplifier occupying an area of 1.3 × 2.1 mm is 70%, while the yield of a vector modulator occupying an area of 13.5 mm^2 is only 30%. This trend is supported by other entries in the table.

The results tabulated are excellent when compared to yields of about 10% in 1982 [4.49]; progress has been significant, considering that the semiconductor community frequently has to derive optimum fabrication techniques for different types of semiconductors.

4.6.4.6 MMIC Cost Minimization Approaches What follows is a brief outline of approaches taken to minimize the costs of MMICs. At first sight some approaches seem contradictory but are in fact complementary to one another. Initially we shall explore the manufacturing approaches taken when option 1 (see beginning of Section 4.6.1) is pursued; before we outline the advantages, problems, and solutions production engineers have defined in the pursuit of option 2.

4.6.4.7 Double- and Single-Chip T/R Modules In 1987 a single-chip T/R module realized at X-band was heralded as a major cost saver for phased arrays [4.61]. It consisted of two FET-based switches, a four-stage power amplifier, a three-stage LNA, and a 4-bit phase shifter. Its total area was 13×4.5 mm or 58.5 mm^2, half of which was occupied by the phase shifter.

Out of the 14 wafers produced, 8 were fully functional and the average chip yield over these 8 wafers was 14%, a very low yield indeed. After considerable efforts by several companies the single chip T/R module was declared a laboratory curiosity in 1992 [4.62]. The optimum semiconductor processes used to fabricate the many different building blocks of the T/R module are slightly different, so compromises have to be made with the one-chip realization. As a consequence, output powers are low, noise figures high, and module yields are low. The essential characteristics of this module are tabulated in the first entry of Table 4.12.

Despite all these drawbacks, it is worth recalling the epoch the first single-chip was realized. More importantly we wish to define how many T/R functions one can pack in the smallest chip area, assuming that production processes performed on the chip can meet the diverse requirements placed upon the module. These considerations are undertaken against a background of digital signal processing chips of phenomenal complexity and significant efforts toward RF-wafer scale integration. In this context it is worth noting that current single-chip digital signal processing chips can have up to 26 million transistors [4.63]. This said, one de-emphasizes neither the frequency difference between chips operating at microwaves and chips operating at relatively low frequencies to perform signal processing functions nor the diverse requirements such as low-noise amplification and high-power generation the former chips have to meet.

An outline of the essential characteristics of representative MMIC T/R modules that were realized in one or two chips follows. The essential characteristics of a system used in portable data collection terminals [4.64] are tabulated in the second entry of the table, while the salient characteristics of a radar used in conjunction with smart munitions [4.65] are tabulated in

TABLE 4.12 Double- and Single-Chip T/R Modules

Application	Frequency (GHz)	MMIC Parts	Chip Area mm × mm	Essential Characteristics	References
T/R module	X-band	2 switches, 1 LNA, 1 power amplifier, and 1 4-bit phase shifter	13 × 4.5	$P_{out} = 501$ mW, NF = 5.5	[4.61]
Portable Data Collection Terminals					
Transceiver Chip	2.4–2.5	LNA, buffer amps, predriver amp, 2 mixers and 1 VCO[a]	1.4 × 2.79	< 5 dB SSB NF, 10–12 dB gain	[4.64]
Power Amplifier Chip		T/R and diversity switches, driver amp, and power amp	1.4 × 2.79	P_{out} 22.5 dBm	
Smart Munitions FM CW Radar	15–25	10 amplifiers, 1 mixer, and a VCO	2.4 × 1.8	$P_{out|1 dB comp} = 12$ dBm	[4.65]
T/R module	2–20 GHz	4-stage power amp, 4-stage LNA, and 2 switches	3.6 × 4.9	Distributed amplification	[4.66]
T/R module	34–36 GHz, Rx; 33–37 GHz, Tx	LNA and power amplifier		NF = 3.5 dB, G = 17 dB	[4.67]
T/R module under development	C-band; X-band; X-Ku-band	Complete modules	8 × 8	$P_{out} = 3.5$ W; $P_{out} = 3.7$ W; $P_{out} = 2$ W	[4.68]

[a] Voltage-controlled oscillator.

the third entry. The chips used in both systems are destined for the mass-production market.

The second system consists of two chips and operates in the industrial scientific and medical band (2.4–2.5 GHz) while the third system consists of one chip and operates at K-band. The characteristics of representative samples of current T/R modules or under development are also tabulated in the same table in entries 4–6.

Initially we can discern a marked trend toward smaller size chips. The chip sizes of 3.9 mm^2 and 4.1 mm^2 (second and third entries) are much smaller than 58.5 mm^2 the chip area of the first T/R module. Admittedly the latter chips do not include a programmable phase shifter, which occupies almost half of the area of the former chip, that is, 29 mm^2. In the fourth and fifth entries the essential characteristics of T/R modules operating over a wide band [4.66] or at mm wavelengths [4.67], respectively, are listed. The last entry, tabulating the essential characteristics of a single chip module [4.68], is important for the following reasons:

1. All RF functions are performed in one chip.
2. Although the size of the chip decreased with the passage of time (170 mm^2, in 1991, 65 mm^2, in 1993 and 64 mm^2, in 1994 [4.68]) it is still considerable.
3. Important issues that we shall consider further in this section are the parts count, the number of die attach or operations, wire or ribbon bonds, and RF tests.
4. While considerable powers are generated by the T/R modules, i.e., 3.5, 3.7, and 2 W at C-, X-, and X-Ku bands, respectively, higher powers can be derived by the use HPAs (high-power amplifiers) connected to the output of the T/R chips. With this approach powers exceeding 10 W can be derived [4.68].

The approach proposed deserves considerable attention because it allows the designer to increase the output power of the array performance by retrofitting the latest HPA to an existing array with the passage of time; similar arguments hold for the LNAs. Indeed one can propose T/R modules that perform all the basic functions required but the low-noise amplification and final power amplification. For these functions separate renewable ICs are utilized. The problems envisaged with this approach will be: (1) the reliability of the interconnections between the MMIC and the two ICs; and (2) the availability of space for arrays operating at mm wavelengths.

In Table 4.13 we can see the resulting economies with modules that are highly integrated when compared to a completed production run that had 8 chips [4.68]. The costs that are strongly related to the number of parts used, RF tests, wire or ribbon bonds, and die (or chip) attach or operations for the

THE SOLID-STATE T/R MODULES 315

TABLE 4.13 Integrated T/R Module Compared with the Previous Design [4.68]

Item	Completed Production	Integrated
MMIC chips	8	2
Parts	250	20
Die attach/Operations	240	18
Wire/ribbon bonds	500	40
RF Tests	30	3

latter modules are typically one-fifth of those corresponding to the former modules. The high integration T/R module is paving the way to low-cost, next generation systems [4.68].

The continuous improvements in materials, the control of many and diverse processes appropriate for the diverse requirements placed upon the chip and computer aided designs coupled with the continuous learning of several talented professionals drawn from diverse backgrounds is now yielding the dividends predicted by production engineers for some time. The ARPA (Advanced Research Projects Agency) programs MIMIC, HDMP (high-density microwave packaging) and MAFET (microwave and analog front-end technology) scheduled to start in fiscal year 1995 will accelerate the process of lowering the costs of MMICs [4.69].

4.6.4.8 RF Wafer Scale Integration: Another High-Yield Approach It is only fair to devote some time and space to the approaches taken to increase the yield of several adjacent modules (realized on one wafer) that constitute a subarray [4.70]; at first it seems contradictory that one can increase the yield of several T/R modules, an approach that implies an increase of the wafer area. While this correlation remains valid, the following important developments have taken place:

1. The realization of nearly damage free GaAs wafer surfaces [4.71]
2. The development of mechanical switches that allow the designer to make full use of built-in redundancy associated with the device and circuit elements within the module

The development of improved wafer polishing techniques has resulted in nearly damage-free wafers; consequently, the yield of small-to-medium-sized devices on these wafers has doubled. Additionally, the module designer builds in double or triple circuit and device redundancy and selects the "best" performing stages with the aid of mechanical switches. After the selection of the best performing stages, the switches are closed to complete the desired amplifier chain by using a suitable wedge bonding tool. The redundancy available [4.72] is illustrated in Table 4.14.

TABLE 4.14 Essential Characteristics of the 6–12-GHz T/R Modules [4.72]

Circuit/Device	Numbers Required	Numbers Available
LNA (NF < 7 dB)	3	5
T/R switch	1	3
Attenuator	1	3
180° analog phase shifter	1	3
180° digital phase shifter	1	2
2-stage driver amplifier	2	4
Power output amplifier ($P_{out} = 0.5$ W)	1	3

The individual circuits are laid out in such a way that the total insertion phase throughout the T/R cell remains the same regardless of which group of circuits is chosen.

The issue of whether high costs are associated with the testing of the many redundant circuits and devices has to be addressed; it is, however, an area that lends itself to automatic testing procedures.

About 30% of started wafers were not completed because of breakages, poor activation, and other causes [4.72]. The RF-wafer scale program has been a high-risk program with potential payoff. It has been highly successful, but additional work is required for larger arrays that satisfy particular system requirements [4.72].

4.6.4.9 Recent Experiences with Multichip T/R Modules We have already outlined the attractive characteristics of option 2 (see beginning of Section 4.6.1), which lead to the multichip module; considerable effort has been expended toward the definition of the problems related to this approach and the implementation of several solutions.

There are three important problems related to multichip modules—two of them have a statistical basis, while the other has a fabrication basis; naturally all of them are interrelated. We alluded to a major problem with multichip modules in the previous section. The number of chips contained in a module dictates how many operations one has to undertake to derive high-yield modules. Costs are therefore correlated to the number of chips a module contains. The next problem is more complex.

In the interest of attaining high-yield modules it is preferable to populate them by known good-dies, (KGDs) or chips instead of untested chips. Hence the drive for testing and burning in all the chips or just the high-power chips prior to module assembly. Experience accumulated with the MMICs used with the COBRA (COunter Battery RAdar) is that most failures occurred prior to 50 h burn in time, so a burn in time of 80 h was accepted [4.73].

The leading defect for the same modules has been wire bond defects; at the start of the production program approximately 80% of the modules had wire bond defects at first visual inspection. After considerable effort the same defects averaged 5% or less [4.73].

Martin Marietta, a participant manufacturer of the above-mentioned modules, is utilizing the microwave high-density interconnect (MHDI) technology, which eliminated the wire bonding and is described in Kole and Ozga [4.73].

What follows are alternative approaches to the solution of the interconnect problems in multichip modules. A conventional T/R module consists of about 5–8 monolithic chips mounted on thick-film printed carrier substrates. With the availability of low-K ceramics, the ability to cofire multiple layers of ceramic and metals to form extremely compact packages with electrical isolation between closely spaced components has emerged [4.74]. Composite materials having low-dielectric constants resulted from the mixing of glasses with ceramics [4.62, 4.75–4.78]. These materials can be fired in air at about 900°C, the temperature range for processing standard thick-film printed circuits [4.75, 4.78]. This new technology is promising for phased array T/R module assemblies, particularly in the higher-frequency bands [4.75]. With these procedures the reliability of the interconnects is maximized, but some processing costs are necessarily added. The comparisons are here made between single-chip and multiple-chip modules.

While the average T/R module consists of 5–8 monolithic chips an EW module was reported to use 72 MMICs [4.79].

4.6.4.10 T/R Realization Issues: Concluding Remarks The key ingredients to understanding the issues in hand is continuous learning and continuous improvement. In the beginning, while the quality of wafers was barely adequate and production engineers could not tightly control the many processes, it was economical to realize T/R modules consisting of several small chips. As the quality of wafers increased and methods of tightly controlling the many processes involved were derived, it is now economical to produce larger chips—so much so that a T/R module consists of one or two large chips. Parallel developments improved the reliability of the interconnects between chips, a major drawback to multichip modules.

There is a widespread consensus that high-level integration is the key to lower module costs [4.70] and programs such as the HDMP program will contribute toward the lowering of the T/R module costs. Similarly, the option of assembling T/R modules from standard off-the-shelf, fully tested, and inexpensive chips is attractive.

At a more fundamental level, Cohen [4.69] suggests the following tasks to complement the two activities described above:

1. The realization of a computer-integrated manufacturing (CIM) facility to perform array architecture and design tradeoff studies.

2. The development of improved array components such as circulators, power supplies, and capacitors needed to meet overall array requirements; in that respect the MAFET program, scheduled to start in 1995, will address these issues.
3. The implementation and demonstration of an adaptable multichip assembly (MCA) and array factory and the subsequent demonstration of the ability of that factory to affordably produce and assemble array hardware suitable for several different defence related applications.

From these considerations it is not hard to draw the conclusion that work on important and challenging phased array problems has just began; this work however would have been impossible to undertake without the experience and knowledge bases already established.

We shall now turn our attention to the constituent parts of the T/R modules.

4.7 THE CONSTITUENT PARTS OF T/R MODULES

In the following sections we shall acquaint the reader with the most essential characteristics of the constituent parts of T/R modules and the technologies used. Whenever necessary, we shall delineate the many the realization approaches taken.

In undertaking this task we are cognizant that our accounts will necessarily be time- and technology-dependent. This being the case, our considerations will serve the purposes of a baseline of specifications and characteristics that newer devices, approaches, and subsystems will have not only to meet but also exceed.

4.7.1 Baseline Characteristics of Power and Low-Noise Amplifiers

As a starting point, we shall begin with the tabulation of the essential characteristics of power and low-noise amplifiers that have been available to the array designer at the date of publication. Tables 4.15 and 4.16 [4.80] (1989), represent typical samples of available chips at that epoch and all entries are for narrowband applications unless otherwise specified.

4.7.1.1 Recent Power Amplifiers Power amplifiers, consisting of medium and final power amplifiers, deliver the array average radiated power to the antenna elements. While the essential characteristics of power and low-noise amplifiers have been tabulated, in Tables 4.15 and 4.16, other important characteristics are worth considering.

THE CONSTITUENT PARTS OF T/R MODULES

TABLE 4.15 Typical Characteristics of Power Amplifier Chips [4.80]

Frequency	P_{out} (W)	PAE (%)	Gain (dB)
L-band	2.5	27	40
L-band	11^a	30	36
S-band	6	22	12
S-band	10^a	17	33
C-band	10^a	30	25
X-band	2^a	15	30
X-band	3	25	12
X-band	6	20	15

^{a} Peak power.

The delivered power should have spectral purity and be stable in amplitude and phase. Typical spectral purity specifications are [4.28]:

| Phase and amplitude additive noise | -105 dBc/Hz at 1 kHz |
| Discrete sidebands | -100 dBc |

(indicated by the stable LO, vibration, power supply ripple and noise).

If all the power amplifiers of an array operate at maximum power output, the array power is maximized but its sidelobes will be high (e.g., 13 and -17 dB for a linear or circular array, respectively). Often the module is required to provide output power control over a range exceeding 20 dB. This requirement conflicts with the requirement for maximum transmitted power; furthermore, the PAE of the amplifier is reduced if the power output is not maximum (with some amplifier designs). One alternative to direct power control is to utilize modules capable of delivering different power levels across the array. The stepped power levels of the array approximate a fixed antenna taper and allow the modules to operate close to their maximum efficiency [4.81].

TABLE 4.16 Typical Characteristics of LNA Chips [4.80]

Frequency	NF (dB)	Gain (dB)
L-band	2.5	32
L-band	2.5	42
L-band	3	30
S-band	3.5	35
S-band	3.6	24
X-band	4	23

If the same number of amplifying stages is used, in a narrowband amplifier and a wideband amplifier, maximum power is attained by the narrowband amplifier because the gain–bandwidth product of active devices is finite.

The theoretical maximum PAE of a class-A amplifier is 50%; Class A operation implies that current flows through the power transistor during the entire electrical cycle. The PAEs of linear GaAs MESFET-based power amplifiers operating at X-band vary between 30 and 35% [4.28]. The reduced efficiency results from device losses and input/output combining losses.

Class AB, B, or C operation implies that the current flows through the power transistor for a period appreciably longer than half, half, and appreciably shorter than half of the entire electrical cycle, respectively.

The PAEs of power amplifiers operating in classes AB and B operation are higher than those operating in class A but have a nonlinear transfer function. Maximum PAE is reached when the amplifier is driven into the nonlinear region of operation; in that regime of operation it is difficult to accurately control the amplitude and phase of each module that will exhibit slightly different saturated output power and input/output transfer characteristics [4.28]. Approaches to cope with these difficulties and increase the PAE of power amplifiers have been proposed in the same reference.

Apart from higher PAEs, class-B operated amplifiers exhibit the following attractive characteristics [4.82]:

1. Negligible power dissipation at no RF power.
2. Under backoff, the efficiency of the Class-B amplifier does not degrade as rapidly as that of the Class-A amplifier.
3. A dynamic range of about 10 dB over which the PAE is greater than 40% and the gain is almost constant.

A PAE as high as 70% has been reported for a single-ended class-B power amplifier operating at C-band and yielding 1.7-W output power. The amplifier employed reactive termination for higher-order harmonics [4.82].

Push–pull amplifiers operating at 0.9 GHz having a PAE of 81% at the 39-dBm level have been reported [4.83]. Tradeoffs between the efficiency and intermodulation in SSPAs (solid-state power amplifiers) have been considered theoretically and experimentally in Duvanaud et al. [4.84].

One unique advantage of MMICs is that power combining can be incorporated within the chip, so that the power requirements for a system can be met by a judicious choice of power combining both within and outside the chip.

In Tables 4.17 and 4.18 we have tabulated the essential characteristics of power amplifiers recently reported in the literature. HEMTs originally considered as low-noise devices can now deliver considerable powers.

As the frequency of operation increases, the power output and the PAE of amplifiers decrease. Low powers translate into shorter radar ranges, while low PAEs result in heat that has to be dissipated. Taking into account the shorter

THE CONSTITUENT PARTS OF T/R MODULES   321

TABLE 4.17 Essential Characteristics of Recent Power Amplifiers Operating below 10 GHz

Frequency (GHz)	P_{out} (W)	PAE (%)	Transistor Type	Discrete/ MMIC	References
1.2–1.4	2000	45–55	Si bipolar	Quad	[4.85]
2.7–2.9	22,000[a]		Si bipolar		[4.2]
3	6	> 30	M	MMIC	[4.32]
3	1.1	61	HBT	D	[4.32]
5–6	1.6	44	M	MMIC	[4.32]
5–6	3	33	M	MMIC	[4.32]
5–6	4	26	M	MMIC	[4.32]
5–6	1.7	70	M	MMIC	[4.32]
5–6	8	32	M	MMIC	[4.32]
6.5–8.5	1.3	42	HBT	MMIC	[4.32]
8	2.5	39	HBT	MMIC	[4.32]
8	5.3	33	HBT	MMIC	[4.32]
8, 12	1.6	38–40	M	MMIC	[4.32]
9	1	40	HBT	MMIC	[4.32]
9.3	12.5	31	HBT	MMIC	[4.32]
9–10	2.5	36	M	MMIC	[4.32]
4, 10	1, 0.5	65–61	M	D	[4.32]
5–10	1	58	HBT	Hybrid	[4.22]
8–10	5	19–22	HBT	MMIC	[4.17]
8.5–11.5	3	30–42	M	MMIC	[4.17]
6–18	1.8	37.5	AlGaAs–GaAsHBT	MMIC	[4.86]

[a] Power combining.
M: MESFET.
D: Discrete.

$\lambda/2$ spacings at mm wavelengths, the designer has formidable problems to manage at these wavelengths.

Given that the PAE of SSDs is strongly dependant on the frequency of operation, it makes sense to obtain as much power as possible at frequency f_1 and multiply the available power to a frequency nf_1, where n is equal to 2 or 3. This approach is taken by Hegazi et al. [4.92] and Ho et al. [4.93] who derived considerable power at 47 GHz, with the view of obtaining useable powers at 94 GHz.

While there are several approaches to spatially combine the power derived from several SSAs, we shall outline one approach because of its uniqueness and the promise it holds for future applications [4.96]. With reference to Figure 4.19a the power generated by a low-power oscillator is fed to a circulator and an orthomode transducer (OMT), the output of which is vertically polarized. A horn-lens combination splits the power received by the OMT to a wide area (the flat surface of the lens) on which an array of

TABLE 4.18 Essential Characteristics of Current Power Amplifiers above ~ 10 GHz

Frequency (GHz)	P_{out} (W)	PAE (%)	Transistor Type	Discrete/ MMIC	References
10	0.7	50	M	D	[4.32]
10	0.5	48–52	HBT	D	[4.32]
10	0.8	58	M	D	[4.32]
10	2	> 40	M	MMIC	[4.32]
10	2	> 40	HBT	MMIC	[4.32]
10	5	> 30	M, HBT	MMIC	[4.32]
16	7, 5	30.8, 35	Pseudomorphic HEMT	MMIC	[4.23]
10–16	0.7	50	AlGaAs–GaAs HBT	D	[4.87]
17.5–24	0.2	> 30	HEMT	MMIC	[4.32][a]
20	0.516	47.1	AlInAs–GaInAs on InP HEMT	D	[4.88]
25.5–27.5	0.1	> 30	HEMT	MMIC	[4.32][a]
28	1.1	10.8	M	MMIC	[4.17]
31	0.235	40	pHEMT	MMIC	[4.89]
32	0.063	40	MBE HEMT	MMIC	[4.17]
33	0.125	21	PHEMT	MMIC	[4.90]
34	0.17	23	MBE MESFET	MMIC	[4.17]
42.5	0.18	14	MBE MESFET	MMIC	[4.17]
44	0.251	33	InP-based HEMT	MMIC	[4.91]
47	0.5		M	MMIC	[4.92]
47	0.18	11.2	M	MMIC	[4.93]
47	0.6		M—4 off	MMIC	[4.93]
57–60.5	0.095	11	MBE MESFET	MMIC	[4.17]
59	0.155	30.1	InP HEMT	D	[4.94]
59	0.288	20.4	InP HEMT—2 off	D	[4.94]
60	0.115	26	PHEMT	MMIC	[4.90]
100.3	0.0064		M-Q-W[b] structures		[4.95]

[a] Predicted
[b] Multiquantum-well.

dual-polarized patch antennas is placed. The vertically polarized probes of the 69 antennas, shown in Figure 4.19b, receive the incoming radiation, amplify it, and feed it to the horizontally polarized probes for retransmission. The resulting power is then received in the HP port of the OMT. This approach eliminates many of the problems associated with the conventional feedthrough spatial power combiner because all the RF power is confined to just one side of the array. This allows the back side of the array to be used for heat sinking and bias connections; additionally the size of this combiner is half that of the feedthrough combiner. At mm wavelengths this approach can be used in a phased array where each array antenna element is fed the

THE CONSTITUENT PARTS OF T/R MODULES 323

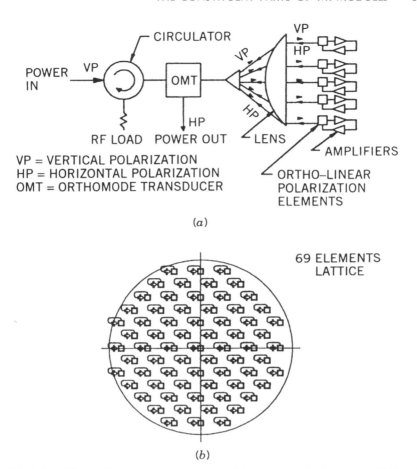

FIGURE 4.19 Illustration of one approach used to generate high power with the aid of a power combining technique. (*Source*: Benet et al. [4.96], © 1993 IEEE.) (*a*) A source generates low power, which reaches the horn–lens combination via a circulator and the orthomode transducer; patch antennas placed at the flat face of the lens receive the incident power and direct it to the orthomode transducer via MMIC amplifiers, their horizontally polarized ports and the horn. The output power is available at the port of the orthomode transducer accommodating the horizontal polarization. (*b*) The 69 patch antennas with their MMICs.

resulting power from an array of elements (69 or more). With this arrangement the phased array radiates considerable power.

4.7.1.2 Recent Low-Noise Amplifiers (LNAs) The function of the LNAs in an array is to define the noise figure of the array and render the other functions that follow them—such as manifolding and beamforming—invisible to the array. Additionally, the same amplifiers should have enough dynamic range determined by the third-order intercept point [4.97]. For a given module and array configuration the designer should ensure that the required noise figure and dynamic range is preserved regardless of the incurred losses along the signal paths.

Receiver dynamic range requirements can range from LNA input third-order intercept values of 0 to 25 dBm [4.28].

The MMIC noise figures tabulated in Table 4.19 are low at cm and mm wavelengths. If operation at mm wavelengths is contemplated the radar designer might opt for an active-on-receive architecture and passive-on-transmit array using tube transmitters. If small-tube amplifiers become available at mm wavelengths, the designer might opt for active phased arrays.

The noise figures of current MMICs are very low at frequencies up to 100 GHz; furthermore, the noise figure of some cooled transistors are exceptionally low. Cooled amplifiers are used routinely in radio-astronomy phased arrays. Pospieszalski et al. [4.105] reported that the noise temperature of HEMT-based LNAs is comparable to those attained by SIS-based receivers at 3 mm.

4.7.1.3 Power and Low-Noise Amplifiers: Concluding Remarks In reviewing the most important elements of the T/R modules, we have outlined the essential characteristics of conventional and recent MMICs. This approach enables the reader to gauge the rate of progress in this area.

There are no lingering doubts that MMICs can perform the low-noise and medium-amplification tasks efficiently and economically. Additionally, the characteristics of the resulting chips can be controlled tightly as production methods are better understood with the passage of time.

Conventional power combining to provide kilowatts of power at L-band and below have been reported; recently the same techniques have been extended to S-band. The application of the same techniques to produce kilowatts of power at C- and X-bands seems possible.

Vacuum tubes can generate significant powers, especially at mm wavelengths. The MPM approach is an important development for it combines the advantages of SSDs and vacuum tubes. In realizing the MPM, great progress has been recorded in the area of miniaturization. Hopefully new devices will be developed from the marriage of microelectronics and vacuum technology.

4.7.2 The Receiver Protector or Limiter

The receiver protector is required to perform several tasks; the most generic one is to protect the LNA from burnout caused by high-power RF energy. The sources of high RF energy are external (an emitter/jammer) or within the module. In the latter case RF power from the final power amplifier is

THE CONSTITUENT PARTS OF T/R MODULES

TABLE 4.19 Essential Characteristics of Current LNAs

Frequency (GHz)	NF (dB)	Gain (dB)	Transistor Type	Discrete/MMIC/ Ambient Temperature	Reference
2.25–2.5	< 0.5	35	HEMT	MMIC	[4.98]
4	0.85	15	GaAs MESFETs	D	[4.99]
7–10.5	≤ 3	22.5	GaAs FETs	MMIC	[4.55]
10	0.6	17	GaAs MESFET	D	[4.100]
9–12.5	≤ 3	22.5	GaAs FETs	MMIC	[4.55]
8–18	1.86	9	PHEMT	MMIC	[4.101]
8–18	0.28	11	PHEMT at $T = 19$ K	MMIC	[4.101]
12	1.67	24	GaAs MESFETs	MMIC	[4.57]
18	0.9	13	GaAs MESFET	D	[4.100]
21–23	< 2	33	InGaAs HEMT	MMIC	[4.102]
18–40	< 4.2	15.6	P-HEMT, GaAsMESFET	Semimonolithic	[4.103]
31–35	4.2	15	Ion-implanted MESFET	MMIC	[4.59]
31–33	2	24–27	InHEMT	Hybrid	[4.104]
31–33	0.3	30–33	InP HEMT, $T = 12$ K	Hybrid	[4.104]
40–45	0.2	33	AlInAs/GaInAs/InP HEMTs, $T = 18$ K	Hybrid	[4.105]
44	3.6	14.4	InGaAs MESFETs	Hybrid	[4.106]
70	0.65	33	AlInAs/GaInAs/InP HEMTs, $T = 18$ K	Hybrid	[4.105]
5–80	4.3[a]	9.3	Pseudomorphic MODFETs	MMIC	[4.107]
56–64	2.7	24.7	PM InP HEMT	MMIC	[4.108]
70–77	6.4[b]	27–21	Pseudomorphic MODFET	MMIC	[4.107]
75–110	5.3–6.8	22	InGaAs/GaAs PHEMT	MMIC	[4.109]
75–110	6	23	InP-based HEMT	MMIC	[4.110]
81	4.3	18.5	P MODFETs	MMIC	[4.111]
91–96	3.2–3.5	17.5	InP HEMT	Hybrid	[4.112]
92–96	5–6	19	PM HEMT	MMIC	[4.113]
94	6.5	49	Pseudomorphic InGaAs HEMT	MMIC	[4.114]
94	4.5–5.5	17	PM HEMT	MMIC	[4.60]
96–100	2	3–19	AlGaAs/InGaAs/GaAs	MMIC	[4.115]
100	4.2	19	HEMT		[4.115]

[a] Below 60 GHz.
[b] At 76 GHz.

reflected from the antenna element (usually because of an existing high VSWR) and leaks into the receiver. The protector therefore must act in an active mode (transmit) or a passive mode of operation. In some cases the protector is used to either attenuate the incoming signals or to blank the receiver.

Given that the protector is between the antenna element and the LNA, its insertion loss has to be minimum. It can be realized in a single-ended or balanced configuration. PIN diodes are commonly used in conjunction with a quarter-wave transmission line [4.28]. Minimum losses of the order of 0.7 dB [4.28] are incurred with the single-ended realization over a relatively narrow band, while the latter realization can be very broad.

PIN diode limiters fabricated using MBE (molecular-beam epitaxy)-grown layers can provide less than 0.2 dB small-signal insertion loss and greater than 15 dB limiting at a frequency of 10 GHz [4.116, 4.117].

A variable attenuator/limiter utilizing planar PIN diode fabrication in GaAs has been developed at X- and Ka-bands; The limiter exhibited an insertion loss of less than 0.5 dB and a 20 dB of variable attenuation. At Ka-band the limiter/attenuator exhibited an insertion loss of typically 1.4 dB and had the same range of attenuation [4.116, 4.118].

4.7.3 Programmable Phase Shifters and Vector Modulators

Narrowband scanning phased arrays utilize programmable phase shifters to steer the resulting array beam to different directions within the array FOV. Usually each antenna element is followed by a phase shifter that operates at either RF or the IF of the array. The phase shifters will take values from 0° to 360° in increments that depend on the required phase tracking accuracy.

The requirements for phase shifters are:

- Phase and amplitude accuracy, repeatability, and resettability
- Minimum amplitude variation between the many phase states
- Meeting the bandwidth requirements
- Low insertion loss
- Power handling capabilities that meet the requirements
- A good impedance match

Often the phase and amplitude errors of the module are determined by the specifications of the programmable phase shifters. The issue of repeatability arises when phase shifters are used in high-power applications. The phase shift introduced by the phase shifter ought to be independent of temperature (both ambient and due to high average RF power). Furthermore, these requirements are preferably met by phase shifters that occupy minimum volume, have minimum weight, and consume low power. If the same phase shifter is used on both the transmit and receive paths, it ought to be reciprocal.

System considerations often dictate what kind of phase shifters the designer uses in a phased array. If the phased array is passive, one selects a phase shifter that has a minimum insertion loss, a requirement that is often satisfied by ferrite phase shifters or solid-state-based analog phase shifters. For an active phased array the insertion loss is still important, but other considerations such as accuracy, the volume it occupies; and its power consumption become as important. Often these requirements are satisfied by digital phase shifters.

Ferrite switches usually have the lowest insertion loss and can handle the highest power; analog solid-state-based phased shifters have comparable or slightly higher insertion loss, can handle significantly lower powers, but are smaller in volume. Digital phase shifters have the highest insertion loss, which is however invisible to the system.

4.7.3.1 The Ferrite Phase Shifter The quest for low-loss phase shifters led to the realization of ferrite phase shifters operating at cm and mm wavelengths; their defining characteristics are considerable bulk, low insertion loss, and substantial power handling capability.

Ferrite phase shifters are two-port devices that are inserted into waveguides to provide variable phase shift to the signals propagating within them by changing the bias field of the ferrite. A ferrite phase shifter consists of an RF structure and an electronic driver under computer control.

Ferrite phase shifters evolved for passive phased arrays that utilized one or a small number of high-power tubes; hence the high-power capability and low insertion loss specification. Useful descriptions of ferrite phase shifters are included in Stark [4.119] and excellent descriptions of currently available ferrite phase shifters are given in Hord [4.120].

In this book we have traced a pronounced trend away from passive phased arrays utilizing one or a small number of high power tubes and toward active arrays utilizing MMIC-based modules that can deliver relatively low powers, tens of watts, to the array antenna elements.

In Chapter 5 we shall trace another trend away from waveguide-based antenna elements for phased arrays and toward printed board antenna elements of various configurations. Lastly we have reflected the widely held view that the total costs, acquisition, and LCC, for active arrays are lower than the costs of passive arrays.

For these reasons we shall refer the reader to References [4.119 and 4.120] for ferrite phase shifters and move over to phase shifters used in MMIC-based modules.

4.7.3.2 Solid-State Phase Shifters Solid-state phase shifters are usually divided into analog and digital depending on how the phase-shift is obtained. In an analog phase shifter a set of voltage settings is used to obtain the required phase shifts. Specific phase shifts (0°, 22.5°, 45°, etc.) are obtained when a digital phase shifter is used. Both types of phase shifters can be monolithic.

4.7.3.2.1 Analog Phase Shifters Analog phase shifters utilize passive components in conjunction with MMIC-compatible Schottky barrier or varactor diodes. Some varactor diodes are used as variable capacitors [4.116]. Passive elements range from capacitors, lumped capacitors, and Lange couplers used as 90° hybrids.

Typically analog phase shifters are of the reflection type, and several of their essential characteristics are tabulated in Table 4.20. Usually the chip area of analog phase shifters is smaller than that occupied by digital phased shifters. Most T/R modules use digital phase shifters; the trend toward digital electronics is manifested here too. Sharma [4.116] is a comprehensive review paper on solid-state control devices.

4.7.3.2.2 Digital Phase Shifters Digital phase shifters and vector modulators are by far the most popular choices for active phased arrays.

We can distinguish the following main types of phase shifters, which are illustrated in Figure 4.20: switched-line (*a*), reflection-line (*b*) and loaded-line (*c*).

Operation of the switched-line phase shifter is straightforward, and the derived phase shift is equal to the phase difference of the two lines used. Operation of the reflection-line phase shifter using an isolator is also straightforward; in most realizations, however, the isolator, which is too expensive and bulky, is substituted by a 90° hybrid. The differential phase shift between the two lines is obtained at the output port of the hybrid.

For small phase shifts ($\phi < 45°$), the loaded-configuration is preferred; it has been shown [4.125] that ϕ is proportional to the susceptance introduced on the IN/OUT line. By switching different susceptances on the line a differential phase shift is introduced. As the characteristic impedance of the line is directly related to the introduced susceptance, a large susceptance or phase shift will introduce unacceptable mismatches.

High-pass/low-pass phase shifters achieve their phase shift by switching the signal path containing either a high- or low-pass circuit. They also provide wide bandwidth and constant phase shift as a function of frequency. This type of phase shifter is easy to realize at the lower microwave frequencies, hence its popularity in these frequency bands. We have tabulated the essential characteristics of typical digital phase shifters in Table 4.21. It is clear that the accuracy obtained when the phase shifter is at IF (first entry in Table

TABLE 4.20 Essential Characteristics of Typical Analog Phase Shifters

Frequency (GHz)	Range	Amplitude Error (dB)	Phase Error (deg)	Chip size (mm × mm)	Insertion Loss (dB)	Reference
8–12.4	0°–105°		±5	1.96 × 2.54	2.5	[4.121]
6–18		±0.7		3.76 × 1.73	2.7	[4.122]
21–26	0°–230°				1.6	[4.123]
16–18	0°–109°	±0.3	±3		1.8	[4.124]

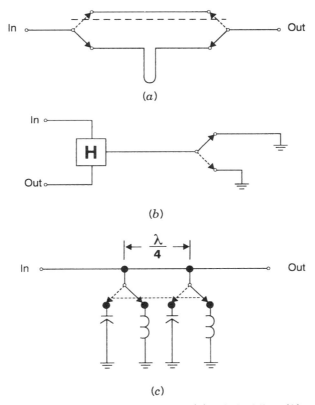

FIGURE 4.20 Conventional phase-shifter types: (a) switched-line; (b) reflection-line; and (c) loaded-line.

4.21) is much higher than that obtained when the same function is performed at RF.

Baluns are used to provide 180° phase changes. With the exception of the baluns, all constituent components of phase shifters are easily fabricated by the use of standard MMIC technology. Conventional off-chip baluns have been used [4.126] because standard monolithic baluns were physically too big [4.127]. Recently a compact monolithic balun consisting of two crossed interdigital couplers has been proposed and realized [4.127]. It has a bandwidth extending from 7 to 19 GHz, an insertion loss of less than 2 dB and occupies an area of 1 × 2 mm.

4.7.3.2.3 Vector Modulators With reference to Figure 4.21, the vector modulator implements the required phase shift by the addition of two vectors aligned with the X and Y coordinates that represent 0°, 90°, 180°, and 270° phase shifts. To derive a given phase, the appropriate quadrant is first chosen;

TABLE 4.21 Essential Characteristics of Typical Digital Phase Shifters

Frequency (GHz)	Bits	Amplitude Error (dB)	Phase Error (deg)	Insertion Loss (dB)	Chip size (mm × mm)	Reference
IF	12	0.02 rms	0.2 rms			[4.48]
L-band	5	0.5 rms	2.5 rms	8.5		[4.80]
2.2–2.3	5	±0.5	7 rms	4.9	5.9 × 2.9	[4.128]
S-band	5	0.4 rms	3 rms	7		[4.80]
3–6	6	±1	< 1 rms	10	3.8 × 3.3	[4.129]
4–18	5	1.5	< 9		10.2 × 12.7[c]	[4.130]
5.5–8.5[a]	5	±1	< 4 rms	13	2.4 × 1.22	[4.126]
X-band	5	2–3		10		[4.131]
X-band	4	0.3 rms	2 rms	7		[4.80]
X-band[a]	6[b]	0.3 rms	3 rms	—	5 × 2.7	[4.44]
17.7–20.2	5	±0.5	±6°	Gain = 16		[4.132]
35–37	4	±1.8	1.5 rms	12.8	3 × 7.8 × 0.1	[4.133]

[a] Vector modulator.
[b] Equivalent to 6-bit digital phase shifter.
[c] Dimensions of the carrier.

then the two vectors of the quadrant are attenuated by the required amount before they are added [4.126]. In another realization the required phase change is attained by simply routing the incoming signal through two cascaded phase shifters; the first phase shifter introduces the 90°, 180°, and 270° shifts, while the second phase shifter introduces the 45°, 22°, and 11° shifts [4.130]. In all realizations the amplitude variations of the output signals are minimized.

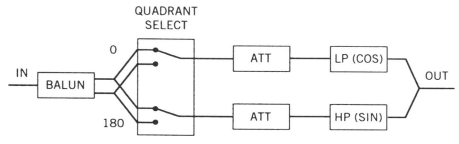

FIGURE 4.21 The block diagram of a vector modulator. The required phase shift is attained by the vector addition of two signals out of phase by 90° appropriately attenuated. First the specific quadrant is selected and the required attenuation is implemented to the two vectors. (*Source*: Ali et al. [4.126], © 1990 IEEE.)

4.7.4 Phase Shifters: At RF or IF

Theoretically there is no difference where the phase shifters are introduced in a phased array system. Sometimes it is convenient to introduce them at the incoming RF, while at other times it is convenient to introduce them at some lower or higher IF. In the extreme, the latter option is used when the incoming RF is converted to some convenient optical wavelength for the sake of attaining substantial true delays and bandwidths in a confined space. Most of the time it is a matter of convenience, which we shall explore.

The discussion here is confined to active phased arrays, where the designer has a choice; this is in contrast to passive arrays, where the phase shifter has to be inserted at the RF. It is recalled here that the attractions of performing the phase-shifting function at a lower IF are lower costs and increased phase and amplitude accuracies.

A couple of issues have to be addressed, however. The IF has to be chosen so that the instantaneous array bandwidth B is easily accommodated. By selecting the IF on the basis that $B/\text{IF} \leq 10\%$, one can easily resolve this issue.

If the down conversion has to take place in the module (a sensible option), LO power can be brought to the module on the transmission line, which is normally dedicated to connect the reference transmitter frequency to the module on transmit. If optical fiber is used for the distribution of signals to and from the modules, a "dedicated line" (which carries a multitude of multiplexed signals) can be used to bring the LO signal to the module.

Figure 4.22 is a block diagram of the module where the phase shifting is performed at the IF [4.48]. The "IF module," as the authors term it, has 12-bit amplitude and phase settings. In passing it is worth noting that the distribution of RF signals to many parts of a phased array constitutes a fundamental problem. It is only recently that photonics enabled designers to solve this problem in a cost effective way. The possibility of using IF phase shifters is only one option systems engineers have to revisit in the light of these developments.

4.7.5 Circulators

Circulators can be used basically to perform the function of duplexers, isolators, and microwave switches (single-pole double-throw). In Figure 4.23 we have illustrated the three basic functions of the circulator. The most prevalent application of circulators is to perform the duplexer function. In the context of phased arrays, circulators are seldom used to perform the isolation function because of their cost and bulk. Circulators used to perform the switching function are used in satellite communications as redundancy switches and in radio astronomy as Dicke switches. There is a continuous and significant research effort to realize circulators, performing the preceding functions, compatible in size to monolithic circuits.

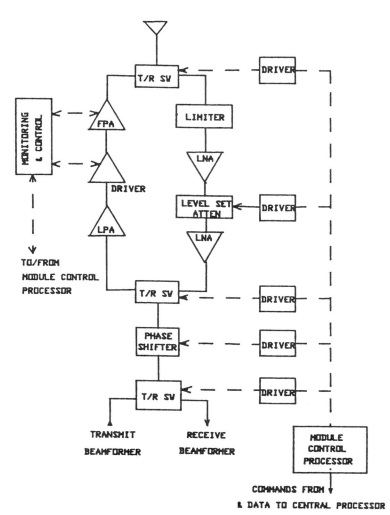

FIGURE 4.22 The block diagram of the module where the phase-shifters are introduced at the intermediate frequency. (*Source*: Aumann and Willwerth [4.48], © 1988 IEEE.)

Essentially there are two types of circulators: the differential phase-shift and the Y-junction circulator. The former circulator can handle very high power (e.g., in the megawatt range) and has a narrow bandwidth, while the latter usually operates over waveguide bandwidths.

Duplexers are used in conventional single-dish radars and phased arrays to provide a convenient, signal route for the receiver and the transmitter to the

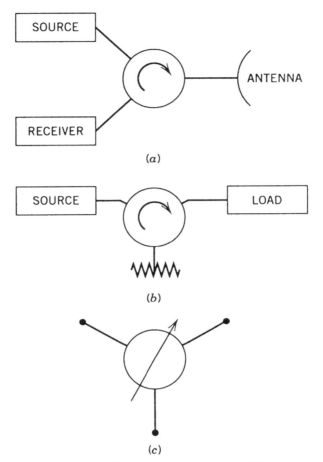

FIGURE 4.23 The three basic functions of circulators: (*a*) the duplexer; (*b*) the isolator; and (*c*) the microwave switch.

antenna. The requirements of the duplexers are:

- High isolation between the transmitter and the receiver
- Low insertion loss and excellent VSWR
- Power handling capability that matches the requirements
- Meeting the bandwidth requirements
- Minimum bulk (consistent with the power handling capability requirement)

The last requirement is important at all wavelengths, especially at mm wavelengths, where the half-wavelength spacing is only a few millimeters wide. Taking into account present technology, a circulator that has to handle high powers is bulky. If the power is not high, the designers would prefer small-size, inexpensive circulators.

Several approaches toward MMIC-based circulators have been reported [4.134–4.136]; additionally, interest in producing a ferrite device that is comparable in size and cost to monolithic microwave circuits has been reported [4.137]. Current research is focused on the intrinsically high Q of YIG resonators and the compatibility of film technology with monolithic circuits [4.137].

4.7.5.1 Ferrite Circulators When a circulator is used in the duplexer mode, it channels the transmitter power to the antenna and the received signals to the LNA and provides high isolation between the LNA and the FPA.

Circulators are typically three-port devices that consist of a ferrite core and permanent magnets to provide the bias field. If an electromagnet is used instead, the circulator can operate in a latching mode as a single-pole double-throw switch. In the absence of a bias field the circulator has a single lowest-order resonant mode with a $\cos \phi$ dependence; when the ferrite is biased, this mode brakes into two resonant modes with slightly different resonant frequencies. The operating frequency of the circulator is then chosen so that the superposition of these two modes add at the output port and cancel at the isolated port [4.97]. Given that the ferrite disks are used as dielectric resonators, the disk diameter is of the order of half a wavelength of the wave propagating through the ferrite. Thus a circulator based on this design is necessarily large [4.138]. Rotrigue [4.139] is an excellent reference for conventional and miniature circulators.

The most popular circulator is the Y-junction circulator, illustrated in Figure 4.24. The three stripline conductors are attached to a center disk at 120° intervals, forming the three circulator ports. These stripline conductors are sandwiched between two ferrite disks and the ground planes; the DC bias field is perpendicular to the ground planes.

Conventional circulators have typical insertion and isolation figures of 0.3–0.5 dB and 20–30 dB, respectively [4.139], depending on the bandwidth required and the frequency of operation. Novel circulator configurations with possible applications in phased arrays have been proposed [4.140].

4.7.5.2 Miniature Circulators In order to decrease the size of the circulators, several approaches have been taken. At UHF and VHF lumped-element circulators have been realized; their volume is one-fifth of their octave bandwidth counterparts, and they are essentially narrowband devices—for example, 4–6% of the center frequency. At frequencies up to 20 GHz, distributed elements are used to realize circulators [4.139].

Conventional ferrite circulators for active phased arrays generally consist of microstrip transmission lines and ferrite "plugs" placed at suitable loca-

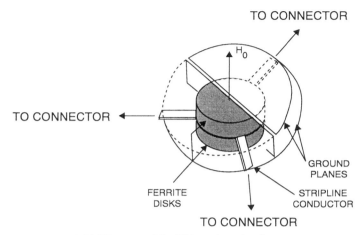
FIGURE 4.24 The Y-junction circulator.

tions; in other realizations ferrite substrates are used. More attractive MMIC compatible methods of fabrication have been proposed [4.138].

A truly miniature circulator operating at X-band has been realized [4.138]. Its bandwidth, insertion loss, and interport isolation are 200 MHz, 0.5 dB, and 15 dB, respectively. The active material of the circulator is a single-crystal disk (or puck) of YIG, the diameter and thickness of which are 0.38 and 0.05 mm, respectively. This is indeed a promising area of research [4.138].

4.7.5.3 MMIC-Based Circulators To further decrease the circulator volume several designers considered MMIC-based circulators. The approach taken by Hara et al. [4.134] is based on the use of an active out-of-phase divider and an active in-phase combiner to realize a quasi-circulator. In other realizations of MMIC-based circulators, distributed amplifiers are used [4.135, 4.136]. Although the bandwidth of these devices is multioctave, their insertion loss is higher than 5 dB and their power handling capability is a few dBm. Despite these limitations, the approaches described above are important.

As isolators and circulators enter the consumer market, their cost, size, and weight decrease; for instance an isolator used for consumer applications at 800 MHz was reported to have dimensions of 6.8 × 6.9 × 4 mm and weighs 0.75 grams [4.141]. Unfortunately no performance details are available.

4.8 CONCLUDING REMARKS

The wide spread adoption of phased arrays for military and civilian applications is based on the thesis that affordable T/R modules are available now. This chapter explores this important thesis in some detail and defines the criteria one should apply before selecting appropriate of T/R modules to meet specific performance specifications and cost constraints.

Up until recently two options were available to the designer: MMIC- or vacuum tube-based T/R modules; now a third option, the microwave power module that consists of a MMIC preamplifier and a mini-TWT is available to the designer.

It is widely accepted that MMIC-based subsystems outperform their vacuum tube counterparts in the performance of all the T/R functions except the generation of high powers; this is especially true at mm wavelengths. Some of the important criteria for the selection procedure are the required module power output, costs (acquisition and LCC), the MTBF issue and the PAE of power amplifiers. While there has been a "benign neglect" of vacuum tube technology, we have witnessed a recent revival of interest in this area.

Although conventional power adding techniques can be used to combine the power output of several MMICs or vacuum tubes, the output power of several solid-state devices can be added at the chip level or with the aid of quasi-optical techniques. The latter approach is bound to be effective at mm wavelengths.

The costs of MMIC modules based on GaAs devices have decreased with the passage of time for the following reasons:

1. The availability of high-quality wafers due to advances in material technology and the evolution of effective quality control methods.
2. The evolution of manufacturing methods that increase the yield and uniformity of devices.
3. Several approaches aimed to decrease manufacturing costs of T/R modules such as RF-wafer scale integration and the multiple chip module, have been pursued.

Most of the above thrusts have been supported by the MIMIC Program, which funded the necessary infrastructure for affordable MMIC modules. Other programs and initiatives such as the development of improved array components and a computer-integrated manufacturing facility to perform array architecture and design tradeoff studies will further decrease costs of phased arrays. From a systems engineering point of view the fun has just begun.

CHAPTER FIVE

Antenna Elements

> Antenna engineering is a field which is bursting with activity, and is likely to remain so in the foreseeable future.
>
> J. R. James, 1990.

Many types of antennas have been used as antenna elements of phased arrays, such as horns, slotted waveguides, and conventional and bent dipoles. Recently researchers and design engineers have shown strong preference for microstrip antennas in the form of patches or printed dipoles. This trend is not surprising given that printed-circuit antennas are lightweight, inexpensive, and can conform to the surfaces of many platforms, such as those for vehicles and airplanes. The comparisons are made with respect to horn antennas and slotted waveguides.

Although printed-circuit antennas (PCAs) were proposed in the early 1950s by Deschamps [5.1], it was not until the 1970s that serious attention was given to these types of antennas. Excellent review papers and books have been written [5.2–5.6] on many theoretical and design aspects of microstrip antennas. However, high-quality microstrip antennas have been realized only recently through the efforts of a diverse group of researchers and designers. It is these aspects of PCAs that we shall explore and emphasize in this chapter.

In the context of active phased arrays, where only low powers exist in and around the array, PCAs are eminently suited for the application. The targets are, however, illuminated by high power, resulting from the combination of all powers transmitted by the many antenna elements.

Another attractive characteristic of microstrip antennas is that they can be integrated with MMICs. In the early literature the integration seemed straightforward; however, considerable work was required before efficient approaches were found. We have already alluded to this integration issue in Section 4.6.2. In this chapter we shall consider some of the successful configurations that have been proposed. Despite these advantages, conventional microstrip antennas (patches and printed dipoles) are bandwidth-limited to a fractional bandwidth of 2–15% and a bandwidth of 40% has been attained by printed dipoles. Their other shortcomings are:

1. Their efficiency can be low because many modes can propagate.
2. The microstrip arrays can suffer from extraneous radiation emanating from feed lines to the antennas and junctions.
3. Power can be dissipated in surface waves.
4. Polarization purity is hard to achieve.
5. Tolerance problems and substrate quality have to be addressed.

While some of these problems are still with us, progress has exceeded conservative expectations. Furthermore, the future of microstrip antennas operating on a stand-alone basis or as antenna elements of phased arrays seems more promising than ever.

In this chapter we shall focus attention on antenna designs that can accommodate two principal polarizations, for this is a generic requirement for many applications.

5.1 OUTLINE OF THE REQUIREMENTS

In this section we shall outline the system requirements related to antenna elements of phased arrays designated for radar or communications applications; these requirements are so stringent that antenna elements for the above-mentioned phased arrays usually meet the requirements for other applications.

Some arrays intended to substitute reflectors or lens apertures operate on a receive-only mode; the signals of interest are emanated from satellites, and one polarization is usually used. For these applications the issues to address are the minimization of losses that are inherent in the transmission lines interconnecting the many antenna elements to a common point, low-cost construction, and ruggedness. To the extent that the designer has to arrange for all transmission lines connecting the antenna elements to one port, one is justified to consider them as phased arrays.

For communications applications, both polarizations are required and a similar requirement exists for single-aperture and phased array-based radars. The high costs of T/R modules prevented the realization of dual-polarization phased array-based radars for some time; suitable architectural approaches to derive the full benefits of polarization information are also needed.

The requirement for polarization purity is satisfied if the cross-polarization fields are of the order of -25 to -35 dB with respect to the maximum main lobe of the copolarization field. It is noted here that the maximum cross-polarization field usually occurs in a direction that subtends a 45° angle with respect to the principal polarizations; Ludwig's third definition of cross-polarization is adopted here [5.7].

For radar applications the minimum requirement is to transmit either of the two principal (vertical and horizontal or left- and right-handed circular polarizations) alternatively on a pulse-to-pulse basis and receive the scattered radiation on two channels accommodating the two principal polarizations. For communications applications the minimum requirement is to transmit and receive two orthogonal or opposite polarizations.

Ideally one would like to have polarization agility on transmit and receive in order to maximize the SNR of the system. For communication and radar applications, polarization agility is required to offset the effects of rainfall and other naturally occurring phenomena on the transmitted/received signals. When polarization is a requirement, the FOV of both polarizations ought to be the same; this requirement is satisfied when the E- and H-plane radiation patterns of the antenna element are coincident, which incidentally is a precondition for the attainment of polarization purity.

The importance of minimizing losses between the antenna elements and the T/R modules has been stressed by many authors and in different parts of this book. Finally, the choice of antenna elements depends on the many considerations already outlined and to some extent on whether scanning in one or two dimensions is required.

5.2 CANDIDATES FOR PHASED ARRAY ANTENNA ELEMENTS

There is no doubt that hybrid-mode and scalar feeds considered in Section 2.6 approximate the ideal feeds. Given their weight and volume, we see them not so much as candidates for phased array antenna elements but as ideal antennas that have optimum characteristics. Thus scalar- and hybrid-mode feeds provide a useful benchmark for other feeds. For applications where weight, volume, and cost are not important, however, these feeds can be used as array antenna elements.

What follows is a brief description of slotted waveguides, which have been used in many phased arrays and extended sections on patch and printed dipole antenna elements. While the latter array elements are suitable for narrowband application, we shall consider antenna elements that are suitable for wideband applications, also.

5.2.1 Antenna Elements for Canonical Arrays, Scanning in One Dimension

While several antenna elements suitable for arrays scanning in two dimensions can be used in arrays where one-dimensional scanning is required, slotted waveguides are particularly suitable for one-dimensional scanning arrays. Given that the latter arrays are less expensive than the former, costs often drive designers to one-dimensional scanning arrays.

Slotted waveguides have been used in a variety of phased arrays, such as the AWACS and ERIEYE already considered. Waveguide slot arrays are among the most important and best understood line-source elements [5.8–5.10]. This being the case, our treatment will be short and confined to an outline of their capabilities and recent applications.

Figure 5.1 illustrates the method of scanning in one dimension, while Figure 5.2 illustrates the three slot waveguide array geometries: (a) the edge slot array, (b) the displaced longitudinal slot array, and (c) the inclined series slot array. Resonant slot arrays are terminated in short circuits to establish a standing wave in the feed waveguide, while traveling-wave slot arrays are terminated in matched loads.

Traveling-wave arrays usually operate over broader bandwidths than their resonant counterparts and have an off-broadside (squinted) pointing angle that is a function of frequency. Resonant arrays are designed to radiate an "in-phase" broadside pattern [5.11]. With the aid of precision manufacturing, traveling-wave arrays can yield radiation patterns having the lowest sidelobes ever reported, such as the AWACS at cm wavelengths. Similarly, waveguide

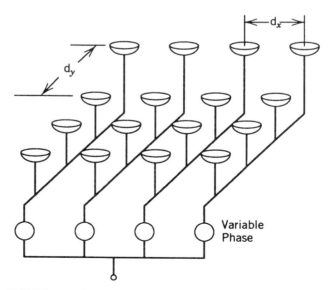

FIGURE 5.1 Array geometry for scanning in one dimension.

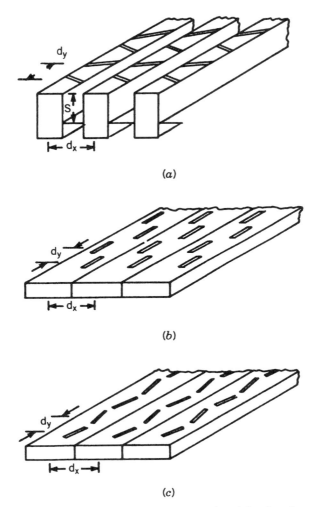

FIGURE 5.2 Typical waveguide slot array geometries: (*a*) edge slot array. (*b*) displaced longitudinal slot array; and (*c*) inclined series slot array.

slot arrays operating at mm wavelengths have been reported [5.12]. Excellent pattern control has been achieved on the basis of design procedures outlined by Yee [5.13] and Oliner [5.14]. Circular polarization can be obtained by different arrangements of slots, including crossed slots [5.15].

The physical size of the waveguide used decreases the degrees of freedom for the designer. It is of interest to note that both the ERIEYE and AWACS arrays are not conformal with the aircraft's skin.

5.2.2 Antenna Elements for Canonical Arrays, Scanning in Two Dimensions

In this section we shall consider antenna elements suitable for canonical phased arrays that are scanned in two dimensions. The most promising candidates for this application are patch antennas, crossed dipoles, and a variety of printed dipoles. While patch antennas and crossed dipoles are used in tile architectures, the printed dipoles are used in brick architectures.

5.3 PATCH ANTENNAS

The patch antenna consists of a metallic ground plane and a thin dielectric on top of which a conducting surface, the patch, is etched; the patch can have several shapes but the rectangular and circular shapes shown in Figure 5.3a and 5.3b, respectively, are commonly used. There are several methods of feeding the patch and some of them are illustrated in the same figure. Figure 5.4a–d illustrates other patch shapes and methods of feeding them, to derive circular polarization.

Typically the longer dimension of a rectangular patch antenna, when it has a single feed point, is less than a wavelength at the frequency of operation in the dielectric, h, substrate material (λ_m), to avoid the excitation of higher order modes; its shorter dimension would be about $\lambda_m/2$ and the thickness of the dielectric would be 0.03–0.05λ. Often λ_m is also designated as λ_g and Teflon or polytetrafluoroethylene, RT Duroid, and other substrates having low relative permittivity are commonly used.

5.3.1 Patch Dimensions

Several methods have been used to analyze patch antennas, including the transmission-line, cavity, and modal expansion models and the Green function method [5.4]. The transmission-line model, applicable only to the rectangular and square patches, views the radiator as a transmission line supporting a quasi-TEM (transverse electromagnetic mode) mode and terminated at its two ends by two radiating apertures, which can be likened to radiating slots. The mode is quasi-TEM because the medium between the conductors is inhomogeneous—that is, the phase velocity is different from the velocity in free space and the velocity in the dielectric. The notion of effective relative permittivity ε_{eff}, which takes into account the inhomogeneity, has been found convenient. The results obtained by using this method are useful but not accurate enough, and the model is not applicable to circular patches.

The cavity model lends deeper physical insights into the radiation mechanisms of patch antennas. More explicitly, the model considers the volume of space under the patch and between the ground plane as leaky cavity and

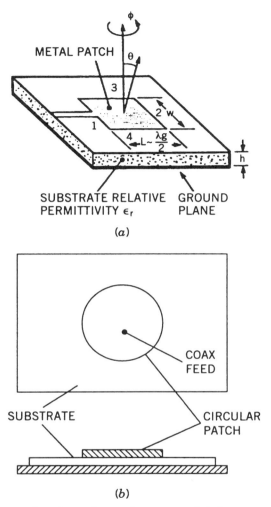

FIGURE 5.3 Geometries of popular patch antennas: (*a*) the rectangular patch [from (5.18)]; and (*b*) the circular patch.

brings out the type of excitation for the different modes (TM$_{mn}$, where m and n can take a range of values) that can provide the radiation field.

Based on this model, the resonant frequency of the rectangular patch illustrated in Figure 5.5, is given by the equation [5.15a]

$$f_\mathrm{r}|_{nm} = \frac{c}{2\sqrt{\varepsilon_\mathrm{r}}} \sqrt{\left[\frac{n}{L}\right]^2 + \left[\frac{m}{w}\right]^2} \qquad (5.1)$$

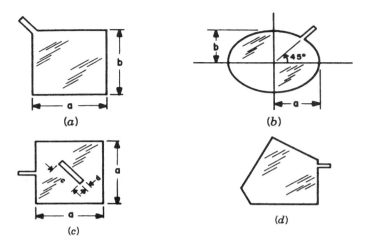

FIGURE 5.4 Geometries of typical patch antennas yielding one circular polarization. (*a*) Almost square; (*b*) elliptical; (*c*) square with a 45° slot; and (*d*) the pentagon patch.

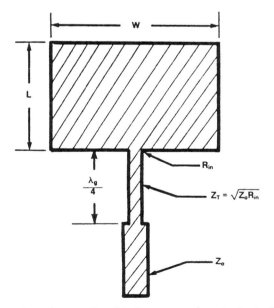

FIGURE 5.5 A rectangular patch with its quarter-length feed. (*Source*: Munson [5.16], © 1988 *Microwave J.*)

where ε_r is the relative permittivity of the substrate, L and W are the length and width of the patch, and m and n are integers with values depending on the mode of excitation The important modes in this case will be $n = 0$ and $m = 1$ or $n = 1$ and $m = 0$. If $n = 0$ and $n = 1$, Equation (5.1) is reduced to

$$L = 0.5 \frac{\lambda_0}{\sqrt{\varepsilon_r}} \qquad (5.2)$$

which is a well-known result [5.16].

To account for the fringing fields at the perimeter of the patch, the following empirical formulas have been used for the affective dimensions [5.17]:

$$L_{eff} = L + \frac{h}{2} \qquad (5.3a)$$

$$W_{eff} = w + \frac{h}{2} \qquad (5.3b)$$

The following more accurate empirical expressions for the resonant frequency and the effective relative permittivity have been proposed in Lo et al. [5.17]

$$F_r|_{nm} = f_r|_{nm} \frac{\varepsilon_r}{\sqrt{\varepsilon_{eff}(L)\varepsilon_{eff}(w)}} \frac{1}{1 + \Delta} \qquad (5.4)$$

where

$$\Delta = \frac{h}{L}\left[0.882 + \frac{0.164(\varepsilon_r - 1)}{\varepsilon_r^2} + \frac{\varepsilon_r + 1}{\pi \varepsilon_r}\left[0.758 + \ln\left(\frac{w}{h} + 1.88\right)\right]\right] \qquad (5.5)$$

$$\varepsilon_{eff}(u) = \frac{\varepsilon_r + 1}{2} + \frac{\varepsilon_r - 1}{2}\left[1 + \frac{10h}{u}\right]^{-1/2} \qquad (5.6)$$

Using the same method of analysis, for a circular patch of radius a, its resonant frequency is given by the equation [5.15a]

$$f_r = \frac{K_{nm}c}{2\pi a_{eff}\sqrt{\varepsilon_r}} \qquad (5.7)$$

and [5.17a]

$$a_{eff} = a\left[1 + \frac{2h}{\pi a \varepsilon_r}\left[\ln\frac{\pi a}{2h} + 1.7726\right]\right]^{1/2} \qquad (5.8)$$

where K_{nm} are the roots of the equation

$$J'_n(x) = 0 \tag{5.9}$$

and the differentiation is with respect to x. The first five nonzero roots of this equation are tabulated in the Table 5.1.

The challenge for designers has been to extract two principal polarizations from the same patch antenna in the presence of a variety of unwanted modes. Several approaches have been taken to meet this challenge; here we shall outline successful approaches that have direct relevance to phased arrays after a formulation of the electric fields at resonance for the rectangular and circular patch antennas.

5.3.2 The Electric Field of the Patch Antennas

The electric fields of the rectangular and circular patch antennas have been derived as [5.15a]

$$E_z = E_0 \cos\frac{m\pi x}{w} \cos\frac{n\pi y}{L} \tag{5.10}$$

$$E_z = E_0 J_n(k_{nm}\rho) \cos n\psi \tag{5.11}$$

respectively; E_0 is an arbitrary constant, ρ, ψ are the radial and azimuthal coordinates of the patch, respectively; and J_n is the Bessel function of the first kind of order n; additionally

$$k_{nm} = \frac{K_{nm}}{a} \tag{5.12}$$

In Lee and Dahele [5.15a] and the references therein methods are described that are used to derive the far-field patterns of patch antennas. Figure 5.6

TABLE 5.1 The First Five Nonzero Roots of $J'_n(x) = 0$

(n, m)	x
(1, 1)	1.841
(2, 1)	3.054
(0, 2)	3.832
(3, 1)	4.201
(1, 2)	5.331

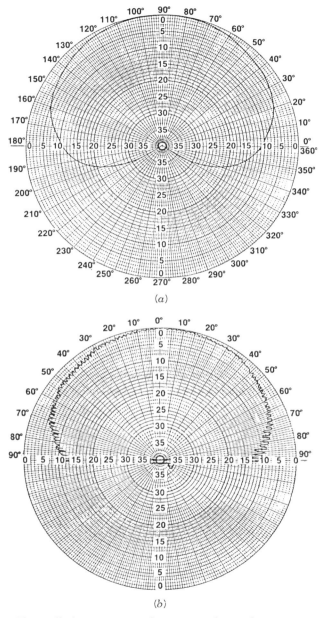

FIGURE 5.6 The radiation patterns of a rectangular and a square patch: (*a*) the theoretical E-plane pattern of a rectangular microstrip antenna element etched on a dielectric with $\varepsilon_r = 2.45$; and (*b*) spinning dipole radiation pattern of a circularly polarized square microstrip antenna element. (*Source*: Munson [5.16], © 1988 *Microwave J.*)

illustrates typical radiation patterns of patch antennas. Figure 5.6a illustrates the theoretical E-plane radiation pattern of a rectangular microstrip element having $\varepsilon_r = 2.45$, while Figure 5.6b shows a spinning dipole radiation pattern of a circularly polarized square microstrip element [5.16]. If we accept the TM_{10} mode as the fundamental mode for a square patches, the TM_{02} mode gives rise to cross-polarization fields, which in turn reduce its polarization purity.

5.3.3 Bandwidth and Efficiency of Patch Antennas

In the context of microstrip antennas, the bandwidth of an antenna element $B (= \Delta f)$ is usually taken to be the range of frequencies over which the input VSWR of the antenna is below a certain value S. The assumption here is that the antenna impedance is matched to the transmission line at resonance. If Q_T is the total Q factor of the antenna, B as a percentage is given by the equation [5.15a]

$$B = \frac{100(S-1)}{Q_T \sqrt{S}}\% \qquad \text{when} \quad S \geq 1 \qquad (5.13)$$

A commonly accepted value for S is 2, Equation (5.13) therefore becomes

$$B = \frac{100}{\sqrt{2}\, Q_T}\% \qquad (5.14)$$

The impedance-matched bandwidth has been adopted by most designers in preference to a bandwidth over which the antenna radiation pattern is acceptable because the antenna radiation pattern continues to be acceptable long after S has reached unacceptable values. While most designers accept this definition of bandwidth, some define the bandwidth as $1/Q_T (\times 100\%)$ [5.18].

To derive Q_T, one has to take into account all the losses normally associated with the microstrip antenna, the main losses are through radiation, surface-wave generation, dielectric losses, and conductor losses. For these losses the corresponding Q factors—Q_r (due to radiation losses), Q_{sw} (due to surface waves), Q_d (due to dielectric losses), and Q_c (due to conductor losses) —can be derived, so that [5.19].

$$\frac{1}{Q_T} = \frac{1}{Q_r} + \frac{1}{Q_{sw}} + \frac{1}{Q_d} + \frac{1}{Q_c} \qquad (5.15)$$

For a rectangular patch, of width W, length $L = \lambda_g/2$, and height h on a substrate of relative permittivity ε_r, illustrated in Figure 5.3a, the following approximate equations have been derived [5.18]:

$$Q_r = \frac{\pi}{4G_R Z_m} \tag{5.16}$$

$$Q_d = \frac{1}{\tan \delta} \tag{5.17}$$

$$Q_c = \frac{78.6\sqrt{f} Z_{m,0} h}{P_m} \quad \text{for copper} \tag{5.18}$$

where $\tan \delta$ is the dielectric loss tangent of the substrate, f is the frequency in GHz, and h is in cm; the conductor was assumed to be copper. Additionally, $Z_{m,0}$, Z_m, P_m, and G_R are the impedance of an air-filled microstrip line of width w and thickness h, the impedance of the same line filled with dielectric, the power dissipated in the latter line, and the radiation conductance of the patch end face, respectively. Z_m is given by the equation [5.18]

$$Z_m = \frac{60\pi}{\sqrt{\varepsilon_r}} \left\{ \frac{w}{2h} + 0.441 + 0.082 \left(\frac{\varepsilon_r - 1}{\varepsilon_r^2} \right) \right.$$

$$\left. + \left(\frac{\varepsilon_r + 1}{2\pi\varepsilon_r} \right) \left[1.451 + \ln\left(\frac{w}{2h} + 0.94 \right) \right] \right\}^{-1} \tag{5.19}$$

for $w/h > 1$ and $Z_{0,m}$, is obtained by setting $\varepsilon_r = 1$. An expression for Z_m when $w/h < 1$ is given in James et al. [5.18]. The equations for P_m for the cases where $h/w \le 2$ and $h/w \ge 2$ are [5.18]

$$P_m = \begin{cases} \left[1 - \left(\frac{w}{4h} \right)^2 \right] \left[1 + \frac{h}{w} \right] & \text{for } \frac{w}{h} \le 2 \\ \dfrac{2\pi[(w/h) + [(w\pi/h)/[(w/2h) + 0.94]]][1 + (h/w)]}{\left\{ \dfrac{w}{h} + \dfrac{2}{\pi} \ln\{2\pi e[(w/2h) + 0.94]\} \right\}^2} & \text{for } \frac{w}{h} \ge 2 \end{cases}$$

$$\tag{5.22}$$

The radiation conductance is given by the following equations [5.18]:

$$G_R = \begin{cases} \dfrac{w_e}{90\lambda_0^2} & \text{for } w_e < 0.35\lambda_0 & (5.23) \\ \dfrac{w_e}{120\lambda_0} - \dfrac{1}{60\pi^2} & \text{for } 0.35\lambda_0 \leq w_e < 2\lambda_0 & (5.24) \\ \dfrac{w_e}{120\lambda_0} & \text{for } w_e \geq 2\lambda_0 & (5.25) \end{cases}$$

where w_e, the equivalent width, is given by the equation

$$w_e = \frac{120\pi h}{\sqrt{\varepsilon_{\text{eff}}}\, Z_m} \tag{5.26}$$

and ε_{eff} is given by Equation (5.6). If we accept the latter definition of antenna bandwidth, the antenna efficiency η is given by the equation

$$\eta = \frac{Q_T}{Q_r} = \frac{1}{Q_r B} = \frac{1}{Q_r \Delta f} \tag{5.27}$$

If we assume that the losses due to surface waves can be neglected, the antenna efficiency η and bandwidth $\Delta f(\%)$ versus the relative permittivity of the substrate ε_r have been illustrated in Figure 5.7a as a function of the ratio h/λ_0. Both the efficiency and the bandwidth of the antenna improve as ε_r decreases. In Figure 5.7b the quantities η and $\Delta f(\%)$ have been plotted against h/λ_0 when the dielectric permittivity ε_r takes the values 1.1–10.3. As the substrate becomes thicker, the efficiency and bandwidth of the antenna increase; similarly, as the relative permittivity is lowered, the efficiency and the antenna bandwidth increase.

These results corroborate the thesis that substrates having low relative permittivities are more suitable than substrates having high relative permittivities. Similarly, thicker substrates are preferred for high-efficiency and relatively wide-bandwidth requirements.

5.3.4 The Worst Case Cross-Polarization Level

With reference to Figure 5.3a, the worst-case cross-polarization level $C_{\max}$ has been calculated by:

1. Estimating the radiation from the radiating apertures 1 and 4.

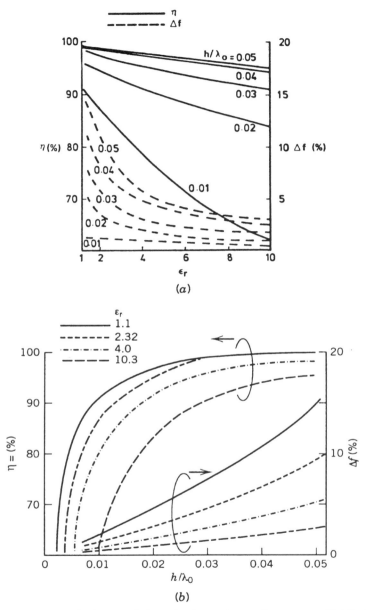

FIGURE 5.7 Essential characteristics of microstrip patch antennas [5.18]. (*a*) The bandwidth Δf (-----) and efficiency η (——) of a rectangular patch antenna as a function of the substrate's relative permitivitty when the substrate thickness h/λ_0 takes the values of 0.01, 0.02, 0.03, 0.04, and 0.05; and (*b*) the bandwidth Δf and efficiency η of a rectangular patch antenna as a function of the substrate's thickness h/λ_0 when the substrate's relative permitivitty takes the values of 1.1, 2.32, 4, and 10.

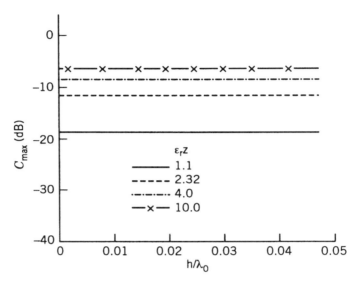

FIGURE 5.8 The maximum cross-polarization radiation C_{max} in decibels, as a function of the substrate thickness h/λ_0, when the substrate's relative permitivitty ε_r takes the values of 1.1, 2.32, 4, and 10 [5.18].

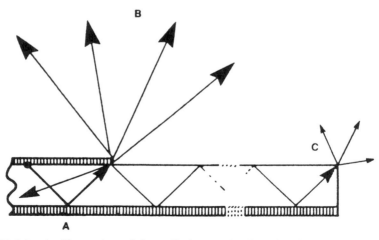

FIGURE 5.9 An illustration of the radiation mechanisms in microstrip patch antennas. (*Source*: Roudot et al. [5.20], © 1988 *Microwave J.*)

2. Using the magnetic wall model for the internal fields of the patch.
3. Accepting Ludwig's third definition of cross-polarization [5.7].

Figure 5.8 illustrates $C_{\max}$ as a function of h/λ_0 when substrates having relative permittivity varying from 1.1 to 10 are used. As can be seen, $C_{\max}$ is only a function of the substrate's relative permittivity and not its thickness; furthermore, $C_{\max}$ improves as the relative permittivity decreases.

The worst-case cross-polarization radiation follows the same trend as the antenna bandwidth and efficiency, illustrated in Figure 5.7. In practice the array factor will decrease the worst-case cross-polarization illustrated in Figure 5.8.

5.3.5 Mutual Coupling Between Microstrip Antennas and Scan Blindness

While one can use simple models to deduce the characteristics of stripline antennas, elaborate and sophisticated procedures are needed to derive the characteristics of patch antennas when they are used as elements of phased arrays.

Figure 5.9 illustrates the complexity of the problem; a point source of current located in the underside of a metallic patch antenna radiates an EM wave [5.20]. Some of the waves designated by A are diffracted, go back under the patch, and store EM energy. The waves designated by B radiate out and contribute to the radiation pattern of the antenna. The waves designated by C remain within the dielectric substrate, trapped by the air–dielectric interface. These are the surface waves that propagate along the two-dimensional interface and decay more slowly than the space waves, which spread into space. In the transmit mode some power fed to the antenna is lost for it does not contribute to the main radiation; the total efficiency of the antenna is therefore reduced. Additionally, surface waves are scattered when they reach the physical boundaries of phased arrays; thus a secondary radiation pattern into the space surrounding the antenna results. To minimize the secondary radiation pattern, designers usually locate dummy antenna elements at the physical boundaries of arrays or place absorbing material along the perimeter of the array.

Surface waves also cause a couple of undesirable effects. When a second patch is in the vicinity of another, currents are induced in it due to coupling to both the space waves and the surface waves of the first patch; thus second patch becomes a secondary radiator. The array designer should therefore aim at minimizing the mutual coupling between antenna elements. Lastly, scan blindness of large phased arrays, toward a scan angle θ, is caused by surface waves propagating in synchrony with a Floquet mode of the structure. At scan angle θ, the array impedance is modified to such an extent that the array does not radiate any power. Consequently the array FOV is defined not by

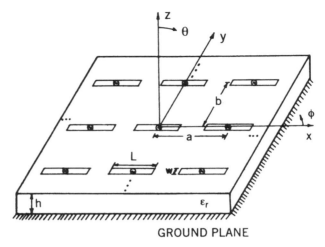

FIGURE 5.10 Geometry of an infinite array of printed dipoles.

the position of the grating lobes, but by the angle at which scan blindness occurs.

It has been shown that the dominant scanning characteristics of an array utilizing microstrip antennas, such as the reflection coefficient, the input resistance trends, scan blindness, and grating lobe effects are dictated by factors such as the element spacing and substrate parameters (h and ε_r) and not by the nature of the microstrip antenna element (patches or dipoles) [5.21].

In Pozar [5.21] and [5.21a] and Pozar and Schaubert [5.22] the problem of scan blindness has been treated in some detail by considering an infinite array of dipole antenna elements printed on a grounded dielectric slab and the derivation of a Green's function for the array scanned to an angle θ, and ϕ. The results summarized here are based on their work.

With reference to Figure 5.10, the spacing between adjacent dipoles is a and b along the x and y directions, respectively, and the coordinates u, v and x_0, y_0 are given by

$$u = \sin\theta \cos\phi \quad \text{and} \quad v = \sin\theta \sin\phi \quad (5.28)$$

$$x_0 = ma \quad \text{and} \quad y_0 = nb \quad (5.29)$$

If we define T_e and T_m by the equations

$$T_e = k_1 \cos k_1 h + jk_2 \sin k_1 h = 0 \quad (5.30)$$

$$T_m = \varepsilon_r k_2 \cos k_1 h + jk_1 \sin k_1 h = 0 \quad (5.31)$$

and

$$k_1^2 = \varepsilon_r k_0^2 - \beta^2 \qquad \text{Im}(k_1) < 0 \qquad (5.32)$$

$$k_2^2 = k_0^2 - \beta^2 \qquad \text{Im}(k_2) < 0 \qquad (5.33)$$

$$\beta^2 = k_x^2 + k_y^2 \quad \text{and} \quad k_0 = 2\pi/\lambda \qquad (5.34)$$

the zeros derived from Equations (5.30) and (5.31) represent the transverse electric (TE) and transverse magnetic (TM) surface waves, respectively, of the unloaded grounded dielectric slab; the loading due to the antenna elements will introduce an error in the calculation of the scan blindness angle not exceeding a few tens of a degree. Either Equation (5.30) or (5.31) is used depending on whether the E- or H-plane scan blindness is required.

Scan blindness can be predicted by comparing the propagation constants of the surface wave of the dielectric slab and the various Floquet modes. The thicker the dielectric material, the larger the number of surface waves it can support. In the interest of limiting the number of modes generated, h, the thickness of the dielectric slab can be limited to values given by the equation [5.23]

$$h < \frac{\lambda}{4\sqrt{\varepsilon_r - 1}} \qquad (5.35)$$

which ensures that only the lowest-order surface wave (TM_0) can propagate. If β_{sw} is the propagation constant of the first (TM) surface-wave mode of the unloaded dielectric slab where $k_0 < \beta_{sw} < \sqrt{\varepsilon_r} k_0$, then a surface wave resonance will occur when β_{sw} matches a particular Floquet-mode propagation constant. Mathematically this condition is satisfied if

$$\left[\frac{\beta_{sw}}{k_0}\right]^2 = \left[\frac{m}{a/\lambda} + u\right]^2 + \left[\frac{n}{b/\lambda} + v\right]^2 \qquad (5.36)$$

from which the scan blindness angle is derived.

Figure 5.11a illustrates the magnitude of the reflection coefficient $|R|$ of an infinite array of printed dipoles on a substrate having a relative permittivity of 2.55 and thickness 0.19λ, as a function of scan angle in the E-, H-, and D- (diagonal) planes; the additional parameters of importance are a and b, which were $a = 0.5774\lambda$ and $b = 0.5\lambda$ for this example. The scan blindness occurs at an angle $\theta = 49.3°$ in the diagonal plane and at angle $\theta = 68.3°$ in the E-plane.

Figure 5.11b illustrates the magnitude of the reflection coefficient $|R|$ of an infinite array of printed dipoles on a substrate having a relative permittivity of 12.8 and thickness 0.06λ, as a function of scan angle in the E-, H-, and D- (diagonal) planes; for this example $a = b = 0.5\lambda$. The scan blindness

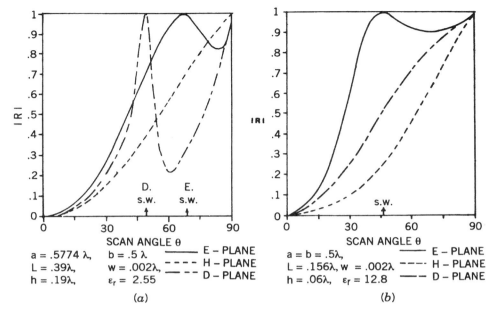

FIGURE 5.11 Scan blindness angles for an infinite array of printed dipoles. (*Source*: Pozar and Schaubert [5.22], © 1984 IEEE.) (*a*) The magnitude of the reflection coefficient of an infinite array of dipoles as a function of scan angle, in the E-, H-, and D- (diagonal) planes, when the relative permitivitty and thickness of the substrate are $\varepsilon_r = 2.55$ and 0.19λ, respectively; and (*b*) the magnitude of the reflection coefficient of an infinite array of dipoles as a function of scan angle, in the E-, H-, and D-planes, when the relative permitivitty and thickness of the substrate are $\varepsilon_r = 12.8$ and 0.06λ, respectively.

occurs at an angle $\theta = 46.1°$ in the E-plane. Usually the array FOV is restricted to a scan angle equal to the scan blindness angle minus 10°.

Figure 5.12 aptly illustrates the array scan blindness dependence on the key substrate parameters h/λ and ε_r and the interelement spacing a. Figure 5.12a illustrates the bandwidth and blindness angle of an array where the spacing is $\lambda/2$ and the substrate is GaAs, as a function of the substrate thickness h/λ [5.24]. As the substrate thickness increases, the fractional bandwidth increases (an observation we are already familiar with), but the blindness angle, measured from the array boresight axis decreases. The array designer therefore has to perform tradeoff studies to meet the array requirements. Figure 5.12b shows the array blindness angle dependence on the relative permittivity of the substrate ε_r for the following cases [5.24]:

1. $a = 0.5\lambda$ and $h/\lambda = 0.06$.
2. $a = 0.52\lambda$ and $h/\lambda = 0.06$.
3. $a = 0.48\lambda$ and $h/\lambda = 0.06$.
4. $a = 0.5\lambda$ and $h/\lambda = 0.02$.

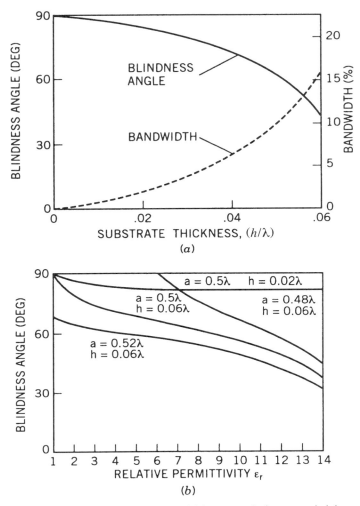

FIGURE 5.12 The impact of substrate thickness, relative permittivity and inter element spacing on the scan blindness angle and percentage bandwidth. (*Source*: Pozar and Schaubert [5.24], © 1986 *Microwave J.*) (*a*) The bandwidth and scan angle blindness of an array as a function of the substrate thickness when the spacing between array elements in $\lambda/2$ and the substrate is GaAs; and (*b*) The scan angle blindness as a function of the substrate's relative permittivity and when the interelement spacing "a" and the substrate thickness h varies.

Cases 1, 2, and 3 illustrate the array blindness dependence on ε_r when a varies and $h/\lambda = 0.06$. Cases 1 and 4 illustrate the array blindness dependence on ε_r when $a = 0.5\lambda$ and h/λ varies. Comparing cases 1 and 2, we observe that the blindness angle decreases as ε_r increases but the blindness angle for case 1 is always greater than that attributed to case 2. Comparing cases 3 and 1, we observe that the blindness angle decreases as ε_r increases but the blindness angle for case 3 is always greater than that attributed to case 1. For cases 4 and 1, the array blindness angle for case 4 is always greater than that attributed to case 1. The four cases demonstrate the interdependence of the array blindness angle on ε_r, h/λ and the interelement spacing a.

5.3.6 Losses Due to Surface Waves

Pozar [5.25] derived approximate and closed-form expressions for the space-wave power P_{sp} and the surface-wave power P_{sw} as well as the space-wave efficiency η, defined as

$$\eta = \frac{P_{sp}}{P_{sp} + P_{sw}} \qquad (5.37)$$

The calculations were obtained by considering an infinite array of dipole antenna elements printed on a grounded dielectric slab. The results, illustrated in Figures 5.13a and 5.13b, are based on the derivation of a Green function for the array of infinitesimal dipoles scanned to an angle θ and ϕ. ε_r for Figures 5.13a and 5.13b is equal to 2.55 and 12.8, respectively. For the latter case the efficiency decreases monotonically for the range of values of h/λ shown. For the former case a similar trend is observed when h/λ takes the values from 0 to 0.1; for values of $h/\lambda > 0.1$ a plateau is reached for the efficiency.

5.3.7 Microstrip-Based Phased Arrays: Design Guidelines

It is useful to draw some general design guidelines for the realization of phased arrays utilizing microstrip antennas in the form of patches or dipoles. To this end we shall treat the space-wave efficiency separately from all the other efficiencies we have defined in Section 5.3.3. The other fundamental distinction is that the guidelines apply only to conventional stripline antennas.

The selection of substrates having low relative permittivity seems mandatory. While the operational bandwidth and efficiency of patch antennas improve as the thickness of the substrate increases, the substrate cannot be too thick, otherwise surface waves will cause scan blindness, which can limit the array FOV.

Apart from the thickness and relative permittivity of the substrate, the other parameter to be considered in the generation of surface waves is the spacing between array elements. While the canonical spacing of half a

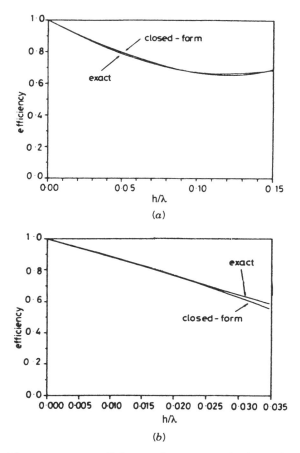

FIGURE 5.13 The space-wave efficiency of an array of microstrip antennas as a function of the substrate thickness (from [5.25]): (a) the case where the relative permitivitty ε_r is equal to 2.55; and (b) the case where the relative permitivitty ε_r is equal to 12.8.

wavelength is accepted so that the grating lobes are placed at the horizon, slight variations from the half-wavelength criterion can be adopted so that the scan blindness angle is shifted outside the required FOV.

5.3.8 Methods of Feeding Patch Antennas

We have already shown several methods of feeding patch antennas in Figures 5.3, 5.4, and 5.5. The feeding point is usually close to the edge of the patch where maximum coupling occurs [5.26].

The simple method of feeding a rectangular patch antenna illustrated in Figure 5.5 has the limitation that some power is radiated from the feeding transmission line. Other more efficient methods of feeding the patches are discussed in Sections 5.5.1.1 and 5.5.1.2. Here, efficiency is equated to maximum power radiated over the largest possible bandwidth.

5.3.9 The Limitations of Conventional Patch Antennas

The basic problem with conventional patch antennas is that there are too many requirements to satisfy simultaneously, including the attainment of optimum input match, maximum efficiency and bandwidth, low cross-polarization fields, and the minimization of surface waves but only a couple of degrees of freedom namely the substrate thickness and relative permittivity. More importantly, the parameters to be optimized are inextricably interrelated. We shall explore the many solutions designers have proposed to overcome these limitations after we consider microstrip dipoles and crossed dipoles.

5.4 MICROSTRIP DIPOLES

Microstrip dipoles have been studied extensively for at least 20 years. The main realizations of microstrip dipoles and the methods of feeding them are shown in Figure 5.14. The dipoles are printed on the surface of a dielectric slab and resemble the free-space dipoles in the sense that radiation results from a harmonically varying dipole moment.

The printed dipoles are easy to manufacture inexpensively, and the electromagnetically coupled (EMC) line is a geometrically attractive method of coupling power into and out of them. For thicker substrates it is possible to alter the radiation properties of printed dipoles by the use of a superstrate

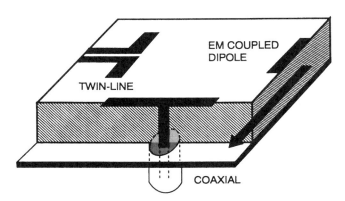

FIGURE 5.14 Microstrip dipoles and three methods of feeding them.

layer so that a substrate–superstrate geometry results. Such a geometry can be used in skins of aircrafts and vehicles.

Schwering [5.27], Katehi et al. [5.27a], and other recent references authored by members of the Alexopoulos group are valuable contributions to the theoretical understanding of printed dipoles. Here we shall use Pozar's work [5.28] to define the regimes of application for microstrip patches and dipoles; the definition of these regimes has been corroborated by authors of the Alexopoulos group [5.27a].

5.4.1 Dipoles versus Patches?

The important criteria for any antenna are its bandwidth of operation and efficiency. Given that both patch and dipole antennas can be easily and inexpensively realized, a comparison of the bandwidths and efficiencies attainable by these antennas is of some importance. We have already considered two definitions of antenna bandwidth; Pozar [5.28] introduces the definition of antenna bandwidth as the bandwidth attainable when the antenna input VSWR is 2.4 and examined the bandwidth attainable by patches and dipoles printed on substrates having ε_r equal to 2.55 and 12.8. The results, the relationship of antenna bandwidth as a percentage to the h/λ ratio, are illustrated in Figures 5.15a ($\varepsilon_4 = 2.55$) and 5.13b ($\varepsilon_r = 12.8$); for both figures $W = 0.3\lambda$ for the patch antenna and the dipole is a half-wave printed dipole.

As can be seen in Figure 5.15a the patch antenna yields a wider bandwidth than the dipole when the substrate thickness, in wavelengths, is lower than about 0.1. The dipole on the other hand yields larger bandwidths when the substrate is thick. The cavity model approximation is seen to be useful for substrates thicknesses $h < 0.045\lambda$. Similar observations can be derived from Figure 5.15b, except that the patch antenna in this case has a larger bandwidth up to about a value for h/λ of 0.07.

Given that commercially available substrates come in certain thicknesses, the dipole yields wider bandwidths when the substrate is thick or when operation at mm wavelengths is required.

In the same paper Pozar considered the power lost to surface waves for both antennas; while there are differences between the two antennas, their power loss to surface waves is comparable.

In Figure 5.15c the dielectric losses of both these antennas are shown [5.28]. ε_r for the substrate is 2.55 and loss tangents of 0.001 and 0.003 are considered. The efficiency due to dielectric losses for the patch antenna is higher for thin substrates. From the foregoing considerations we can draw the following conclusions:

1. At mm wavelengths, where the substrate is thick, the dipole is a better choice; by contrast, the patch antenna is the preferred choice for arrays designed to operate at cm wavelengths.

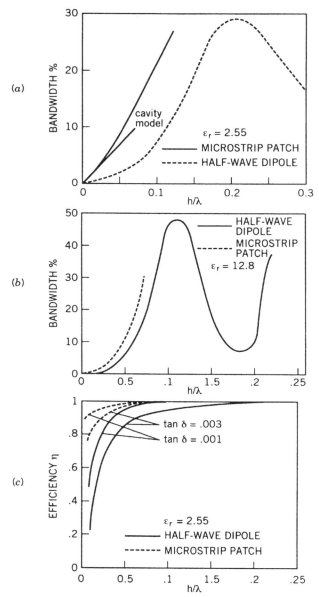

FIGURE 5.15 The percentage bandwidth and efficiency of arrays consisting of microstrip patches (———) and half-wave dipoles (-----) as a function of substrate thickness. (*Source*: Pozar [5.28], © 1983 IEEE.) (*a*) The bandwidth versus substrate thickness when the relative permittivity ε_r is equal to 2.55. (*b*) Same as (*a*) but the relative permittivity ε_r is equal to 12.8; and (*c*) the percentage efficiency versus substrate thickness when the substrate has a relative permittivity ε_r of 2.55 and when the loss angle tan δ takes the values of 0.001 and 0.003.

2. These findings nicely define the regime of application for the two kinds of antennas.

5.4.2 Applications: Wideband Printed Dipoles Suitable for Brick Array Architectures

Edward and Rees [5.29] reported the realization of a printed dipole, illustrated in Figure 5.16, with its coupling structure (also known as an integrated balun); the dipole is suitable for the brick array architecture and its VSWR-2 bandwidth is 40%. Guidelines for the design of the integrated dipole and coupling structure are also presented by the authors.

While the radiation patterns of antenna elements, including dipoles and patches, are well known and have been measured by many researchers, the same patterns and other essential parameters have to be measured when the antenna element is connected to a coupling structure located between the antenna element and its T/R module.

It is instructive to consider the effects of the coupling structure, used in Edward and Rees [5.29], on the radiation pattern of the printed dipole qualitatively here, as a sample of the interaction effects.

The radiation patterns of the printed dipole shown in Figure 5.16 were measured and calculated by using the electric-field integral equation (EFIE) approach suitably modified to compute the resulting fields of arbitrarily

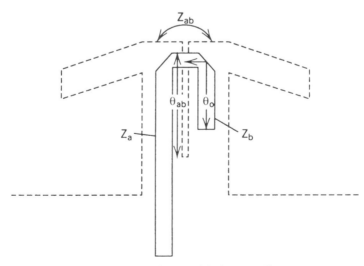

FIGURE 5.16 The printed-circuit dipole with its coupling structure, realized by Edward and Rees. (*Source*: [5.29], © 1987 *Microwave J.*)

364 ANTENNA ELEMENTS

shaped finite conducting/dielectric composite structures embedded in an infinite homogeneous medium such-as free space [5.30].

The resulting radiation patterns when the angle ψ takes the values of 10°, 30°, 45°, 60°, 75°, and 90° are shown in Figure 5.17. When $\psi = 30°$, the radiation pattern of the printed dipole does not show any undesirable pattern asymmetries, which are obvious when ψ takes the other values. It is clear that the interaction between the antenna element and its coupling structure is important for it affects the radiation pattern of the antenna element.

5.4.3 Applications: An Experimental Array Utilizing Crossed Printed Dipoles

An experimental 8×8 element array, shown in Figure 5.18, utilizing crossed printed dipoles has been reported by Wolfson and Sterns [5.31]. The authors outline the significant consequences of arrays having microstrip radiating elements that have poor input impedance characteristics:

1. Operation over narrow bandwidths (e.g., a few percent)
2. Scan angles not wider than $\pm 20°$

Furthermore, if the radiators are dual-polarized elements, the polarization purity is often less than 18 dB. The authors reported the following results

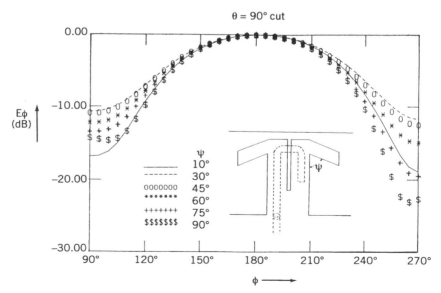

FIGURE 5.17 A printed-circuit dipole when ψ takes the values of 10°, 30°, 45°, 60°, 75°, and 90° and the resulting radiation patterns. (*Source*: Rao et al. [5.30], © 1991 IEEE.)

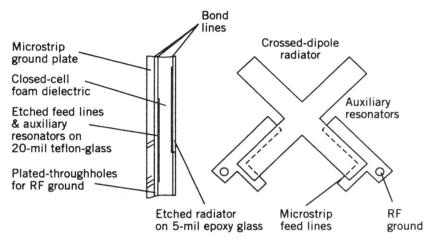

FIGURE 5.18 One of the 64 EMC cross-dipoles with its auxiliary resonators used in an array reported by Wolfson and Sterns. (*Source*: [5.31], © 1984 IEEE.)

derived from their experimental array utilizing the cross-dipole array elements shown in Figure 5.18 with auxiliary resonators:

1. An input match of better than 2.3:1 for all polarizations over a 4.3% bandwidth and $\pm 45°$ azimuth scan and an input match of 1.65:1 over $\pm 20.7°$ azimuth scan.
2. Polarization purity of about 25 dB for linear polarizations over the frequency of 5.65–5.9 GHz and $\pm 45°$ azimuth scan. For circular polarization the axial ratio was 1 dB over the scan of $\pm 20.7°$ and about 3 dB at a scan of $\pm 45°$, in azimuth.

The results are impressive and indicative of the difficulties of obtaining good input impedances and polarization purity over wide scan angles.

5.5 THE QUEST FOR HIGH-QUALITY DUAL-POLARIZED ANTENNAS

Many strategies have been pursued to realize high-quality dual-polarized antennas suitable for phased array applications. While we are interested in antennas having excellent polarization purity, we should not ignore other vital characteristics such as low-insertion loss, the attainment of as many design degrees of freedom as possible, cost and complexity. In this section we shall broadly outline the diverse solutions several designers reported [5.32–5.45] before we examine in detail some key measurements and approaches.

Let us limit our field of view to conventional patch antennas yielding the two linear polarizations pro tem. Also recall that patch antennas can support several modes which cause high cross-polarization fields. The use of two feed points on one patch contributes to the cancellation of the undesirable cross polarization fields.

Figure 5.19 shows a simple and effective way of deriving the two linear polarizations from patch antennas in an array context. In Figure 5.20 the two linear polarizations are derived with the aid of a hybrid ring or a ratrace connected between a patch having two probes and the T/R module.

In Figures 5.4a–5.4d we have seen how one circular polarization can be derived from different patches when only one probe is used. In Figures 5.21a and 5.21b, two probes are used on a patch to derive the circular polarizations with the aid of 90° hybrid and in Figure 5.21c four probes are used on a patch to derive the circular polarizations.

If one wants to persevere with conventional patch antennas, Figure 5.22 illustrates four ways of forming 2×2 subarrays and derive the two linear polarizations.

It is now time to recall that the conventional patch antenna is required to have many desirable characteristics but has only two degrees of freedom. By contrast the EMC patch, shown in Figure 5.23 offers several degrees of freedom. It consists of a patch and a parasitic element, also referred to as the feeding patch, etched on substrates separated by a dielectric slab; thus the designer has the freedom to select the relative permittivity and thickness of the two substrates and of the dielectric slab as well as the selection of the dimensions of the patch and parasitic/feeding element. These extra degrees of freedom allow the designer to meet a variety of requirements.

When the circular polarizations are required, sequential subarrays offer low cross-polarization fields over a larger bandwidth and wider scan angles; additionally the mutual coupling between the subarray antenna elements is diminished. While a 2×2 sequential subarray is considered in Section 5.5.2.2, it is only a special case of sequential subarrays having more elements.

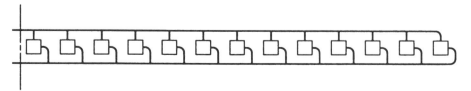

FIGURE 5.19 The derivation of the two linear polarizations in an array context (from [5.33]).

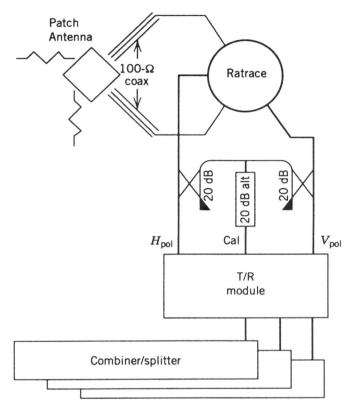

FIGURE 5.20 The derivation of two linear polarizations with the aid of a hybrid ring (ratrace) connected between a square patch having two probes and the T/R module. (*Courtesy*: TNO, Paquay et al. [5.34].)

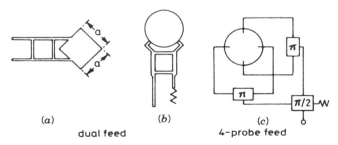

FIGURE 5.21 Three approaches to derive one circular polarization from patch antennas: (*a*) and (*b*) dual feed approaches; and (*c*) four feed approach.

368 ANTENNA ELEMENTS

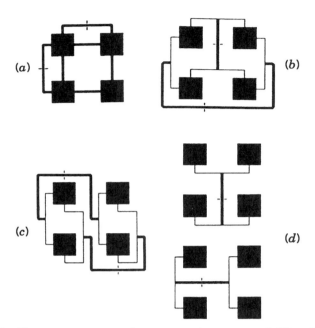

FIGURE 5.22 Four arrangements of square patch antennas yielding the two linear polarizations. (*Source*: Levine and Shtrikman [5.43], © 1990 J. Wiley & Sons.)

5.5.1 Single-Element Approaches and Applications

The antenna array elements can be either conventional patches or EMC patches and Daniel et al. [5.32] is a useful and up-to-date source of information on antenna elements used in phased arrays.

5.5.1.1 Conventional Patches In an array context one can simply derive the two principal linear polarizations by the method illustrated in Figure 5.19, adopted by Johansson et al. [5.33]. The cross-polarization radiation of the array of patches was between -27 (V-pol) and -30 dB (H-pol) in the two principal planes, at $\pm 0.15°$, with respect to the array boresight axis while the maximum cross-polarization levels were between -10 dB and -15 dB. The array operated at 5.331 GHz $\pm$ 10 MHz and was developed for the European Space Agency for the advanced SAR (ASAR). Simplicity and low-cost are the attractive characteristics of this approach.

Several approaches to obtain circular polarization by utilizing the two linear polarizations available from patch antennas are shown in Figure 5.21 [5.33a]. The circular polarization is simply obtained from a square (*a*) or

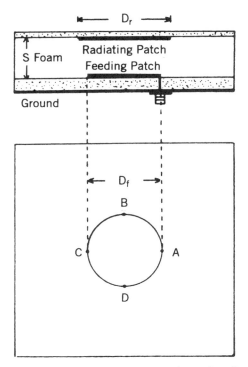

FIGURE 5.23 The ECM patch (from [5.36]).

circular (*b*) patch by introducing a 90° phase shift between the two linear polarizations before combining them.

For wideband applications, four probes are used to derive one circular polarization from a circular patch, shown in Figure 5.21*c*. If the two linear polarizations are required, the 90° phase shift between the two linear polarizations is omitted.

While a patch antenna having four feed points has a wider impedance and axial-ratio bandwidth (typically $\geq 10\%$), than a conventional patch [5.33b], it is complicated and more prone to RF losses in a phased array context. Let us explore the performance of a dual-polarization patch antenna utilizing two feed points.

The following antenna element was developed for the PHased ARray SAR (PHARUS), which is designed to operate at 5.3 GHz. Paquay et al. [5.34] realized a square patch antenna that exhibited some -25-dB cross-polarization level in the D plane, which is a remarkable result. Figure 5.20 is the block diagram for the arrangement the authors adopted; the ratrace provides high isolation between the two probes and the two inputs to the T/R module. Coaxial lines having a characteristic impedance of about 100 Ω are used to

370 ANTENNA ELEMENTS

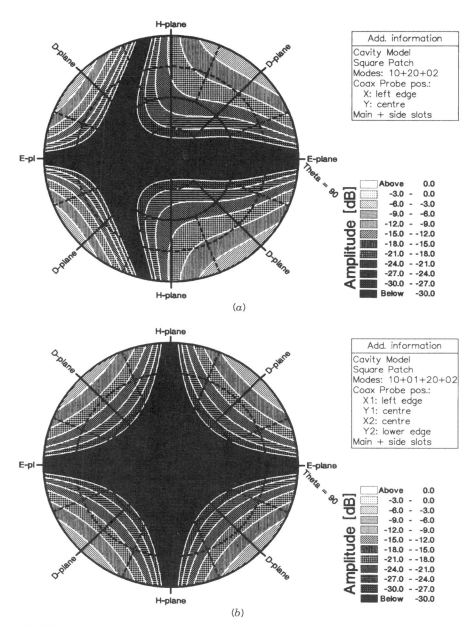

FIGURE 5.24 The co- and cross-polarization radiation of a single and dual feed patch antenna (*Courtesy*: TNO, Paquay et al. [5.34]. (*a*) The cross-polarization of a single feed patch antenna. (*b*) The cross-polarization of a dual feed patch antenna.

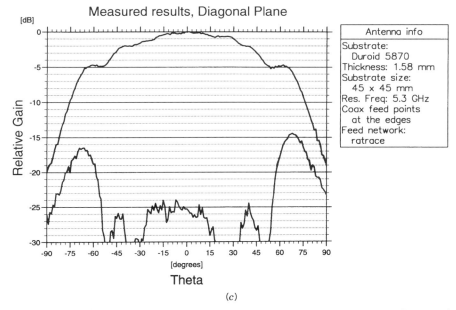

FIGURE 5.24 (c) The co- and cross-polarization radiation patterns of a patch antenna illustrated in Figure (5.20) having a dual feed.

connect the patch to the ratrace. These lines act as impedance transformers between the patch and the ratrace which is made on a ceramic substrate and occupies less than 18×18 mm. Figures 5.24a and 5.24b show the cross-polarization radiation with one and two probes separated by 90°, respectively; the probes here are fed with identical powers. The fields located around the boresight axis of the patch, calculated by using the cavity model, are low with this arrangement because the TM_{02} mode that generates cross-polarized fields is heavily attenuated. The measured radiation pattern of the patch in the D-plane is shown in Figure 5.24c. An isolation between the polarizations was about 30 dB. The cross-polarization radiation over ± 45 ° is about -25 dB, which is an exceptional result.

While the method of connecting the patch to the T/R module is unique to the patch reported by Paquay et al., other authors reported a different approach. Zurcker et al. [5.35] reported measurements taken when the coupling to a single patch was implemented with the aid of two slots. The frequency of operation was 2.4 GHz, the VSWR-2, bandwidth was 5%, and the measured cross-polarization level was comparable to that reported by Paquay et al.

5.5.1.2 The Electromagnetically Coupled (EMC) Patch The EMC patch, depicted in Figure 5.23, consists of a radiating patch and a feeding patch separated by foam; the patch feeding points are A, B, C, and D. The EMC patch has a wider bandwidth than its single patch counterpart [5.36] and this attractive feature is directly attributed to its dual-resonant structure. The geometry of the EMC patch is mechanically convenient for the radiating patch is protected from the weather and can conform to the skin of the platform.

A broadband EMC patch with a VSWR = 1.7 bandwidth of 14.4% around a central frequency of 1.67 GHz has been reported [5.37]; its cross-polarization level was better than -27 dB over the same frequency range. Similar performance was obtained by another patch operating at S-band [5.38].

An EMC patch was reported having a 13% bandwidth and a maximum cross-polarization level of -30 dB at L-band [5.39]; design guidelines for the EMC patch are also given.

Recent theoretical work that corroborated the experimental work related to the ECM patches has appeared in the literature [5.40, 5.41].

5.5.2 Subarrays of Antenna Elements

One of the key questions related to forming contiguous subarrays is how the array scan blindness is affected as $N = 2, 3, 4, \ldots$ antenna elements are connected to form a subarray. Pozar [5.42] reported the results of his analysis of infinite arrays composed of arbitrary subarrays of microstrip antenna elements. He used a full-wave moment method solution that included mutual coupling between elements in the subarray as well as the coupling between subarrays. The magnitude of the reflection coefficient as a function of scan angle was derived for subarrays consisting of two or three antenna elements. The other assumptions were that the interelement spacing is $\lambda/2$, that each element is fed inphase with equal amplitude, and that a uniform phase progression is applied across the subarrays.

The most significant results of the study are:

1. The magnitude of the reflection coefficient of the radiating elements significantly decreases when $N = 2, 3$, over scan angles of 0–90°, as Figure 5.25a illustrates. In the same figure the case where subarraying is not implemented is shown and the scan blindness angle occurs at 76°.
2. As the intersubarray spacing is λ and 1.5λ for the cases when $N = 2, 3$ some power is dissipated into the grating lobes and the beam efficiency defined as the ratio of the main-beam power to the total power radiated decreases. Figure 5.25b shows the beam efficiency of the array as a function of the scan angle for the cases where $N = 1, 2$, and 3.

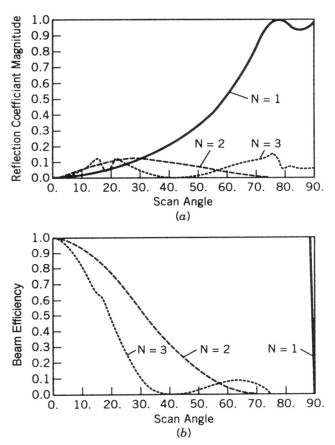

FIGURE 5.25 The effects of subarraying on two critical array parameters as a function of scan angle. (*Source*: Pozar [5.42], © 1992 IEEE.) (*a*) the magnitude of the reflection coefficient of the radiating elements as a function of the array scan angle for the case were there is no subarraying $N = 1$ and when N is equal to 2 or 3; and (*b*) the array beam efficiency as a function of scan angle for the cases when N is equal to 1, 2, and 3.

The first result is of considerable import, for the designer can guarantee that the array FOV is determined solely by the subarray spacing. The designer, however, should not select high values of N, for the beam efficiency decreases and the reflection coefficient does not improve as N takes the values higher than 2.

In what follows we shall discern subarrays of similar antenna elements having the same orientation shown in Figure 5.22 from subarrays of antenna elements that have a sequential orientation. Although we shall explore 2 × 2 subarrays of patch antennas, larger subarrays consisting of other elements are possible [5.33a].

5.5.2.1 Subarrays of Similar Patches Having the Same Orientation and Applications

An experimental comparison of four sets of dual-polarized square patch antennas, shown in Figure 5.22, has been undertaken [5.43]; the lowest measured cross-polarization radiation along the principal planes of the antennas was about -28 dB, corresponding to the set where the two polarizations were derived from two different sets of patches illustrated in Figure 5.22d. This case is impractical, for the area the patches occupy is twice that occupied by the other three sets; the case is, however, illustrative of the lowest possible cross-polarization level one can expect with this geometric configuration.

The next lowest cross-polarization level obtained by the arrangement illustrated in Figure 5.22b is at the level of -24 dB and an interpolarization isolation of -33 dB. These cross-polarization levels are indicative of what one can expect from patch antennas having the geometries shown. The cross-polarization levels reached by the patches illustrated in Figures 5.22a and 5.22c were -18 and -15 dB, respectively. The bandwidths ranged between 3 and 8%.

Huang [5.44] slightly modified the connections between the patches shown in Figure 5.22b and reported an interpolarization isolation figure of -40 dB and a cross-polarization level of -28 dB.

It is of interest to explore whether EMC patches can yield circular polarizations. With reference to Figure 5.23, the measured input VSWR was 1.22 : 1 over the frequency range of 4.01–4.47 GHz and 1.92 : 1 over 3.85–4.58 GHz [5.45] when linear polarization is used. However, when the EMC is fed at two points, such as A and B, it generated highly elliptical polarization because of the asymmetrical feed structure. With reference to Figure 5.26, a 2 × 2 subarray consisting of EMC patches and the associated circuitry, yielded the two circular polarizations and had the required symmetry [5.36] and [5.45]. The axial ratio was therefore better than 0.5 dB over the bandwidth of interest. This low figure for the axial ratio is obtained because of the symmetrical arrangement of the subarray elements—see the next section also.

Apart from the excellent electrical characteristics this subarray has, it is lightweight and volumetrically attractive. The unit is estimated to be 1 in. thick and to weigh 100 g; the corresponding antenna element used in INTELSAT VI is over 12 in. long and weighs approximately 300 g [5.36].

"Clearly this technology holds the potential for tremendous weight savings when large feed arrays are required," wrote R. M. Sorbello [5.36]. At C-band a small penalty in additional insertion loss may be paid; at Ku-band, however the additional loss can be prohibitive.

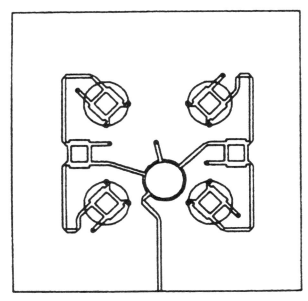

FIGURE 5.26 Circularly polarized array of four EMC patch antennas, one of which is shown in Figure 5.23 [5.36].

5.5.2.2 Sequential Subarrays and Applications With reference to Figure 5.27a, two rectangular patches are used to derive the two linear polarizations. If the two circular polarizations are required, a 90° phase is inserted between the two polarizations before they are combined. At angles off the boresight axis, however, a spatial delay is introduced between the two elements as shown in Figure 5.27b. This delay, in turn, introduces high cross-polarization levels as shown in Figure 5.27c.

With reference to Figure 5.28a the spatial delay in one row or column is opposite that of the other row or column and consequently cancel each other; hence a very low cross-polarization field results, as illustrated in Figure 5.28b [5.46]; measurements taken corroborate the calculations [5.46]. The concept can be extended to larger subarrays utilizing other antenna elements [5.33a, 5.46]. The 2×2 subarray considered here is but a special case of subarrays having many elements [5.33a].

5.6 WORK IN PROGRESS

In this section we shall briefly outline work in progress in several directions before we consider wideband antenna elements suitable for phased arrays.

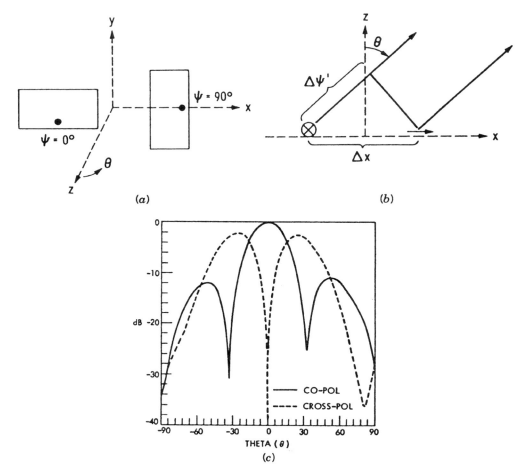

FIGURE 5.27 A two-element subarray arranged to yield a circular polarization. (*Source*: Huang [5.46], © 1986 IEEE.) (*a*) Subarray geometry; (*b*) the geometry illustrating the spatial phase delay $\Delta\psi$; and (*c*) the calculated radiation pattern of the subarray when the circular polarization is derived.

Given that subarrays are important, several authors have explored the possibility of improving the interconnections between the antenna elements of subarrays. Microstrip antenna subarrays, consisting of a central aperture coupled microstrip antenna element feeding additional elements via coplanar microstrip lines, have been described along with their relative advantages and a basic design procedure by Duffy and Pozar [5.47]. Legay and Shafai [5.47a] describe a 2×2 subarray that consists of 4 identical patches electromagnetically by a driven patch etched on a lower substrate.

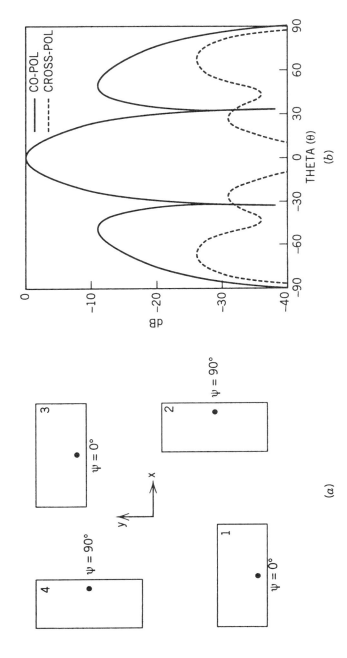

FIGURE 5.28 A four-element sequential subarray. (*Source*: Huang [5.46], © 1986 IEEE.) (*a*) The subarray geometry; and (*b*) the calculated radiation pattern of the subarray.

378 ANTENNA ELEMENTS

Two references explored the realization of polarization-agile patch antennas [5.48, 5.49], a promising area for future research.

Reference [5.50] summarized work related to aperture antennas fabricated on ferrite substrates with and without the application of a magnetic field. The same paper includes references that explore the beam scanning, antenna pattern, and the RCS control of antennas by the application of a magnetic field. When a uniform magnetic field was applied to a circular patch antenna on a ferrite material, the ferrite microstrip antenna (FMA) yielded a circularly polarized radiation pattern in accord with theoretical predictions. Again, this is a field for promising research.

We have already explored aperture-coupled patch antennas [5.35]; Haddad and Pozar [5.51] explored methods to increase the efficiency of the process described above using a thick ground plane; the thick ground is useful as a heat sink for active MMIC circuitry or as a mechanical support for thin substrates, a case that is particularly advantageous in mm-wave phased array applications.

5.7 WIDEBAND ANTENNA ELEMENTS

A case for phased arrays having wide instantaneous bandwidths has been made in Chapter 1. The required bandwidths, it is recalled, is over several octaves.

While the dimensions of narrowband antennas are defined by lengths, the dimensions of frequency-independent antennas are defined by angles. There are two fundamental characteristics related to wideband (or frequency-independent) antennas [5.52]:

1. Their beamwidths are invariant with frequency. In a phased array context this requirement will ensure that the array FOV remains unaltered within the band of operation.
2. Their input VSWRs over the required frequency range is acceptable (viz., ≤ 2).

When the two polarizations are required, the differences between the E- and H-plane radiation patterns should be negligible.

Antennas approximating these characteristics are naturally acceptable. If the same antennas have low profile and are conformal like the microstrip antennas, their use expands.

5.7.1 Log-Periodic, Patch-Based Antenna

Unfortunately the dimensions of microstrip antennas are determined by lengths; hence their frequency of operation is limited to 10–20% bandwidth.

The first choice for wideband antennas is therefore the log-periodic array, which utilizes patches of different sizes, depicted in Figures 5.29 and 5.30 [5.53]. Figure 5.29a shows the log-periodic array utilizing several patches of different sizes, and the feed line coupled to the patches; the radiating action at different frequencies f_1, f_2, and f_3 is also shown, while the array topology is shown in Figure 5.29b. The performance characteristics are summarized in Table 5.2, drawn from Hall [5.53]. The antenna has many attractive characteristics: it is conformal and has a low profile and wide bandwidth.

TABLE 5.2 Summary of Performance of 36-Element Microstrip Log-Period Array [5.53]

Bandwidth	4–16 GHz
Gain	> 8 dB
Efficiency	> 79%
Input return loss	> 7 dB
Beamwidths	40° × 92° at 4 GHz
	30° × 84° at 16 GHz

5.7.2 Spiral-Mode Microstrip (SMM) Antenna

It is well known that the planar spiral-mode antennas, such as the equiangular and sinuous antennas, have multioctave bandwidths [5.52, 5.54]. Unfortunately, these antennas radiate to both sides of the spiral plane. The placement of a lossy cavity on one side of the spiral or sinuous structure absorbs all the undesired radiation in that direction; with this arrangement the bandwidth of the antennas typically extends from 2 to 18 GHz. The lossy cavity, however, is deeper than the radius of the spiral and therefore not suitable for low-profile surface mounting.

If the lossy cavity is substituted by a lossless cavity [5.55] or a conducting plane [5.56], the achievable bandwidth is 40 and 20%, respectively. Wang and Tripp [5.57] reported the design and realization of an experimental spiral-mode antenna, depicted on Figure 5.31; they achieved good input matching even when the spiral was 0.1 in. above the ground plane; additionally, the absorbing material in the periphery of the spiral contributed greatly to the quality of the radiation patterns over the 2–18-GHz frequency range.

This important broadband and low-profile antenna will find many uses, some of which have been explored in Tripp et al. [5.58]. The SMM antennas can be used in phased arrays having a tile architecture.

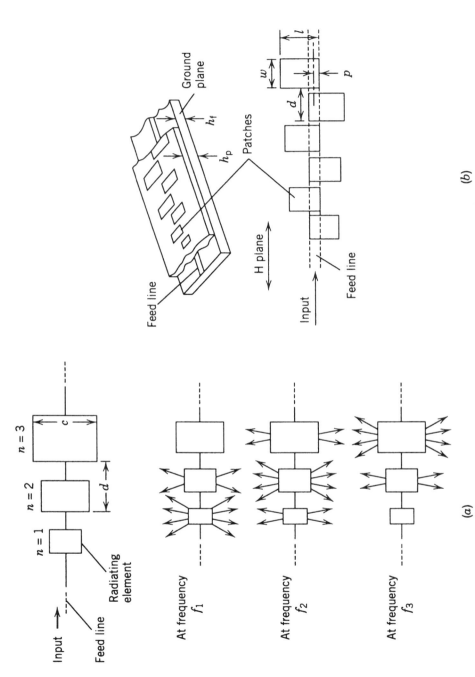

FIGURE 5.29 A log-periodic array utilizing patch antennas. (*Source:* [5.53], © 1986 *Microwave J.*) (*a*) The radiating elements of

FIGURE 5.30 Log-periodic array of electromagnetically microstrip patches. (*Source*: Hall [5.53], © 1986 *Microwave J.*)

5.7.3 *Planar and Antipodal Tapered Slotline Antennas*

The planar tapered slotline antenna (TSA), shown in Figure 5.32a, is also known as the *Vivaldi or slotline antenna* [5.59, 5.60]. Other versions of the same antenna are the antipodal [5.61, 5.62] and symmetrical antipodal [5.63] shown in Figures 5.31b and 5.31c, respectively. Additionally, the taper can take many forms, such as linear, exponential or the tangent of an angle [5.62]. These antennas are members of the class of aperiodic, continuously scaled, slow leaky endfire traveling-wave antennas. These antennas can have:

1. Good input return loss over several octaves.
2. Equal E- and H-plane radiation patterns; this is a remarkable characteristic considering that the antennas are planar.
3. Reasonable cross-polarization characteristics over the same bandwidth.

With these properties, the antennas can be easily considered as elements for wideband phased arrays having the brick architecture.

TSAs have been used in the focal plane of a reflector for imaging of a scene [5.64, 5.65]. Fourikis et al. [5.62] summarizes work done prior to 1993 and is a comprehensive parametric study of planar and antipodal TSAs; design guidelines to realize TSAs having equal E- and H-plane HPBWs are also outlined.

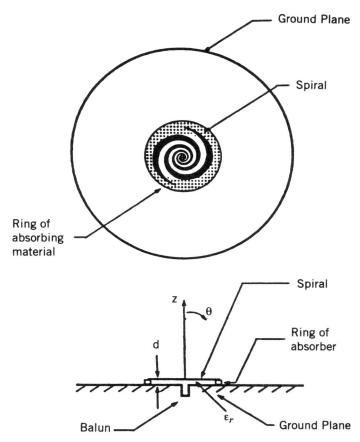

FIGURE 5.31 The spiral-mode microstrip developed by Wang and Tripp and Tripp et al. [5.57, 5.58]. (*Source*: Tripp et al. [5.58], © 1993 IEEE.)

The gain of these antennas varies between 8 and 12 dB over the frequency range 3–12 GHz [5.62].

The measured input return loss of the planar, antipodal, and symmetrical antipodal TSAs are shown in Figure 5.33a [5.63]. As can be seen, the input return loss of the antipodal TSA is superior to that corresponding to the planar TSA [5.62, 5.63]. The additional problem with planar TSAs is the difficulty of maintaining the accuracy of very thin slots or the dimension, W, shown in Figure 5.32a. [5.62]. The input return loss of the antipodal and the balanced antipodal are comparable.

The measured cross-polarization level of the same set of antennas is shown in Figure 5.33b [5.63]. The cross-polarization level of the planar TSA is superior to that of the antipodal [5.62, 5.63]. The balanced TSA exhibits the lowest cross-polarization levels of the three TSAs. A slight E-plane squint has

WIDEBAND ANTENNA ELEMENTS 383

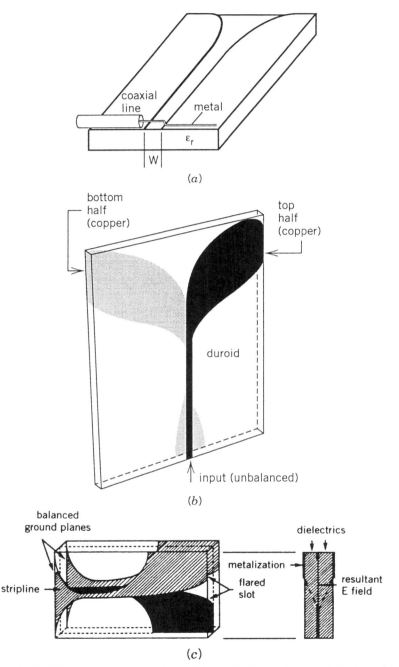

FIGURE 5.32 Three important versions of MIC slotline antennas: (*a*) planar [5.62]; (*b*) antipodal [5.62]; and (*c*) balanced antipodal [5.63].

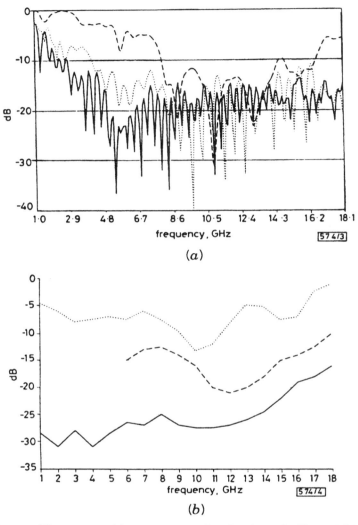

FIGURE 5.33 The measured key parameters for the three significant TSAs: planar (---); antipodal ($\cdots$); and symmetrical antipodal (—) (from [5.63]). (a) The input return loss; and (b) the cross-polarization level.

been observed between 10° and 20° for high-gain balanced symmetrical antipodal [5.63]. It is felt that this squint will become less significant for low-gain balanced symmetrical antipodal antennas suitable for phased arrays having a FOV of, say, 140° [5.63]. The balanced antipodal TSA requires a suitable transition before it is connected to the input of the T/R module that

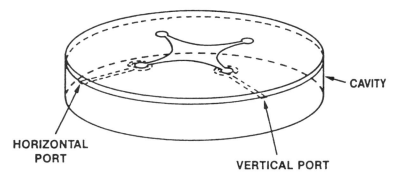

FIGURE 5.34 The dual-polarized microstrip flared slot antenna. (*Source*: Povinelli [5.67], © 1987 IEEE.)

is unbalanced. For dual-polarization applications the eggcrate geometry proposed in Povinelli and Johnson [5.66] is appropriate.

Although the antennas described above are suitable for phased arrays having a brick architecture, the planar microstrip flared slot antenna [5.67, 5.68], shown in Figure 5.34 has similar characteristics to the flared slot antennas and is a suitable candidate for phased arrays having a tile architecture. Given the current popularity of the tile architecture and the fact that the two linear orthogonal polarizations can be derived from the same phased-array real estate, the planar microstrip flared slot antennas has a promising future.

5.8 CONCLUDING REMARKS

The modern thrusts in antenna elements suitable for phased arrays, are toward inexpensive, lightweight printed circuit board or microstrip antenna elements that can conform to the skin of platforms such as airplanes and away from heavy slotted waveguide structures and metal horns; these solutions allow the airplanes to have an aerodynamic shape and accommodate phased arrays. Additionally the same antennas can be easily connected to MMIC-based T/R modules.

The characterization of patch antennas has been undertaken in this chapter; while patches are efficient at cm wavelengths, dipoles are preferred for operation at mm wavelengths.

For a long time microstrip antenna elements such as patches suffered from the presence of several modes, so the migration to the promised land was slow. Recently methods, examined in this chapter, to minimize the presence of other modes and the resulting cross-polarization radiation have been successfully implemented and the rate of migration toward microstrip antennas will accelerate. Approaches and benefits resulting from subarraying (conventional and sequential) are also considered.

The chapter ends with a detailed examination of wide-band antenna elements suitable for the tile or brick architecture. Future work toward antenna systems than can provide polarization agility and adaptability is important.

References

CHAPTER 1

1.1 C. A. Federer, Jr., Solar thermal test facility. *Sky Telescope*, April, p. 286 (1978).

1.2 Encyclopedia Britannica: Archimedes Macropaedia, vol. 13, p. 872, 1993.

1.3 I. G. Sakas, Archimedes burnt Roman's fleet by means of flat mirrors, *Technical Chronicles*, Official Publication of the Technical Chamber of Greece, 9/567, p. 779, September 1973 (in Greek with English abstract) and references therein.

1.4 P. R. Jaszka, Aegis system still gets high marks. *Def. Electron.*, October, p. 48 (1988).

1.5 B. Sweetman, Leading-edge technology for Swedish AEW. *Int. Def. Rev.* **21** (No. 3), 277 (1988).

1.6 R. J. Mailloux, J. F. McIlvenna, and N. P. Kernweis, Microstrip array technology. *IEEE Trans. Antennas Propag.* **AP-29** (1), 2 AP-29(1)25 (1981).

1.7 L. Murphy, SEASAT and SIR—A microstrip antennas. *Proc. Workshop Printed Circuit Antenna Technol. 1979*, Pap. No. 18 (1979).

1.8 A. J. Parrish, Electronically steerable arrays: Current and future technology and applications. MSc. Project Report, Loughborough University of Technology and Royal Air Force College, Cranwell, UK, 1989.

1.9 K. Coriolis, The true cost aircraft in service. *Jane's Def. Wkly.*, September 27, p. 680 (1986).

1.10 P. J. Napier, A. R. Thomson, and R. D. Ekers, The very large array: Design and performance of a modern synthesis radio telescope. *Proc. IEEE* **71** (11), 1295 (1983).

1.11 E. Brookner, Phased array radars. *Sci. Am.* **252** (2), 94 (1985).

1.12 A. Blondel, Belg. Pat. 163,516 (1902); Br. Pat. 11,427 (1903).

1.13 S. G. Brown, Br. Pat. 14,449 (1899).

1.14 R. M. Foster, Directive diagrams of antenna arrays. *Bell Syst. Tech. J.* **5**, 292 (1926).

1.15 P. S. Carter, Circuit relations in radiating systems and applications to antenna problems. *Proc. IRE* **20**, 1004 (1932).

1.16 H. T. Friis and C. B. Feldman, A multiple unit steerable antenna for short-wave reception. *Proc. IRE* **25**, 841 (1937).

1.17 H. A. Wheeler, The radiation resistance of an antenna in an infinite array or waveguide. *Proc. IRE* **36**, 478 (1948).

1.18 R. M. Scudder and W. H. Sheppard, AN/SPY-1 phased array antenna. *Microwave J.* **17**, 51 (1974).

1.19 W. T. Patton, Compact constrained feed phased array for AN/SPY-1. *Microwave J. Intens. Course Notes, Prac. Phased-Array Syst. Lect.*, April, No. 8 (1975).

1.20 L. Stark, Theory of phased arrays. *Proc. IEEE* **62** (12), 1661 (1974).

1.21 R. H. Dicke and R. Beringer, Microwave radiation from the sun and moon. *Astrophys. J.* **103**, 373 (1946).

1.22 L. L. McCready, J. L. Pawsey, and R. Payne-Scott, Solar radiation at radio frequencies and its relation to sunspots. *Proc. R. Soc. London, Ser. A* **190**, 357 (1947).

1.23 J. G. Bolton and G. J. Stanley, Variable source of radio frequency radiation in the constellation of Cygnus. *Nature (London)* **161**, 312 (1948).

1.24 J. G. Bolton, G. J. Stanley, and O. B. Slee, Positions of three discreet sources of galactic radio-frequency radiation. *Nature (London)* **164**, 101 (1949).

1.25 W. Baade and R. Minkowski, Identification of the radio sources in Cassiopeia, Cygnus A and Puppus A. *Astrophys. J.* **119**, 206 (1954).

1.26 W. Baade and R. Minkowski, On the identification of radio sources. *Astrophys. J.* **119**, 215 (1954).

1.27 W. N. Christiansen and J. A. Warburton, The distribution of radio brightness over the solar disk at a wavelength of 21 cm. *Aust. J. Phys.* **6**, 190 (1953).

1.28 A. E. Covington and N. W. Broten, Brightness of the solar disk at a wavelength of 10.3 cm. *Astrophys. J.* **119**, 569 (1952).

1.29 B. Y. Mills and A. G. Little, A high resolution system of a new type. *Aust. J. Phys.* **6**, 272 (1953).

1.30 W. N. Christiansen and D. S. Mathewson, Scanning the sun with a highly directional antenna. *Proc. IRE* **46** (1), 127 (1958).

1.31 J. D. Kraus, R. T. Nash, and H. C. Ko, Some characteristics of the Ohio State 360-foot radio telescope. *IRE Trans. Antennas Propag.* **AP-9** (1), 4 (1961).

1.32 E. J. Blum, A. Boischot, and J. Lequeux, Radio astronomy in France. *Proc. IREE Aust.* **24** (2), 208 (1963).

1.33 G. A. Pinchuk et al. Multibeam operational mode at the RATAN 600 radio telescope. *IEEE Antennas Propag. Mag.* **APM-35** (5), 18 (1993).

1.34 J. P. Wild, Special Edition on The Culgoora radioheliograph. *Proc. IREE Aust.* **28**, No. 9 (1967).

1.35 K. I. Kellermann, Radio astronomy: The next decade. *Sky Telescope*, September, p. 247 (1991).

1.36 S. A. Hovanessian, *Radar Design and Analysis*. Artech House, Dedham, MA, 1984.

1.37 M. E. Skolnik, *Introduction to Radar Systems*. McGraw-Hill, New York, 1980.

1.38 N. C. Currie and C. E. Brown, *Principles and Applications of Millimeter-wave Radar*. Artech House, Dedham, MA, 1987.

1.39 J. Preissner, The influence of the atmosphere on passive radiometric measurements. *AGARD Conf. Repr.* No. 245 (1978).

1.40 K. Miya, ed., *Satellite Communications Technology*. KDD Engineering and Consulting, Tokyo, p. 106 (1981).

1.41 J. I. Marcum, A statistical theory of target detection by pulsed radars. *Rand Corp Res. Memo* **RM-754**, December (1947); reprinted: *IRE Trans. Inf. Theory* **IT-6**, No. 2, 269 (1960).

1.42 J. L. Lawson and G. E. Uhlenbeck, eds., *Threshold Signal*. Dover, New York, 1965.

1.43 S. O. Rice, Mathematical analysis of random noise. *Bell Syst. Tech. J.* **23**, 282 (1944); **24**, 46 (1945).

1.44 P. Swerling, Probability of detection for fluctuating targets. *Rand Corp. Res. Memo* **RM-1217**, March (1954); *IRE Trans. Inf. Theory* **IT-6**, April, p. 269 (1960).

1.45 D. P. Meyer and H. A. Mayer, *Radar Target Detection*. Academic Press, New York, 1973.

1.46 S. Chandrasekhar, *Radiative Transfer*. Oxford Univ. Press, London, 1955.

1.47 M. H. Cohen, Radio astronomy polarization measurements. *Proc. IRE* **46** (1), 172 (1958).

1.48 R. D. Hayes and J. L. Eaves, *Study of Polarization Techniques for Target Enhancement*, AF33(615)-2523 AD. 31670 Report A-871. Georgia Institute of Technology, Atlanta, 1966.

1.49 G. A. Ioannides and D. E. Hammers, Optimum antenna polarization for target discrimination in clutter. *IEEE Trans. Antennas Propag.* **AP-27** (3), 357 (1979).

1.50 A. A. Swartz et al., Optimal polarizations for achieving maximum contrast in radar images. *J. Geophys. Res.* **91** (B12), 15332 (1988).

1.51 A. J. Poelman, Virtual polarisation adaptation: A method of increasing the detection capability of a radar system though polarisation-vector processing. *IEE Proc. Part F* **130** (5), 261 (1981).

1.52 A. J. Poleman, Polarisation-vector translator in radar systems. *IEE Proc., Part F* **130** (4), 383 (1983).

1.53 A. J. Poelman and J. R. F. Guy, Multinotch logic-product polarisation suppression filters: A typical design example and its performance in a rain-clutter environment. July, *IEEE Proc. Part F* **131** (4), 383 (1984).

1.54 E. Hammers, M. Fugita, and A. Klein, Applications of adaptive polarisation. *Mil. Microwaves Conf. Rec.*, London, 1980, p. 383 (1980).

1.55 D. Giuli, M. Fossi, and M. Cherardelli, A technique for adaptive polarization filtering in radars. *Record of the IEEE 1985 Int. Radar Conf.*, p. 213 (1985).

1.56 R. D. Brown and H. Wang, Adaptive multiband polarization processing for surveillance radar. *Proc. IEEE Int. Syst. Eng. Conf. 3rd, 1991*, p. 17 (1991).

1.57 R. D. Brown and H. Wang, *Improved Radar Detection Using Adaptive Multiband Polarization Processing*, Natl. Telesyst. Conf. George Washington University, Virginia Campus, Washington, DC, 1992.

1.58 T. S. Chu, Rain induced cross-polarization at cm and mm-wavelengths. *Bell Syst. Tech. J.* **53** (8), 1557 (1977).

1.59 M. Yamada et al., Compensation techniques for rain depolarisation in satellite communications. *Radio Sci.* **17** (5), 1220 (1982).

1.60 W. J. Vogel, Terrestrial rain depolarization compensation experiment at 11.7 GHz. *IEEE Trans. Commun.* **COM-31** (11), 1241 (1983).

1.61 N. Fourikis, Future directions for millimeter-wave systems. *Invited Keynote Pap., Int. Conf. Infrared Millimeter Waves, 16th*, Lausanne, Switzerland, *1991*, p. 424 (1991).

1.62 A. J. Blanchard and B. J. Jeans, Antenna effects in depolarization measurements. *IEEE Trans. Geosci. Remote Sens.* **GE-21** (1), 113 (1983).

1.63 F. T. Ulaby et al., *Michigan Microwave Canopy Scattering Model*, 22486-T-1 7/88. University of Michigan, Ann Arbor, 1988.

1.64 W. D. White, Circular radar cuts rain clutter. *Electronics* **27**, 158 (1954).

1.65 L. M. Novak et al., Optimal processing of polarimetric synthetic-aperture radar imagery. *Lincoln Lab. J.* **3** (2), 133 (1990).

1.66 W. A. Holm, Applications of polarimetry to target/clutter discrimination in millimeter wave radar systems. *Proc. SPIE—Polarimetry: Radar, IR Visible, UV X-Ray* **1317**, 148 (1990).

1.67 M. R. B. Dunsmore, Bistatic radars, Chapter 11. In *Advanced Radar Techniques and Systems*, G. Galati, Ed., IEE Peregrinus, 1993.

1.68 I. Groger, W. Sander, and W.-D. Wirth, Experimental phased array radar ELRA with extended flexibility. *IEEE AES Mag.*, November, p. 26 (1990).

1.69 S. A. Hovanessian, *Introduction to Synthetic Array and Imaging Radars*. Artech House, Dedham, MA, 1980.

1.70 D. R. Wehner, *High Resolution Radar*. Artech House, Dedham, MA, 1987.

1.71 F. Johansson, L. Rexberg, and N. O. Petersson, Theoretical and experimental investigation of large microstrip array antenna. *IEEE Colloq. Recent Developments in Microstrip Antennas*, 4/1 (1993).

1.72 J. Schindall, Aircraft detection and identification uses passive electronic-support measures. *Microwave Syst. News*, October, p. 78 (1986).

1.73 R. D. Curtis, On overview of surface navy ESM/ECM development. *J. Electron. Def.*, March, p. 31 (1982).

1.74 M. Simpson, High ERP phased array ECM systems. *J. Electron. Def.*, March, p. 41 (1982).

1.75 D. B. Hoisington, Low-probability of detection radar designs challenge tactical ESM capabilities. In *The International Countermeasures Handbook*, 11th ed., EW Communications, Inc. Palo Alto, CA, p. 350, 1986.

1.76 R. C. Dixon, *Spread Spectrum Systems*. Wiley, New York, 1977.

1.77 J. Munday and M. C. Pinches, Jaguar-V frequency hopping radios system. *IEE Proc., Part F* **129** (3), 213 (1982).

1.78 N. Fourikis, Novel receiver architectures for ESM facilities. *J. Electr. Electron. Eng., Aust.* **6** (3), 341 (1986).

1.79 G. L. Verschuur, Interstellar molecules. *Sky Telescope*, March, p. 379 (1992).

1.80 A. A. Penzias and R. W. Wilson, A measurement of excess temperature at 4080 Mc/s. *Astrophys. J.* **142**, 419 (1965).

1.81 J. L. Pawsey and R. N. Bracewell, *Radio Astronomy*. Oxford Univ. Press, Oxford, 1955.

1.82 J. W. M. Baars, The measurement of large antennas with cosmic radio sources. *IEEE Trans. Antennas Propag.* **AP-21** (4), 461 (1973).

1.83 M. L. Perlman, E. M. Rowe, and R. E. Watson, Synchrotron radiation—light fantastic. *Phys. Today* **27**, 30 (1974).

1.84 I. S. Shklovsky, *Cosmic Radio Waves*. Harvard Univ. Press, Cambridge, MA, 1960.

1.85 J. H. Oort, The Crab Nebula. *Sci. Am.* **196** (3), 59 (1957).

1.86 G. Bekefi and A. H. Barrett, *Electromagnetic Vibrations, Waves and Radiation*. MIT Press, Cambridge, MA, 1977.

1.87 R. H. Dicke, The measurement of thermal radiation at microwave frequencies. *Rev. Sci. Instrum.* **17**, 268 (1946).

1.88 A. Gagliano and R. H. Platt, An upward looking airborne millimeter wave radiometer for atmospheric water vapor sounding and rain detection. *Proc. SPIE—Int. Soc. Opt. Eng.* **544**, 112 (1985).

1.89 M. Ryle and A. Hewish, The synthesis of large radio telescopes. *Mon. Not. R. Astron. Soc.* **120**, 220 (1960).

1.90 A. Hewish, The realisation of giant radio telescopes by synthesis techniques. *Proc. IREE Aust.* **24**, (2), 225 (1963).

1.91 D. M. Le Vine et al., Initial results in the development of a synthetic aperture microwave radiometer. *IEEE Trans. Geosci. Remote Sens.* **28**, 614 (1990).

1.92 B. Y. Mills, Cross-type radio telescopes. *Proc. IREE Aust.* **24** (2), 132 (1963).

1.93 K. V. Sheridan, M. Morimoto, and J. P. Wild, The Culgoora radioheliograph—18. Initial observations. *Proc. IREE Aust.* **28** (9), 380 (1967).

1.94 N. Fourikis, The Culgoora radioheliograph—7. The branching network. *Proc. IREE Aust.* **28** (9), 315 (1967).

1.95 J. P. Wild, Matthew Flinders Lecture 1974—Radio Investigations of the Sun. *Records Austral. Acad. Science* **3** (1), November (1974).

1.96 B. Vowinkel et al., Airborn imaging system using a cryogenic 90 GHz receiver. *IEEE Trans. Microwave Theory Tech.* **MTT-20**, 535 (1981).

1.97 B. Vowinkel, K. Gruener, and W. Reinert, Cryogenic all solid-state millimeter-wave receivers and airborne radiometry. *IEEE Trans. Microwave Theory Tech.* **MTT-31**, 996 (1983).

1.98 H. A. Malliot, A cross beam interferometer radiometer for high resolution microwave sensing. *IEEE Aerospace Applications Conf. Digest*, p. 77 (1993).

1.99 A. S. Milman, Sparse-aperture microwave radiometers for Earth remote sensing. *Radio Sci.* **23** (2), 193 (1988).

1.100 C. S. Ruf et al., Interferemetric synthetic aperture microwave radiometry for remote sensing of the Earth. *IEEE Trans. Geosci., Remote Sens.* **26** (5), 596 (1988).

1.101 C. T. Swift, D. M. Le Vine, and C. S. Ruf, Aperture synthesis concepts in microwave remote sensing of the Earth. *IEEE Trans. Microwave Theory Tech.* **39** (12), 1931 (1991).

1.102 N. W. Broten et al., Long base line interferometry: A new technique. *Science* **156**, p. 1592 (1967).

1.103 W. K. Klemperer, Long baseline radio interferometry with independent frequency standards. *Proc. IEEE* **60**, 602 (1972).

1.104 A. R. Thompson, J. M. Moran, and G. W. Swenson, *Interferometry and Synthesis in Radio Astronomy*. Wiley, New York, 1986; reprinted by Krieger Press, Melbourne, FL, 1991.

1.105 K. I. Kellerman and A. R. Thomson, The very long baseline array. *Science* **229**, 123 (1985).

1.106 K. I. Kellerman and A. R. Thomson, The very long baseline array. *Sci. Am.* **258**, 44 (1988).

1.107 P. J. Napier et al., The very long baseline array. *Proc. IEEE* **82** (5), 658 (1994).

1.108 R. S. Sade and L. Deerkoski, Tracking and data relay satellite operations in the 1980's. *Proc. AIAA/NASA Symp. Space Tracking Data Syst.*, p. 77 (1981).

1.109 G. S. Levy et al., Very long baseline interferometric observations made with an orbiting radio telescope. *Science* **234**, 117 (1986).

1.110 G. S. Levy et al., VLBI using a telescope in Earth orbit. I. The observations. *Astrophys. J.* **336** (No. 2, Part 1), 1098 (1989).

1.111 R. P. Linfield et al., 15 GHz space VLBI observations using an antenna on a TDRSS satellite. *Astrophys. J.* **358** (No. 1, Part 1), 350 (1990).

1.112 A. W. Rudge, Skyhooks, fish-warmers and hub-caps: Milestones in satellite communications. *IEE Proc., Part F* **132** (1), 1 (1985).

1.113 K. Miya, ed., *Satellite Communications Technology*. KDD Engineering and Consulting, Tokyo, p. 339, 1981.

1.114 W.-D. Wirth, Omnidirectional low probability of intercept radar. *Proc. Int. Radar Conf.*, Paris, *1989* (1989).

1.115 E. R. Billam, System aspects of multiple beams in phased array radar. *IEE Colloq. Multiple Beam Antennas Beamformers, 1989*, p. 1/1 (1989).

1.116 W.-D. Wirth, Radar signal processing with active receiving array. *IEE Proc. RADAR-77*, p. 219 (1977).

1.117 C. Latham, Martello—A long range 3-D radar. *Mil. Microwaves Conf. Proc.*, London, *1984*, p. 79 (1984).

1.118 W. P. M. N. Keizer, New active phased array configurations. Conf. Proc. Mil. Microwaves, *1990*, p. 564 (1990).

1.119 L. I. Ruffe and G. F. Scott, LPI considerations for surveillance radar. *Int. Conf. Radar 1992*, p. 200 (1992).

1.120 B. J. Reits and A. Groenenboom, FMCW signal processing for a pulse radar. *Int. Conf. Radar 1992*, p. 332 (1992).

1.121 A. G. Stove, Linear FMCW radar techniques. *IEE Proc., Part F (Radar and Signal Processing)*, *1992* **139** (5), p. 343 (1992).

1.122 P. D. L. Beasley et al., Solving the problems of a single antenna frequency modulated CW radar, *Proc. IEEE 1990 Int. Radar Conf.*, p. 391 (1992).

1.123 C. Shannon and W. Weaver, *The Mathematical Theory of Communications*. Univ. of Illinois Press, Urbana, 1964.

1.124 F. H. Perrin, Methods of appraising photographic systems. *J. Soc. Pict. Telev. Eng.* **69**, 152 (1960).

1.125 D. R. Wehner, *High Resolution Radar*. Artech House, Dedham, MA, 1987.

1.126 N. Fourikis and D. R. Wehner, High resolution long range ISAR imaging systems. *Int. J. Infrared Millimeter Waves* **14** (2), 405 (1993).

1.127 N. Fourikis, Novel power combining circular arrays operating at mm-wavelengths. *Proc. Int. Conf. Infrared Millimeter Waves*, *16th*, Lausanne, Switzerland, *1991*, p. 503 (1991).

1.128 N. Fourikis, D. R. Wehner, and K. W. Eccleston, A proposal for a novel surveillance and imaging phased array radar. *Proc. Int. Conf. Infrared Millimeter Waves*, *17th*, *1992*, p. 146 (1992).

1.129 E. Briemle, Passive detection and location of noise jammers. *AGARD Conf. Proc.* **AGARD-CP-488**, p. 33-1 (1990).

1.130 W. Xu, The challenges and the ways to deal with-where is airborne fire control radar going. *Proc. IEEE Natl. Aerosp. Electron. Conf.*, *1993*, Vol. 1, p. 303 (1993).

1.131 D. Richardson, *Stealth Warplanes*. Salamander Books, London, 1989.

1.132 H. F. Harmuth, Use of ferrites for absorption of electromagnetic waves. *IEEE Trans. Electromagn. Compat.* **EMC-27**, (2), 100 (1985).

1.133 R. T. Petty, ed., *Jane's Weapon Systems*, 1987–1988. Jane's Publishing, London (1987).

1.134 N. Fourikis, Antenna elements and architectures for wide-band multifunction active phased arrays. *Invited Keynote Pap., Int. Conf. Millimeterwave Infrared Technol.*, Beijing, *1992*, p. 24 (1992).

1.135 N. Fourikis and N. Lioutas, Novel wideband multifunction phased arrays. *Proc. Int. Conf. Infrared Millimeter Waves*, *16th*, Lausanne, Switzerland, *1991*, 499 (1991).

1.136 M. A. Burke, Multifunction/shared aperture systems or smart skins now. *J. Electron. Def.* **14** (1), 29 (1991).

1.137 Z. Zhang, Tactical advantages of quasi-wide-band phased array radar. *Acta Sinica* **21** (3), 86 (1993).

1.138 S. J. Fu, J. L. Vizu, and D. L. Grose, Determination of ground emitter location. *IEEE AES Mag.* **3** (12), December, p. 15 (1988).

1.139 P. V. Noah, Background characterization techniques for target detection using scene metrics and pattern recognition. *Opt. Eng.* **30** (3), 154 (1991).

1.140 V. G. Nebabin, *Methods and Techniques of Radar Recognition* (D. K. Barton, ed./transl.). Artech House, Boston and London, 1995.

1.141 T. F. Brukiewa, Active array radar systems applied to air-traffic control. *IEEE Natl. Telesyst. Conf.*, *1994*, p. 27 (1994).

1.142 M. Michelson, W. W. Shrader, and J. C. Wieler, Terminal Doppler weather radar. *Microwave J.* February, p. 139 (1990).

1.143 R. J. Bowles and J. Goodson, Dual-use technology and the national airspace system. *IEEE Natl. Telesyst. Conf.*, *1994*, p. 1A-4 (1994).

1.144 D. S. Sego, Ultrawide band active radar array antenna for unmanned air vehicles. *IEEE Natl. Telesyst. Conf.*, *1994*, p. 13 (1994).

1.145 R. N. Longuemare, Advanced phased arrays redefine radar. *Def. Electron.*, April, p. 63 (1990).

1.146 P. F. Goldsmith, Focal plane arrays for millimeter-wave astronomy. *IEEE MTT-S Int. Microwave Symp. Dig.*, p. 1255 (1992).

1.147 K. I. Kellerman, Radio astronomy: The next decade. *Sky Telescope*, September, p. 247 (1991).

1.148 J. Davis, The quest for high resolution. *Sky Telescope*, January, p. 29 (1992).

1.149 A. Bottcher et al., LEO satellite networks: Fundamental system parameters and their mutual dependence. *ITG-Fachberichte, Mobile Commun. Conf.*, 124, September 1993, p. 165, (in German).

1.150 J. J. Schuss et al., Design of the Iridium phased array antennas. *IEEE Antennas Propag. Soc. Int. Symp.*, *1993*, Vol. 1, p. 218, 1993.

CHAPTER 2

2.1 R. C. Spencer, *Paraboloid Diffraction Patterns from the Standpoint of Physical Optics*, Rep. No. T-7. MIT Radiat. Lab., Massachusetts Institute of Technology, Cambridge, MA, 1942.

2.2 L. J. Chu, *Theory of Radiation from Paraboloidal Reflectors*, Rep. No. V-1. MIT Radiat. Lab., Massachusetts Institute of Technology, Cambridge, MA, 1942.

2.3 J. F. Ramsay, Tubular beams from radiating apertures. *Advances in Microwaves*. Vol. 3, L. Young, ed. Academic Press, New York, p. 127, 1968.

2.4 J. F. Ramsay, Lambda functions describe antenna/diffraction patterns. *Microwaves, Antenna Des. Suppl.*, June, p. 69 (1967).

2.5 J. F. Ramsay, An introduction to Lambda functions and transforms. *PTGAP Int. Symp. Prog. Dig.*, July, p. 205 (1963).

2.6. D. B. Steinberg, Relationships and fundamentals examined for phased-array design. In *The Microwave System Designer's Handbook*. EW Communications, Palo Alto, CA, 3rd ed., p. 93, 1985.

2.7 K. K. Miya, ed., *Satellite Communications Technology*. KDD Engineering and Consulting, Tokyo, 1981.

2.8 J. Ruze, The effect of aperture errors on the antenna radiation pattern. *Nuovo Cimento, Suppl.* **9** (3), 364 (1952).

2.9 J. Ruze, Antenna tolerance theory—A review. *Proc. IEEE* **54** (4), 633 (1966).

2.10 S. Zarghamee, On antenna tolerance theory. *IEEE Trans. Antennas Propag.* **AP-15**, 777 (1967).

2.11 G. Katow, 85 foot Az-el antenna structure deformation from gravity loads. *JPL Space Programs Summ.* **33-44** (3), 106 (1968).

2.12 H. G. Weiss, The Haystack microwave research facility. *IEEE Spectrum* **2**, 50 (1965).

2.13 J. H. Davis, *The Evaluation of Reflector Antennas*, Tech. Rep. No. NGL-006-70-1. University of Texas at Austin, 1970.

2.14 J. R. Cogdell and J. H. Davis, On separating aberrant effects from random scattering effects in radio telescopes. *Proc. IEEE* **61** (9), 1344 (1973).

2.15 S. von Hoerner, Radio-telescopes for millimeter wavelengths. *Astron. Astrophys.* **41**, 301 (1975).

2.16 P. W. Hannan, Microwave antennas derived from the Cassegrain telescope. *IRE Trans. Antennas Propag.* **AP-9**, 140 (1961).

2.17 S. Silver, *Microwave Antenna Theory and Design*. McGraw-Hill, New York, 1949.

2.18 E. A. Ohm, A proposed multiple-beam microwave antenna for earth stations and satellites. *Bell Syst. Tech. J.* **53**, 1657 (1974).

2.19 C. Dragone and D. C. Hogg, The radiation pattern and impedance of offset and symmetric near field Cassegrain and Gregorian antennas. May *IEEE Trans. Antennas Propag.* **AP-22**, 472 (1974).

2.20 J. S. Cook, E. M. Elam, and H. Zucker, The open Cassegrain antenna. Part 1. Electromagnetic design and analysis. *Bell Syst. Tech. J.* **44**, 1255 (1965).

2.21 T. S. Chu and R. H. Turrin, Depolarization properties of offset reflector antennas. *IEEE Trans. Antennas Propag.* **AP-21**, 339 (1973).

2.22 R. Graham, The polarisation characteristics of offset Cassegrain aerials. *Proc. Inst. Electr. Eng. Int. Conf. Radar Present Future Conf. Pub.* **105**, 23 (1973).

2.23 H. Tanaka and M. Mizusawa, Elimination of crosspolarization on offset dual-reflector antennas. *Electron. Commun. Jpn.* **58**, 71 (1975).

2.24 Y. Mizugutch, M. Akagawa, and H. Yokoi, Offset dual reflector antenna. *Dig. Antennas Propag. Soc. Int. Symp.*, 1976, p. 2 (1976).

2.25 C. Dragone, Offset multireflector antennas with perfect pattern symmetry and polarization discrimination. *Bell Syst. Tech. J.* **57**, 2663 (1978).

2.26 R. A. Shore, A simple derivation of the basic design equation for offset dual reflector antennas with rotational symmetry and zero cross-polarization. *IEEE Trans. Antennas Propag.* **AP-33** (3), 114 (1985).

2.27 N. Fourikis, A parametric study of the constraints related to Gregorian/Cassegrain offset reflectors having negligible cross polarization. *IEEE Trans. Antennas Propag.* **36** (1), 144 (1988).

2.28 J. Ruze, Lateral feed displacement in a paraboloid. *IEEE Trans. Antennas Propag.* **AP-13**, 660 (1965).

2.29 W. V. T. Rusch and A. C. Ludwig, Determination of the maximum scan-gain contours of a beam-scanning paraboloid and their relation to the Petzval surface. *IEEE Trans. Antennas Propag.* **AP-21**, 141 (1973).

2.30 C. Dragone, A first-order treatment of aberrations in Cassegrain and Gregorian antennas. *IEEE Trans. Antennas Propag.* **AP-31**, 331 (1982).

2.31 P. F. Goldsmith, Focal plane arrays for millimeter wavelength astronomy. *IEEE MTT-S Int. Microwave Symp. Dig.*, p. 1255 (1992).

REFERENCES

2.32 W. B. Wetherell and M. P. Rimmer, General analysis of aplanatic Cassegrain, Gregorian and Schwarschild telescopes. *Appl. Opt.* (11), 2817 (1972).

2.33 P. F. Goldsmith et al., Focal plane imaging systems for millimeter wavelengths. *IEEE Trans. Microwave Theory Tech.* **41** (10), 1664 (1993).

2.34 N. Fourikis and B. MacA. Thomas, Beam separation constraints in a beam-switched radio-telescope. *Proc. IREE Aust.* **35** (7), p. 199 (1974).

2.35 S. Stein, On cross coupling in multiple beam antennas. *IRE Trans. Antennas Propag.* **AP-20**, 548 (1962).

2.36 K. S. Yngvensson, Realizable feed-element patterns and optimum aperture efficiency in multibeam antenna systems. *IEEE Trans. Antennas Propag.* **AP-36**, 1637 (1988).

2.37 N. R. Erickson et al., A 15 element imaging array for 100 GHz. *IEEE Trans. Microwave Theory Tech.* **40** (1), 1 (1992).

2.38 D. N. Black and J. C. Wiltse, Millimeter-wave characteristics of phase-correcting Fresnel zone plates. *IEEE Trans. Microwave Theory Tech.* **MTT-35** (12), 1122 (1987).

2.39 M. Shoucri et al., A passive millimeter wave camera for landing under low visibility conditions. *Proc. Natl. Telesystems Conf.*, p. 109 (1993).

2.40 G. S. Dow et al., W-band MMIC direct detection receiver for passive imaging system. *IEEE MTT-S Int. Microwave Symp. Dig.*, p. 163 (1993).

2.41 P. S. Hall and S. J. Vetterlein, Review of radio frequency beamforming techniques for scanned and multiple beam antennas. *IEEE Proc., Part H: Microwaves, Opt. Antennas* **137** (5), 293 (1990).

2.42 R. J. Mailloux, Hybrid antennas. In *The Handbook of Antenna Engineering* (A. W. Rudge et al., eds.), Chapter 5. IEE Peregrinus, 1986.

2.43 E. Brookner, ed., *Aspects of Modern Radar*. Artech House, Boston and London, 1988.

2.44 E. Brookner, ed., *Radar Technology*. Artech House, Norwood, MA, 1977.

2.45 A. W. Rudge, Current trends in antenna technology and prospects for the next decade. *IEEE Antennas Propag. Soc. Newsl.*, December, p. 5 (1983).

2.46 R. M. Sorbello et al., 20 GHz phased-array-sed antennas utilizing distributed MMIC modules. *COMSAT Tech. Rev.* **16** (2), 339 (1985).

2.47 R. M. Sorbello, Advanced satellite antenna development for the 1990s. *AIAA Int. Commun. Satellite Syst. Conf., 12th*, Arlington, VA, *1988*, Collec. Tech. Pap., p. 322 (1988).

2.48 R. M. Sorbello et al., MMIC: A key technology for future communications satellite antennas. *Proc. SPIE: Monolithic Microwave Integr. Circuits Sensors, Radar Commun. Syst.* **1475**, 175 (1991).

2.49 F. T. Assal, A. I. Zaghoul, and R. M. Sorbello, Multiple spotbeam systems for satellite communications, *AIAA Int. Commun. Satellite Syst. Conf., 12th*, Arlington, VA, *1988*, Collect. Tech. Pap., p. 322 (1988).

2.50 G. Cortez-Medellin and P. F. Goldsmith, Analysis of segmented reflector antennas for a large millimeter wave radio telescope. *IEEE AP-S Int. Symp. Dig.* **4**, 1886 (1992).

2.51　T. S. Mast and J. E. Nelson, Fabrication of the Keck ten meter telescope primary mirror. *Proc. SPIE: Opt. Fabr. Test. Workshop: Large Telescope* **542**, 48 (1985).

2.52　W. T. Sullivan, III, Radio astronomy's golden anniversary. *Sky Telescope*, December, p. 544 (1982).

2.53　A. Molker, High-efficiency phased array antenna for advanced multibeam; multiservice mobile communication satellite. *Third Int. Conf. Satellite Systems Mobile Commun. Navagat.*, p. 75, 1983.

2.54　N. Fourikis, A new class of millimetre wave telescopes. *Astron. Astrophys.* **65**, 385 (1978).

2.55　N. Fourikis, Several aspects related to the realisation of annular synthesis telescopes operating at millimetre wavelengths. *J. Electr. Electron. Eng. Aust.*, p. 193 (1982).

2.55a　N. Fourikis, Single structure, steerable synthesis telescopes utilizing offset reflectors. Ph.D dissertation, University of NSW, School of Electrical Engineering & Computer Sciences, September 1984.

2.56　N. Fourikis, Novel power combining circular arrays at mm-wavelengths. *Proc. Int. Conf. Infrared Millimeter Waves*, *16th*, Lausanne, Switzerland, *1991*, p. 503 (1991).

2.57　A. C. Ludwig, Antenna feed efficiency. *JPL Space Programs Summ.* **37-26** (4), 200 (1965).

2.58　A. W. Love, *Electromagnetic Horn Antennas*. IEEE Press, New York, 1976.

CHAPTER 3

3.1　A. Krienski, Equivalence between continuous and discrete radiating arrays. *Can. J. Phys.* **39**, 35 (1961).

3.2　R. J. Mailloux, *Phased Array Handbook*. Artech House, Norwood, MA, 1994.

3.3　C. Hemmi, in Hal Schrank's Antenna Designer's Notebook. Bandwidth of the array factor for phased-steered arrays, *IEEE Antennas Propag. Mag.* **35** (1), 72 (1993).

3.4a　R. S. Elliott, Beamwidth and directivity of large scanning arrays. *Microwave J. Part I*, December, p. 53 (1963).

3.4b　R. S. Elliott, Beamwidth and directivity of large scanning arrays. *Microwave J., Part II*, January, p. 74 (1964).

3.5　J. S. Stone, U.S. Pats. 1,643,323 and 1,715,433.

3.6　C. A. Balanis, *Antenna Theory—Analysis and Design*. Harper & Row, New York, 1982.

3.7　C. L. Dolph, A current distribution for broadside arrays which optimizes the relationship between beam width and side-lobe level. *Proc. IRE Waves Electrons* **34**, June, p. 335 (1946).

3.8　H. J. Riblet, Discussion on 'A current distribution for broadside arrays which optimizes the relationship between beam width and side-lobe level' by C. L. Dolph. *Proc. IRE* **35**, p. 489 (1947).

3.9 D. Barbiere, A method for calculating the current distribution of Tschebyscheff arrays. *Proc. IRE* **40**, January, p. 78 (1952).

3.10 J. Stegen, Excitation coefficients and beamwidths of Tschebyscheff arrays. *Proc. IRE* **41**, November, p. 1671 (1953).

3.11 J. Drane, Jr., Useful approximations for the directivity and beamwidth of large scanning Dolph-Chebyshev arrays. *Proc. IEEE* **56**, November, p. 1779 (1968).

3.12 W. W. Hansen and J. R. Woodyard, A new principle in directional antenna design. *Proc. IRE* **26** (3), 333 (1938).

3.13 S. A. Shelkunoff, A mathematical theory of linear arrays. *Bell Syst. Tech. J.* **22**, 80 (1943).

3.14 R. S. Elliott, *Pattern Synthesis for Antenna Arrays*, IEEE Lect. Notes, 1988.

3.15 R. S. Elliott, Improved pattern synthesis for equispaced linear arrays. *Alta Freq.* **51** (6) 296 (1982).

3.16 R. C. Hansen, Array pattern control and synthesis. *Proc. IEEE* **80** (1) 141 (1992).

3.17 T. T. Taylor, The design of line-source antennas for narrow beamwidth and low-sidelobes. *IRE Trans. Antennas Propag.* **AP-3** (1), 16 (1955).

3.18 T. T. Taylor, *One Parameter Family of Line Sources Producing Modified $Sin(\pi u)/\pi u$ Patterns*, Tech. Memo. 324, Contract AF 19(604)-262-F-14. Hughes Aircraft Co., Culver City, CA, 1953.

3.19 R. S. Elliott, On discretizing continuous aperture distributions. *IEEE Trans. Antennas Propag.* **AP-25** (5), 617 (1977).

3.20 R. C. Hansen, A one-parameter circular aperture distribution with narrow beamwidth and lowside lobe. *IEEE Trans. Antennas Propag.* **AP-14**, 477 (1976).

3.21 T. T. Taylor, Design of circular apertures for narrow beamwidth and low sidelobes. *IRE Trans. Antennas Propag.* **AP-18**, 17 (1960).

3.22 R. C. Hansen, *Array Chapters in Handbook of Antenna Design*, Vol. 2. A. W. Rudge, et al., Eds., IEE Peregrinus, 1983.

3.23 R. C. Hansen, *Microwave Scanning Antennas*, Vol. 1. Academic Press, New York, 1964. Ch. 1; Peninsula Publ., 1985.

3.24 E. T. Bayliss, Design of monopulse antenna difference patterns with low sidelobes. *Bell Syst. Tech. J.* **47**, 623 (1968).

3.25 H. J. Orchard et al., Optimizing the synthesis of shaped beam antenna patterns. *IEE Proc., Part H: Microwaves, Opt. Antennas* **132**, 63 (1985).

3.26 Y. U. Kim and R. S. Elliott, Shaped-pattern synthesis using pure real distributions. *IEEE Trans. Antennas Propag.* **36**, 1645 (1988).

3.27 J. Powers, Utilisation of the Lambda functions in the analysis and synthesis of monopulse difference patterns. *IEEE Trans. Antennas Propag.* **AP-15** (6), 771 (1967).

3.28 W. H. Von Aulock, Properties of phased arrays. *IRE Trans. Antennas Propag.* **AP-9**, 1715 (1960).

3.29 T. C. Cheston and J. Frank, Phased array radar antennas. *In Radar Handbook* (M. Skolnik, ed.), Chapter 7. McGraw-Hill, New York, 1990.

3.30 E. D. Sharp, A triangular arrangement of planar-array elements that reduces the number needed. *IRE Trans. Antennas Propag.* **AP-9**, 126 (1961).

3.31 B. Y. Mills, The Sydney University cross-type radio telescope. *Proc. IREE Aust.* **24** (2), 156 (1963).

3.32 A. E. Convington, A compound interferometer. *J. Royal Astron. Soc. Canada* **54**, February, 67 (1960).

3.33 N. R. Labrum et al., A compound interferometer with a 1.5 minute of arc fan beam. *Proc. IRE (Aust.)* **24** (2), February, 148 (1963).

3.34 J. Frank and K. W. O'Haver, Phased array antenna development at the Applied Physics Laboratory. *Johns Hopkins APL Tech. Dig.* **14** (4), 339 (1993).

3.35 F. P. Irzinski, A coaxial waveguide commutator feed for a scanning circular phased array antenna. *IEEE Trans. Microwave Theory Tech.* **MTT-29**, 266 (1981).

3.36 J. P. Wild, A new method of image formation with annular aperture and an application in radio astronomy. *Proc. R. Soc. London, Ser. A* **286**, 499 (1965).

3.37 J. P. Wild, The Culgoora radioheliograph. 1. Specification and general design. *Proc. IREE Aust.* **28** (9), 279 (1967).

3.38 N. Goto and Y. Tsunoda, Sidelobe reduction of circular arrays with a constant excitation amplitude. *IEEE Trans. Antennas Propag.* **AP-25** (6), 890 (1977).

3.39 J. J. Lee, Sidelobes control of solid-state array antennas. *IEEE Trans. Antennas Propag.* **36** (3), 339 (1988).

3.40 R. E. Collin, Aperture-type antennas. In *Antennas and Radiowave Propagation*, Ch. 4. McGraw-Hill, New York, p. 164, 1985.

3.41 R. E. Willey, Space tapering of linear and planar arrays. *IRE Trans. Antennas Propag.* **J-AP62**, 369 (1962).

3.42 P. S. Hacker, *in* Hal Schrank's Antenna Designer's Notebook 'Thinned arrays: Some fundamental considerations.' *IEEE Antennas Propag. Mag.* **34** (3), 43 (1992).

3.43 E. Brookner, ed., *Aspects of Modern Radar*, Chapter 2, p. 28. Artech House, Dedham, MA, 1988.

3.44 M. Skolnik et al., Statistically designed density-tapered arrays. *IEEE Trans. Antennas Propag.* **AP-12**, 408 (1964).

3.45 R. J. Mailloux and E. Cohen, Statistically thinned arrays with quantized element weights. *IEEE Trans. Antennas Propag.* **39** (4) 4 (1991).

3.46 J. Frank and R. Coffman, Hybrid active arrays. *IEEE, Proc. Natl. Telesyst. Conf.*, p. 19 (1994).

3.47 R. J. Mailloux, Array grating lobes due to periodic phase, amplitude and time delay quantization. *IEEE Trans. Antennas Propag.* **AP-32** (12), 1364 (1984).

3.48 D. K. Barton and H. R. Ward, *Handbook of Radar Measurements*. Prentice-Hall, Englewood Cliffs, NJ, 1969.

3.49 C. J. Miller, Minimizing the effects of phase quantization errors in an electronically scanned array. *Proc. 1964 Symp. Electronically Scanned Phased Arrays and Applications*, RADC-TDR-64-225, RADC Griffiss AFB, **1**, 17.

3.50 J. Ruze, The effect of operature errors on the antenna radiation pattern. *Nuovo Cimento, Suppl.* **9** (3), 361 (1992).

3.51 R. E. Elliott, Mechanical and electrical tolerances for two-dimensional scanning antenna arrays. *IRE Trans. Antennas Propag.* **AP-6**, 114 (1958).

3.52 J. L. Allen, *The Theory of Array Antennas*, MIT Lincoln Lab. Tech. Rep. No. 323. Massachusetts Institute of Technology, Cambridge, MA, 1963.

3.53 M. I. Skolnik, Non-uniform arrays. In *Antenna Theory* (R. E. Collin and F. J. Zucker, eds.), Chapter 6. McGraw-Hill, New York, 1969.

3.54 H. J. Moody, *A Survey of Array Theory and Techniques*, Res. Labs Rep. No. 6501.3. RCA Victor Co., 1963.

3.55 T. C. Cheston, Effect of random errors on sidelobes of phased arrays. *IEEE APS Newsl.-Antennas Des. Notebook*, April (1985).

3.56 J. K. Hsiao, *Array Sidelobes, Error Tolerances, Gain and Beamwidth*, RL Rep. 8841. Naval Research Laboratory, Washington, DC, 1984.

3.57 J. K. Hsiao, Design of error tolerance of a phased array. *Electron. Lett.* **21** (19), 834 (1985).

3.58 R. D. Kaplan, Predicting antenna sidelobe performance. *Microwave J.*, September, p. 201 (1986).

3.59 S. O. Rice, Mathematical analysis of random noise. In *Selected Papers on Noise and Stochastic Processes* (N. Wax, ed.). Dover, New York, 1954.

3.60 B. D. Steiberg, *Principles of Aperture and Array Systems Design*. Wiley, New York, 1976.

3.61 M. T. Borkowski and D. G. Leighton, Decreasing cost of GaAs MMIC modules is opening up new areas of application. *Electron. Prog.* **29** (2), 32 (1989).

3.62 D. N. McQuiddy, Jr. et al., Transmit/receive module technology for X-band active array radar. *Proc. IEEE* **79** (3), 308 (1991).

3.62a R. Tang and R. W. Burns, Array technology. *Proc. IEEE* **80** (1), 173 (1992).

3.63 R. J. Mailloux, Antenna array architecture. *Proc. IEEE* **80** (1), 163 (1992).

3.64 E. Levine et al., A study of microstrip array antennas with the feed network. *IEEE Trans. Antennas Propag.* **AP-37** (4), 426 (1989).

3.65 P. M. Esker et al., Composite stripline phase shifter with low loss and minimum weight. *IEEE MTT-S Int. Microwave Symp. Dig.*, p. 1239 (1990).

3.66 D. R. Carey and W. Evens, The Patriot radar in tactical air defense. *Microwave J.*, May, p. 325 (1988).

3.67 E. D. Cohen, Active electronically scanned arrays. *IEEE Natl. Telesyst. Conf.*, *1994*, p. 3 (1994).

3.68 T. F. Brukiewa, Active array radar systems applied to air traffic control. *IEEE Natl. Telesyst. Conf.*, *1994*, p. 27 (1994).

3.69 E. R. Billam, Phased array radar and the detection of low cross section targets. *Proc. 19th European Microwave Conf.*, *1989*, p. 55 (1989).

3.70 D. E. Lingle, D. P. Mikszan, and D. Mukai, Advanced technology ultrareliable radar. *Proc. IEEE Natl. Radar Conf.*, *1989*, p. 1 (1989).

3.71 R. M. Lockerd and G. E. Crain, Airborne active element array radars come of age. *Microwave J.*, January, p. 101 (1990).

3.72 R. H. Abrams, Jr. and R. K. Parker, Introduction to the MPM: What it is and where it might fit. *IEEE MTT-S Int. Microwave Symp. Dig.*, p. 107 (1993).

3.73 J. A. Christensen et al., MPM technology developments: An industry perspective. *IEEE MTT-S Int. Microwave Symp. Dig.*, p. 115 (1993).

3.74 B. Edward and D. Rees, A broadband printed dipole with integrated balun. *Microwave J.*, May, p. 339 (1987).

3.75 N. Fourikis, N. Lioutas, and N. V. Shuley, Parametric study of the co- and cross-polarisation of tapered planar and antipodal slotline antennas. *IEE Proc., Part H: Microwaves, Opt. Antennas* **140** (1), 17 (1993).

3.76 L. D. S. Langley, P. S. Hall, and P. Newham, Novel ultrawide-bandwidth Vivaldi antenna with low cross polarisation. *Electron. Lett.* 19 (23), 2005 (1993).

3.77 M. J. Povinelli and J. A. Johnson, Design and performance of wideband dual-polarized stripline notch arrays. *Proc. IEEE AP-S Int. Symp.*, 1988, **1**, p. 200 (1988).

3.78 I. L. Newberg and J. J. Wooldridge, An affordable low-profile multifunction structure (ALMS) for an optoelectronic (OE) active array. *IEEE MTT-S Int. Microwave Symp. Dig.*, p. 509 (1993).

3.79 I. L. Newberg and J. J. Wooldridge, Revolutionary active array radar using solid state "modules" and fiber optics. *Record 1993 IEEE National Radar Conf.*, p. 88 (1993).

3.80 G. J. Scalsi, J. P. Turtle, and P. H. Carr, MMIC's for airborne phased arrays. *Proc. SPIE—Monolithic Microwave Integ. Circuits Sensors, Radar Commun. Syst.* **1475**, 2 (1991).

3.81 J. K. Kinzel, B. J. Edward, and D. Rees, V-band, space-based phased arrays. *Microwave J.*, January, p. 89 (1987).

3.82 J. J. H. Wang and V. K. Tripp, Design of multioctave spiral-mode microstrip antennas. *IEEE Trans. Antennas Propag.* **39** (3), 332 (1991).

3.83 P. S. Hall, Microstrip antenna array with multi-octave bandwidth. *Microwave J.*, March, p. 133 (1986).

3.84 M. J. Povinelli, A planar broad-band flared microstrip slot antenna. *IEEE Trans. Antennas Propag.* **AP-35**, (8) 968 (1987).

3.85 N. Fourikis et al., Microprocessor tuned antennas. *2nd Aust. (ATERB/CSIRO) Symp. Antennas*, Sydney, Australia, *1989*, Paper 21 (1989).

3.86 M. P. Purchine, J. T. Aberle, and C. R. Birtcher, A tunable L-band circular microstrip patch antenna. *Microwave J.*, October, p. 80 (1993).

3.87 M. Priolo et al., Transmit/Receive modules for the 6-18 GHz multifunction arrays. *IEEE MTT-S Int. Microwave Symp. Dig.*, Vol. 3, p. 1227 (1990).

3.88 R. Tang, A. Popa, and J. J. Lee, Applications of photonic technologies to phased array antennas. *IEEE MTT-S Int. Microwave Symp. Dig.*, p. 758 (1990).

3.89 Y. T. Lo, A. mathematical theory of antenna arrays with randomly spaced elements. *IRE Trans. Antennas Propag.* **AP-12**, May, 257 (1964).

3.90 B. D. Steinberg, *Principles of Aperture and Array Design*. Wiley, New York, 1976.

3.91 B. D. Steinberg, The peak sidelobes of the phased array having randomly located elements. *IEEE Trans. Antennas Propag.* **AP-20**, March, 129 (1972).

3.92 B. D. Steinberg, *Microwave Imaging with Large Antenna Arrays—Radio Camera Principles and Techniques*. Wiley-Interscience, New York, 1983.

3.93 H. K. Schuman and B. J. Strait, On the design of unequally spaced arrays with nearly equal sidelobes. *IEEE Trans. Antennas Propag.* **AP-16**, 493 (1968).

3.94 W. A. Sandrin et al., Design of arrays with unequal spacing and partially uniform amplitude taper. *IEEE Trans. Antennas Propag.* **AP-14**, 642 (1969).

3.95 E. Mattsson, Designing thinned phased arrays. *Proc., Eur. Microwave Conf., 1971*, B2/3:1 (1971).

3.96 A. P. Goffer, M. Kam, and P. R. Herczfeld, Wide-bandwidth phased arrays using random subarraying. *Proc., Euro. Microwave Conf., 20th*, p. 241 (1990).

3.97 G. J. Laughlin, E. V. Byron, and T. C. Cheston, Very wideband phased-array antenna. *IEEE Trans. Antennas Propag.* **AP-20**, 699 (1972).

3.98 J. E. Boyns and J. H. Provencher, Experimental results of a multifrequency array antenna. *IEEE Trans. Antennas Propag.* **AP-20**, 106 (1972).

3.99 D. G. Shively and W. L. Stutzman, Wideband arrays with variable element sizes. *IEE Proc., Part H: Microwaves, Opt. Antennas* **137** (4), 138 (1990).

3.100 W. T. Stutzman, Wide bandwidth antenna array design. *Proc. IEEE Southeast. Reg. Meet.*, Raleigh, NC, p. 97 (1985).

3.101 R. D. Curtis, On overview of surface navy ESM/ECM developments. *J. Electron. Def.*, March, p. 37 (1982).

3.102 R. L. Moynihan, Phased arrays for airborne ECM—The rest of the story. *Microwave J.*, January, p. 34 (1987).

3.103 M. Simpson, High ERP phased array ECM systems. *J. Electron. Def.*, March, p. 41 (1982).

3.104 L. Josefsson, Designing a broadband phased array antenna suitable for use in ECM systems. *Def. Electron.*, February, p. 85 (1980).

3.105 V. Rossi and G. Damen, Solid state jamming antenna. *Conf. Proc. Mil. Microwaves, 1990*, p. 446 (1990).

3.106 J. D. Sparno, PAPPORT tactical self protection systems design. *Conf. Proc. Military Electron.*, 1980, p. 28 (1980).

3.107 D. J. Rice and A. S. Kaufman, Airborne self-protection jammer—ASPJ/ALQ-165. *Conf. Proc.* Military Microwaves, 1982, p. 389 (1982).

3.108 J. G. Teti et al., Wideband airborne early warning (AEW) radar. *Rec. IEEE Natl. Radar Conf., 1993*, p. 239 (1993).

3.109 K. Kalbasi, R. Plumb, and R. Pope, An analysis and design tool for a broadband dual feed circles array antenna. *IEEE AP-S Int. Symp.*, Vol. 4, p. 2085 (1992).

3.110 M. J. Povinelli and C. E. Grove, Wideband apertures for active planar multifunction phased arrays. *Proc. IEEE Natl. Radar Conf., 1989*, p. 125 (1989).

3.111 N. Fourikis, Antenna elements and architectures for wide-band multifunction active phased arrays. *Invited Keynote Pap., Int. Conf. Millimeter Wave Infrared Technol., 2nd*, Beijing, *1992*, p. 24 (1992).

3.112 N. Fourikis and N. Lioutas, Novel wideband multifunction phased arrays. *Proc. Int. Conf. Infrared Millimeter Waves*, *16th*, Lausanne, Switzerland, *1991*, p. 499 (1991).

3.113 M. A. Burke, Multifunction/shared aperture systems or smart skins now. *J. Electron. Def.* **14** (1), 29 (1991).

3.114 H. V. Cottony, Wide-frequency band array system. *IEEE Trans. Antennas Propag.* **AP-18**, 774 (1970).

3.115 M. N. Cohen, A tabular synthesis technique for broadband/thinning linear phased arrays. *Proc. Int. Conf. Infrared Millimeter Waves*, *16th*, Lausanne, Switzerland, *1991*, p. 501 (1991).

3.116 J. Arsac, Nouveau reseau pour l'observation radiostronomique de la brillance sur le soleil a 9350 Mc/s. *C. R. Hebd. Seances Acad. Sci.* **240**, 942 (1955).

3.117 A. T. Moffet, Minimum redundancy linear arrays. *IEEE Trans. Antennas Propag.* **AP-16**, 172 (1968).

3.118 M. Felli and P. Pampaloni, The information of a minimum redundancy linear antenna array. *Alta Freq.* **44** (5), 240 (1975).

3.119 K. A. Blanton and J. H. McClellan, New search algorithm for minimum redundancy linear arrays. *Int. Conf. Acoust., Speech, Signal Process.*, *1991*, Vol. 2, p. 1361 (1991).

3.120 D. A. Linebarger, I. H. Sudbough, and I. G. Tollis, A unified approach to design of minimum redundancy arrays. *Asilomar Conf. Signals, Syst. Comput. 24th*, *1990*, Vol. 1, p. 143 (1990).

3.121 C. S. Ruf, Numerical annealing of low-redundancy linear arrays. *IEEE Trans. Antennas Propag.* **41** (1), 85 (1993).

3.122 R. N. Bracewell et al., The Stanford five element radio telescope. *Proc. IEEE* **61**, 1249 (1973).

3.123 P. J. Napier, A. R. Thompson, and R. D. Ekers, The very large array: Design and performance of a modern synthesis radio telescope. *Proc. IEEE* **71** (11) 1295 (1983).

3.124 A. S. Milman, Sparse-aperture microwave radiometers for Earth remote sensing, *Radio Sci.* **23** (2), 193 (1988).

3.125 C. S. Ruf et al., Interferemetric synthetic aperture microwave radiometry for remote sensing of the Earth. *IEEE Trans. Geosci. Remote Sens.* **26** (5), 596 (1988).

3.126 C. T. Swift, D. M. Le Vine, and C. S. Ruf, Aperture synthesis concepts in microwave remote sensing of the Earth. *IEEE Trans. Microwave Theory Tech.* **39** (12), 1931 (1991).

3.127 D. M. Le Vine et al., Initial results of the development of a synthetic aperture microwave radiometer. *IEEE Trans. Geosci. Remote Sens.* **28** (4), 614 (1990).

3.128 Y. Lee and S. U. Pillai, An algorithm for optimal placement of sensor elements. *Int. Conf. Acoust., Speech Signal Process.*, *1988*, Vol. 5, p. 2674 (1988).

3.129 W.-L. Chen and B.-N. Yeheskel, Minimum redundancy array structure for interference cancellation. *IEEE AP-S Int. Symp.*, London, Ontario, Canada, *1991*, p. 121 (1991).

REFERENCES

3.130 X. Huang, J. P. Reilly, and M. Wong, Optimal design of linear array of sensors. *Proc. Int. Conf. Acoust., Speech Signal Process.*, *1991*, Vol. 2, p. 1405 (1991).

3.131 M. B. Jorgenson, M. Fattouche, and S. T. Nichols, Applications of minimum redundancy arrays in adaptive beamforming. *IEE Proc., Part H: Microwaves, Antennas Propag.* **138** (5), 441 (1991).

3.132 R. N. Bracewell, Radio astronomy techniques. In *Handbuch der Physik*, S. Flügge, ed., Vol. 54. Springer-Verlag, Berlin, 1962.

3.133 P. S. Hall and S. J. Vetterlein, Review of radio frequency beamforming techniques for scanned and multiple beam antennas. *IEE Proc., Part H: Microwaves, Opt. Antennas* **137** (5), 293 (1990).

3.134 W. D. White, Pattern illuminations in multiple beam antennas. *IRE Trans. Antennas Propag.* **T-AP62**, 430 (1962).

3.135 J. L. Allen, A theoretical limitation on the formation of lossless multiple beams in linear arrays. *IRE Trans. Antennas Propag.* **T-AP62**, 350 (1961).

3.136 S. Stein, On cross coupling in multiple beam antennas. *IRE Trans. Antennas Propag.* **T-AP62**, 548 (1962).

3.137 J. M. Chambers, R. Passmore, and J. Ladbrooke, Beamforming for a multi-beam radar. *Int. Radar Conf.* **82**, 390, 1982.

3.138 J. R. Wallington, Beamforming options for phased array radar. *Conf. Proc. Military Microwaves*, 379, 1986.

3.139 *IEE Colloq. Multiple Beam Antennas Beamformers* (1989).

3.140 N. Fourikis, The Culgoora radioheliograph—7. The branching network. *Proc. IREE Aust.* **28** (9), 315 (1967).

3.141 E. J. Blum, Le reseux Nord-Sud a lobes mortiples. Complement au grant interferometre de la station de Nancay. *Ann. Astrophys.* **24** (4), 359 (1961).

3.142 L. Cardone, Ultra-wideband microwave beamforming technique. *Microwave J.*, April, 21 (1985).

3.143 K. E. Alameh, R. A. Minasian, and N. Fourikis, Photonics-based beamforming network for active phased arrays. *Proc. Aust. Conf. Opt. Fibre Technol., 18th 1993*, p. 360 (1994).

3.144 K. E. Alameh, R. A. Minasian, and N. Fourikis, Hardware compressed photonic beamformer architecture for multi-beam active arrays. *Proc. Aust. Conf. Opt. Fibre Technol., 19th, 1994*, p. 41 (1995).

3.145 K. E. Alameh, R. A. Minasian, and N. Fourikis, High capacity optical interconnections for phased array beamforming. *IEEE J. Lightwave Technol., Spec. Issue* **13** (6), 1116 (1995).

3.146 R. A. Minasian, K. E. Alameh, and N. Fourikis, Wavelength multiplexed photonic beamformer architecture for microwave phased array. *Microwave Opt. Technol. Lett.*, Oct. 5, 1995, p. 84.

3.147 N. Fourikis, D. R. Wehner, and K. W. Eccleston, A proposal for a novel surveillance and imaging phased array radar. *Invited Pap. Int. Conf. Infrared Millimeter Waves, 17th, 1992*, p. 146 (1993).

3.148 A. S. Daryoush, Interfaces for high-speed fiber optic links: Analysis and Experiment. *IEEE Trans. Microwave Theory Tech.* **39** (12), 2031 (1991).

3.149 S. Banerjee et al., A wide band microwave/photonic distribution network for an X-band active phased array antenna. *IEE Colloq. Microwave Opto-Electron. Dig.*, 1994, p. 46.

3.150 A. Goutzoulis and K. Davies, Compressive 2-D delay line architecture for the time-steering of phased-array antennas. *Proc. SPIE—Optoelectron., Signal Processing Phased-Array Antennas II*, 270, 1990.

3.151 N. A. Riza, Multiple-simultaneous phased array antenna beam generation using an acousto-optic system. *Proc. SPIE—Analog Photon.* **1790**, 95 (1992).

3.152 B. Y. Mills et al., The Sydney University cross-type radio telescope. *Proc. IREE Aust.* **24**, 156 (1963).

3.153 M. I. Large and R. H. Frater, The beam forming system for the Molonglo radio telescope. *Proc. IREE Aust.* **30**, 227 (1969).

3.154 T. W. Clarke, H. S. Murdoch, and M. I. Large, The delay line system for the Molonglo radio telescope. *Proc. IREE Aust.* **30**, 236 (1969).

3.154a N. J. Easton, F. C. Bennett, and C. W. Miller, Analogue beamformers. *IEE Colloq. Multiple Beam Antennas Beamformers*, p. 12/1 (1989).

3.155 A. I. Zaghloul, F. T. Assal, and R. M. Sorbello, Multibeam active phased array system configurations for communication satellites. *1987 IEEE Military Commun. MILCOM'87*, p. 289, Vol. 1 (1987).

3.156 J. Blass, Multidirectional antenna—new approach top stacked beams. *IRE Int. Conv. Rec.*, Part 1, p. 48 (1960).

3.157 M. Fassett, L. J. Kaplan, and J. H. Pozgay, Optimal synthesis of ladder network array antenna feed systems. *Antennas Propag. Symp.*, Amherst, Mass., p. 58 (1976).

3.158 N. Inagaki, Synthesis of beam forming networks for multiple beam array antennas with maximum feed efficiency. *IEE Int. Conf. Antennas Propagation 1987*, p. 375 (1987).

3.159 P. J. Wood, An efficient matrix feed for an array generating overlapped beams. *IEE Int. Conf. Antennas Propag., 1987*, p. 371 (1987).

3.160 P. S. Hall and S. J. Vetterlein, Advances in microstrip multiple beam arrays. *Int. Conf. Antennas Propag., 7th, 1991*, Vol. 1, p. 129 (1991).

3.161 P. S. Hall and S. J. Vetterlein, Integrated multiple beam microstrip arrays. *Microwave J.* **35** (Issue 1), 103 (1992).

3.162 J. Butler and R. Howe, Beamforming matrix simplifies design of electronically scanned antennas. *Electron. Des.* **9**, 170 (1961).

3.163 L. MacNamara, Simplified design procedure for Butler matrices incorporating 90° or 180° hybrids. *IEE Proc., Part H: Microaves, Antennas Propag.* **134** (1), 50 (1987).

3.164 J. Shelton and J. Hsiao, Reflective Butler matrix. *IEEE Trans. Antennas Propag.* **AP-27** (5), 651 (1979).

3.165 J. R. F. Guy, Proposal to use reflected signals through a single Butler matrix to produce multiple beams from a circular array antenna. *Electron. Lett.* **28** (4), 209 (1985).

3.166 N. A. Blokhia and B. A. Mishustin, Design of planar beam-shaping circuits. *Radio Electron. Commun. Syst. (Engl. Transl.)* **27** (2), 45 (1984).

3.167 P. E. K. Chow and D. E. N. Davis, Wide bandwidth Butler matrix network. *Electron. Lett.* **3**, 252 (1967).

3.168 M. J. Withers, Frequency insensitive phase shift networks and their application in a wide-bandwidth Butler matrix. *Electron.* **5** (20), 496 (1969).

3.169 R. Levy, A high power X-band Bulter matrix. *Microwave J.*, April, p. 153 (1984).

3.170 J. R. Wallington, Analysis, design and performance of a microstrip Butler matrix. *Proc. Eur. Microwave Conf. 6th*, Brussels, *1973*, p. A 14.3.1 (1973).

3.171 W. Charczenco et al., Integrated optical Butler matrix for beamforming in phased array antennas. *Proc. SPIE—Optoelectron. Signal Process. Phased-Array Antennas II*, p. 196 (1990).

3.172 T. P. Waldron, S. K. Chin, and R. J. Naster, Distributed beamsteering control of phased array radars. *Microwave J.*, September, p. 133 (1986).

3.173 P. R. Herczfeld et al., Optical control of MMIC-based T/R modules. *Microwave J.*, May, p. 309 (1988).

3.174 N. A. Riza, A compact electrooptic controller for microwave phased-array. *Proc. SPIE—Photon. Processors, Neural Networks and Memories* **2026**, 286 (1993).

3.175 W.-D. Wirth, Signal processing for target detection in experimental phased array radar ELRA. 1981 *IEE Proc., Part F* **128**, 311 (1981).

3.176 W. Ng et al., Wideband fiber-optic delay network for phased array antenna steering. *Electron. Lett.* **25**, (21), 1456, October (1989).

3.177 E. Ackerman, S. Wanuga, and D. Kasemset, Integrated 6-bit photonic true-time delay for lightweight 3-6 GHz radar beamformer. *IEEE MTT-S Int. Microwave Symp. Dig.*, p. 681 (1992).

3.178 L. H. Gessell and T. M. Turpin, True time delay beam forming using acousto-optics. *Proc. SPIE—***1703**, *Opt. Technol. Microwave Appl./Optoelectron. Signal Process. Phased Array Antennas*, Vol. 3, p. 592 (1992).

3.179 M. L. VanBlaricum, Photonic systems for antenna applications. *IEEE Antennas Propag. Mag.* **36** (5), 30 (1994).

3.180 A. Seeds, Microwave optoelectronics. Tutorial review. *Opt. Quantum Electron.* **25**, 219 (1993).

3.181 A. Seeds, Optical technologies for phased array antennas. Invited paper. *IEICE Trans. Electron.* **E76-C** (2), 198 (1993).

3.182 A. S. Glista, Airborne photonics, A technology whose time has come. *AIAA/IEEE Digital Avion. Syst. Conf., 12th, DASC*, p. 336 (1993).

3.183 R. Benjamin, Optical techniques for generating multiple agile antenna beams. *IEE Colloq. Multiple Beam Antennas Beamformers, 1989*, p. 11/1 (1989).

3.184 D. K. Paul, Optical beam forming and steering for phased array antenna. *IEEE Proc. Natl. Telesyst. Conf. 1993, Commercial Applicat. Dual-Use Technol.*, p. 7 (1993).

3.185 D. D. Curtis and R. J. Mailloux, A critical look at photonics for phased array systems. *IEEE AP-S Int. Symp./URSI Radio Sci. Meet. Nucl. EMP Meet., 1992 Dig.*, Vol. 2, p. 717 (1992).

3.186 H. Steyskal, Digital beamforming antennas: An introduction. *Microwave J.*, January, p. 107 (1987).

3.187 H. Steyskal and J. F. Rose, Digital beamforming for radar systems. *Microwave J.*, January, p. 121 (1989).

3.188 J. F. Rose, Digital beamforming receiver technology. *IEEE AP-S Int. Symp.*, *1990*, Vol. 1, p. 380 (1990).

3.189 J. Herd, Experimental results from a self-calibrating digital beamforming array. *IEEE AP-S Int. Symp.*, *1990*, p. 384 (1990).

3.190 J. L. Langston and K. Hinman, A digital beamforming processor for multiple beam antennas. *IEEE AP-S Int. Symp.*, *1990*, p. 383 (1990).

3.191 A. K. Prentice, A digitally beamformed phased array receiver for tactical bistatic radar. *IEE Colloq. Active Passive Components Phased Array Syst.* 1992, p. 11/1.

3.192 M. J. D. Powell, Radial basis functions for multivariable interpolation: A review. *IMA Conf. Algorithms Approx. Funct. Data*, RMCS Shrivenham (1985).

3.193 D. S. Broomhead and D. Lowe, Multivariable interpolation and adaptive networks. *Complex Syst.* **2**, 321 (1988).

3.194 T. Lo, H. Leung, and J. Litva, Nonlinear beamforming. *Electron. Lett.* **27** (4), 350 (1991).

3.195 S. C. Liu, A fault correction technique for phased array antennas. *IEEE AP-S Int. Symp.*, *1992*, p. 1612 (1992).

3.196 K. M. Lee, R. S. Chu, and S. C. Liu, A performance monitoring/fault isolation and correction system of a phased array antenna using transmission-line signal injection with phase toggling method. *IEEE AP-S Int. Symp.*, *1992*, p. 429 (1992).

3.197 R. A. Shore, Sidelobe sector nulling with minimized weight perturbations. **RADC-TR-85-40**, AD-A 157057 (1985).

3.198 R. Tang and R. Brown, Cost reduction techniques for phased arrays. Invited paper. *Microwave J.*, January, p. 139 (1987).

3.199 T. Harper, W. Kennan, and N. K. Osbrink, In search of the $300 T/R module. *Microwave J.*, March, p. 48 (1986).

3.200 W. P. M. N. Keizer, New active phased array configurations. *Conf. Proc. Mil. Microwaves*, *1990*, p. 564 (1990).

3.201 T. F. Brukiewa, Active array radar systems applied to air-traffic control. *IEEE Natl. Telesyst. Conf.*, *1994*, p. 27 (1994).

3.202 M. Michelson, W. W. Shrader, and J. C. Wieler, Terminal Doppler weather radar. *Microwave J.*, February, p. 139 (1990).

3.203 D. J. Sego, Ultrawide band active radar array antenna for unmanned air vehicles. *IEEE Natl. Telesyst. Conf.*, *1994*, p. 13 (1994).

CHAPTER 4

4.1 B. C. Considine, Solid-state transmitters take advantage of continuing advances in device technology. *Electron. Prog.* **29** (2), 24 (1989).

4.2 D. J. Rivera, An S-band solid-state transmitter for airport surveillance radars. *IEEE 1993 Natl. Radar Conf. Rec.*, p. 197 (1993).

4.3 Millimeter wave extended interaction klystrons (amplifiers and oscillators) and transmitter subsystems. Varian Booklet, November (1990).

4.4 J. W. Hansen, US TWTs from 1 to 100 GHz. State of the Art Reference. Microwave J., p. 179 (1989).

4.5 N. C. Currie and C. E. Brown, *Principles and Applications of Millimeter-Wave Radar*. Artech House, Dedham, MA 1987.

4.6 L. Sivan, *Microwave Tube Transmitters*. Chapman & Hall, London, 1994.

4.7 A. Freiley et al., The 500kW CW X-band Goldstone solar system radar. *IEEE MTT-S Int. Microwave Symp. Dig.*, p. 125 (1992).

4.8 R. H. Abrams, Jr. and R. K. Parker, Introduction to the MPM: What it is and where it might fit. *IEEE MTT-S Int. Microwave Symp. Dig.*, p. 107 (1993).

4.9 A. Staprans and R. S. Symons, The 1990 Microwave Power Tube Conference. *Microwave, J.*, December, p. 26 (1990).

4.10 H. Bierman, Microwave tubes reach new power and efficiency levels. *Microwave J.* **30** (2), 26 (1987).

4.11 W. Manheimer, Application of gyrotrons to radar and atmospheric sensing. *Proc. Int. Conf. Infrared Millimeter Waves, 17th, 1992*, p. 142 (1993).

4.12 R. M. Lhermitte, Cloud and precipitation remote sensing at 94 GHz. *IEEE Trans. Geosci. Remote Sensing* **29** (3), May, 207 (1988).

4.13 H. Li, A. J. Illingworth, and J. Eastment, A simple method of Dopplerizing a pulsed magnetron radar. *Microwave J.* **37** (4), 226 (1994).

4.14 D. J. Christie, Advances in solid-state magnetron modulation. *Microwave J.* **36** (1), 111 (1993).

4.15 Y. C. Shih and H. J. Kuno, Solid-state sources from 1 to 100 GHz. State of the Art Reference. *Microwave J.*, p. 145 (1989).

4.16 P. T. Greiling and L. Nguyen, Ultrahigh-frequency InP-based HEMTs for millimeter wave applications. *Proc. SPIE—Monolithic Microwave Integr. Circuits Sensors, Radar Commun. Syst.* **1475**, (1991).

4.17 H. Q. Tserng and P. Saunier, Advances in power MMIC amplifier technology in space communications. *Proc. SPIE—Monolithic Microwave Integr. Circuits Sensors, Radar Commun. Syst.* **1475**, 74 (1991).

4.18 P. M. Smith and A. W. Swanson, HEMTs low-noise and power transistors for the 1 to 100 GHz. *Appl. Microwave*, May, p. 63 (1989).

4.19 R. K. Parker, Recent advances in high frequency vacuum electronics. *Proc. Int. Conf. Infrared Millimeter Waves, 16th*, Laussanne, Switzerland, *1991*, p. 1 (1991).

4.20 R. Mallavarpu and M. P. Puri, High power CW with multi-octave bandwidth from power-combined mini-TWTs. *IEEE MTT-S Int. Microwave Symp. Dig.*, p. 1333 (1990).

4.21 J. W. Mink and D. G. Rutledge, Guest Editors of the Special Issue on Quasi-Optical Techniques. IEEE Trans. on MTT, Oct. 1993, Vol. 41, Number 10.

4.22 M. Salib et al., A 5-10 GHz, 1-Watt HBT amplifier with 58% peak power-added efficiency. *IEEE Microwave Guided Wave Lett.* **MGWL-4** (10), 321 (1994).

4.23 H. Q. Tserng and P. Saunier, A highly-efficient 7-Watt 16 GHz monolithic pseudomorphic HEMT amplifier. *IEEE MTT-S Int. Microwave Symp. Dig.*, p. 87 (1993).

4.24 R. Mallavarpu and G. MacMaster, 100 W peak/30 W average broadband X-band solid-state amplifier. *Conf. Proc. Military Microwaves '86*, p. 354, 1986.

4.25 *IEEE Standard Dictionary of Electrical and Electronics Terms*, 1984, p. 537.

4.26 R. L. Woods, Microwave power source overview. *Mil. Microwaves Conf. Proc.*, London, *1986*, p. 349 (1986).

4.27 R. H. Giebeler and J. E. Huddart, Specifying, designing and interfacing tubes in the modern radar. *Milit. Microwaves Conf. Proc.*, Brighton, *1986*, p. 366 (1986).

4.28 D. N. McQuiddy, Jr. et al., Transmit/Receive module technology for X-band active array radar. *Proc. IEEE* **79** (3), 308 (1991).

4.29 J. Berenz, InP MMICs: Coming of age. *Microwave J.* **36** (8), 113 (1993).

4.30 Y. Saito et al., Reliability testing of state-of-the-art PM HEMT MMICs three-stage low-noise amplifier. *14th Ann. IEEE GaAs IC Symp., Tech. Dig.*, p. 153 (1992).

4.31 A. R. Francoeur, Naval space surveillance system (NAVSPASUR) solid state transmitter modernization. *Proc. 1989 IEEE Natl. Radar Conf.*, p. 147 (1989).

4.32 C. R. Green et al., GaAs MMIC power amplifiers for radar and communication systems. *Conf. Proc. Military Microwaves*, p. 132, 1992.

4.33 R. L. Ernst, R. L. Camisa, and A. Presser, Graceful degradation properties of matched n-port power amplifier combiners. *IEEE MTT-S Int. Microwave Symp. Dig.*, p. 174 (1977).

4.34 M. S. Gupta, Degradation of power combining efficiency due to variability among signal sources. *IEEE Trans. Microwave Theory Tech.* **40** (5), 1031 (1992).

4.35 M. Belna, D. Henry, and B. L. Smith, High peak-power TWT with very wide bandwidth. *Military Microwaves—Spec. Sess. Microwave Tubes*, London, *1988*, p. 22 (1988).

4.36 H. G. Kosmahl, Tubes vs. solid-state: Reality vs. promises. *Microwaves & RF* **24** (5), 167 (1985).

4.37 W. C. Morchin, *Early Warning Radar*. Artech House, Dedham, MA, 1989.

4.38 G. Ewell, *Radar Transmitters*. McGraw-Hill, New York, 1981.

4.39 L. Cosby, MPM applications: A forecast of near- to long-term applications. *IEEE MTT-S Int. Microwave Symp. Dig.*, Vol. 1, p. 111 (1993).

4.40 A. Brees et al., Microwave power module (MPM) development and results. *Proc. IEEE Int. Electron Devices Meet.*, Washington, DC, *1993*, p. 145.

4.41 J. A. Christensen et al., MPM Technology developments: An industry perspective. *IEEE MTT-S Int. Microwave Symp. Dig.*, p. 115 (1993).

4.42 J. Peterson, MIMIC Program spawns GaAs infrastructure. *Microwaves RF*, January, p. 55 (1995).

4.43 E. D. Cohen, MIMIC from the Department of Defense perspective. *IEEE Trans. Microwave Theory Tech.* **38** (9), 1171 (1990).

4.44 F. L. M. van den Bogaart and J. G. Bij de Vaate, Production results of a transmit/receive-MMIC chip set for a wide band active phased array radar at X-band. Conference Proceedings, Military Microwaves 92, p. 138.

4.45 R. Goyal and S. S. Bharj, MMIC: On-chip tunability. *Microwave J.* **30** (4), 135 (1987).

4.46 S. F. Paik, Microwave integrated circuits. *Electron. Eng.* **26** (2), 29 (1985).

4.47 J. McIlvenna, Monolithic phased arrays for EHF communication terminals. *Microwave J.*, March, p. 113 (1988).

4.48 H. M. Aumann and F. G. Willwerth, Intermediate frequency transmit/receive modules for low-sidelobe phased array application. *Proc. 1988 IEEE Natl. Radar Conf.*, p. 33 (1988).

4.49 R. S. Pengelly, Broadband monolithic microwave circuits for military applications. *Military Microwaves*, Oct., p. 244 (1982).

4.50 A. Podell, D. Lockie, and S. Moghe, Practical GaAs ICs designed for microwave subsystems. In *The Microwave System Designer's Handbook*, 4th ed., Vol. 16, No. 7, p. 327. EW Communications Inc., Palo Alto, CA, 1986.

4.51 I. J. Bahl et al., GaAs IC's fabricated with the high-performance, high-yield multifunction self-aligned gate process for radar and EW applications. *IEEE Trans. Microwave Theory Tech.* **38** (9), 1232 (1990).

4.52 I. J. Bahl et al., Multifunction SAG process for high-yield, low-cost GaAs microwave integrated circuits. *IEEE Trans. Microwave Theory Tech.* **38** (9), 1175 (1990).

4.53 D. G. Fisher, GaAs IC applications in electronic warfare, radar and communications systems. *Microwave J.*, May, p. 275 (1988).

4.54 A. Cetronio and R. Graffitti, A reproducible high yield technology for GaAs MMIC production. *Alta Freq.* **55** (3), 173 (1986).

4.55 C. Mayousse et al., 'Design and Technology' optimisation for high yield monolithic GaAs X-band low noise amplifiers. *Proc. 17th European Microwave Conf.*, p. 267 (1987).

4.56 P. H. Landbrooke, *MMIC Design: GaAs FETs and HEMTs*, Chapter 1. Artech House, Norwood, MA, 1989.

4.57 N. Shiga et al., X-band MMIC amplifier with pulse-doped GaAs MESFET's. *IEEE Trans. Microwave Theory Tech.* **39** (12), 1987 (1991).

4.58 S. X. Bar et al., Manufacturing technology development for high yield pseudomorphic HEMT. *IEEE GaAs IC Symp.*, p. 173 (1993).

4.59 J. Mondal, High performance and high-yield Ka-band low-noise MMIC using $0.25\text{-}\mu\text{m}$ ion-implanted MESFET's. *IEEE Microwave Guided Wave Lett.* **1** (7), 167 (1991).

4.60 H. Wang et al., High-yield W-band monolithic HEMT low-noise amplifier and image rejection downconverter chips. *IEEE Microwave Guided Wave Lett.* **3** (8), 281 (1993).

4.61 W. R. Wisseman et al., X-band GaAs single-chip T/R radar module. *Microwave J.* **30** (9), 167 (1987).

4.62 R. Tang and R. W. Burns, Array technology. *Proc. IEEE* **80** (1), 173 (1992).

4.63 J. E. Brewer et al., A single-chip digital signal processing system. *Proc. Sixth Ann. IEEE Int. Conf. Wafer Scale Integration*, p. 265, 1994.

4.64 B. Khabbaz et al., A high performance 2.4 GHz transceiver chip-set for high volume commercial applications. *IEEE Microwave Millimeter Wave Monolithic Circuits Symp. Dig., 1994*, p. 11 (1994).

4.65 H. Fudem et al., A highly integrated MMIC K-band transmit/receive chip. *IEEE MTT-S Int. Microwave Symp. Dig.*, p. 137 (1993).

4.66 M. J. Schindler et al., A single chip 2-20 GHz T/R/ module. *IEEE Microwave Millimeter Wave Monolithic Circuits Symp. Dig., 1990*, p. 99 (1990).

4.67 A. Kurdoghlian et al., The demonstration of Ka-band multi-functional MMIC circuits fabricated on the same PHEMT wafer with superior performance. *IEEE Microwave Millimeter Wave Monolithic Circuits Symp. Dig., 1993*, p. 97 (1993).

4.68 T. F. Brukiewa, Active array radar systems applied to air traffic control. *Proc. IEEE Natl. Telesyst. Conf.*, San Diego, CA, *1994*, p. 27 (1994).

4.69 E. D. Cohen, Active electronically scanned arrays. *Proc. IEEE Natl. Telesyst. Conf.*, San Diego, *1994*, CA, p. 3 (1994).

4.70 L. R. Whicker et al., A new approach to active phased arrays through RF-wafer scale integration. *IEEE MTT-S Int. Microwave Symp. Dig.*, p. 1223 (1990).

4.71 Manufacturing technology for microwave monolithic components. *Dig. Tech. Pap., U.S. Conf. GaAs Manuf. Technol., 1989* (1989).

4.72 L. R. Whicker and J. D. Murphy, RF-wafer scale integration: A new approach to active phased arrays. *Proc. Int. Conf. Wafer Scale Integration*, IEEE Comput. Soc. Press, p. 291 (1992).

4.73 R. F. Kole and S. E. Ozga, Jr., COBRA: Lessons learnt. *Proc. IEEE Natl. Telesyst. Conf.*, San Diego, CA, *1994*, p. 3 (1994).

4.74 R. Johnson, Multichip modules: Next generation packages. *IEEE Spectrum* **27**, 34 (1990).

4.75 Y. Shimada, Y. Yamashita, and H. Takamizawa, Low dielectric constant glass-ceramic substrate with AgPd wiring for VLSI package. *IEEE Trans. Components, Hybrids Manuf. Technol.* **11**, 163 (1988).

4.76 W. E. Pence et al., New TCE-matched glass-ceramic multi-chip module, electrical design and characterization. *IEEE Electron. Components Conf. Proc., 39th, 1989*, p. 647 (1989).

4.77 K. Kato, Y. Shimada, and H. Takamizawa, Low dielectric constant new materials for multilayer ceramic substrate. *IEEE Trans. Components, Hybrids Manuf. Technol.* **13**, 448 (1990).

4.78 W. A. Vitriol et al., Use of low temperature cofired ceramics (LTCC). In *Ceramic Substrates Packages for Electronic Applications*. American Ceramic Society, 1990.

4.79 G. Garbe et al., A. 6.8–10.7 GHz EW module using 72 MMICs. *1993 IEEE MTT-S Dig.*, p. 1329.

4.80 M. T. Borkowski and D. G. Leighton, Decreasing cost of GaAs MMIC modules is opening up new areas of application. *Electron. Prog.* **29** (2), 32 (1989).

4.81 J. J. Lee, Sidelobes control of solid-sate array antennas. *IEEE Trans. Antennas Propag.* **36** (3), 339 (1988).

4.82 I. J. Bahl et al., Class-B power MMIC amplifiers with 70 percent power-added efficiency. *IEEE Trans. Microwave Theory Tech.* **37** (9), 1315 (1989).

4.83 S. Toyoda, High efficiency single and push-pull power amplifiers. *IEEE MTT-S Int. Microwave Symp. Dig.*, Vol. 1, p. 277 (1993).

4.84 C. Duvanaud et al., Optimization of trade-offs between efficiency and intermodulation in SSPAs based on experimental and theoretical considerations. *IEEE MTT-S Int. Microwave Symp. Dig.*, p. 285 (1993).

4.85 Microwave Modules and Devices, A 2kW, L-band radar module using 550 W "quad" building blocks. *Microwave J.* **28** (6), 157 (1985).

4.86 M. Salib et al., A 1.8-W, 6-18-GHz HBT MMIC power amplifier with 10-dB gain and 37.5 peak power-added efficiency. *IEEE Microwave Guided Wave Lett.* **3** (9), 325 (1993).

4.87 N. L. Wang, W. J. Ho, and J. A. Higgins, 0.7 W X-Ku-band high gain, high-efficiency common base power HBT. *IEEE Microwave Guided Wave Lett.* **1** (9), 258 (1991).

4.88 M. Matloubian, 20 GHz high efficiency AlInAs-GaInAs on InP power HEMT. *IEEE Microwave Guided Wave Lett.* **3** (5), 142 (1993).

4.89 P. Saunier and H.-Q. Tserng, Fabrication and performances of pHEMT Ka-band 3-stage amplifiers for phased array applications. *Proc. SPIE—Int. Conf. Millim. & Submillim. Waves* **2211**, 187 (1994).

4.90 G. L. Lan et al., Millimeter-wave pseudomorphic HEMT MMIC phased array components for space communications. *Proc. SPIE—Monolithic Microwave Integr. Circuits Sensors, Radar Commun. Syst.* **1475**, 184 (1991).

4.91 W. Lam, High efficiency InP-based HEMT MMIC power amplifier for Q-band applications. *IEEE Microwave Guided Lett.* **3** (11), 420 (1993).

4.92 G. Hegazi et al., A 0.5-Watt 47-GHz power amplifier GaAs monolithic circuits. *IEEE Microwave Guided Wave Lett.* **2** (2), 61 (1992).

4.93 T. Ho et al., A 0.6-Watt U-band monolithic MESFET power amplifier. *IEEE MTT-S Int. Microwave Symp. Dig.*, p. 531 (1993).

4.94 M. Matloubian et al., V-band high efficiency high power AlInAs/GaInAs/InP HEMTs. *IEEE MTT-S Int. Microwave Symp. Dig.*, p. 535 (1993).

4.95 C. Meng et al., GaAs/AlGaAs multiquantum well structure applied to high frequency IMPATT devices. *IEEE MTT-S Int. Microwave Symp. Dig.*, p. 539 (1993).

4.96 J. A. Benet et al. Spatial power combining for millimeterwave solid state amplifiers. *1993 IEEE MTT-S Dig.*, p. 619.

4.97 D. M. Pozar, *Microwave Engineering*. Addison-Wesley, Reading, MA, 1990.

4.98 S. E. Rosenbaum et al., A 1-GHz three stage AlInAs-GaInAs-InP HEMT MMIC low-noise amplifier. *IEEE Microwave Guided Wave Lett.* **3** (8), 265 (1993).

4.99 P. J. Apostolakis et al., Microwave performance of low-power ion implanted 025-micron gate GaAs MESFET for low-cost MMIC's applications. *IEEE Microwave Guided Wave Lett.* **3** (8), 278 (1993).

4.100 M. Feng, J. Laskar, and J. Kruse, Ultra low-noise performance of 0.15-micron gate GaAs MESFET's made by direct ion implantation for low-cost MMIC's applications. *IEEE Microwave Guided Wave Lett.* **2** (5), 194 (1992).

4.101 C. C. Yang et al., A cryogenically-cooled wide-band HEMT MMIC low-noise amplifier. *IEEE Microwave Guided Wave Lett.* **2** (2), 58 (1992).

4.102 P. D. Chow et al., InGaAs HEMT LNA and doublers for EHF SATCOM ground terminals. *Proc. SPIE—Monolithic Microwave Integr. Circuits Sensors, Radar Commun. Syst.* **1475**, 43 (1991).

4.103 M. Kimishima and T. Azhizuka, 18-40 GHz semi-monolithic balanced cascade amplifiers using AlGaAs/InGaAs P-HEMT and HaAS MESFET. *IEEE MTT-S Int. Microwave Symp. Dig.*, p. 523 (1993).

4.104 R. Lai et al., An ultra-low noise cryogenic Ka-band InGaAs/InAlAs/InP HEMT front-end receiver. *IEEE Microwave Guided Wave Lett.* **4** (10), 3 (1994).

4.105 M. W. Pospieszalski et al., Millimeter-wave, cryogenically coolable amplifiers using AlInAs/GaInAs/InP HEMT's. *IEEE MTT-S Int. Microwave Symp. Dig.*, p. 515 (1993).

4.106 G. W. Wang, R. Kaliski, and Y. Chang, InGaAs MESFET's for millimeter-wave low-noise applications. *IEEE Microwave Guided Wave Lett.* **1** (4), 76 (1991).

4.107 J. Braunstein et al., High performance narrow and wide bandwidth amplifiers in CPW-technology up to W-band. *15th Ann. GaAs IC Symp.*, p. 277 (1993).

4.108 R. Lai et al., A high performance and low DC power V-band MMIC LNA using 0.1 μm InGaAs/InAlAs/InP HEMT technology. *IEEE Microwave Guided Wave Lett.* **3** (12), 447 (1993).

4.109 S. M. Joseph et al., 75-110 GHz InGaAs/GaAs HEMT high gain MMIC amplifier. *15th GaAs IC Symp.*, p. 273 (1993).

4.110 H. Wang et al., A monolithic 75-110 GHz balanced InP-based HEMT amplifier. *IEEE Microwave Guided Wave Lett.* **3** (10), 381 (1993).

4.111 M. Trippe et al., mm-Wave MIMIC receiver components. *IEEE Microwave and Millimeter Wave Monolithics Circuits Symp. Dig., 1991*, p. 51 (1991).

4.112 K. H. G. Duh et al., A super low-noise 0.1 μm T-gate InAlAs-InGaAs-InP HEMT. *IEEE Microwave Guided Wave Lett.* **1** (5), 114 (1991).

4.113 D. C. W. Lo, A monolithic 1X2 W-band four-stage low-noise amplifier array. *15th Ann. GaAs IC Symp., 1993*, p. 281 (1993).

4.114 H. Wang et al., High performance W-band monolithic pseudomorphic InGaAs HEMT LNA's and design/analysis methodology. *IEEE Trans. Microwave Theory Tech.* **40** (3), 417 (1992).

4.115 H. Wang et al., A high gain low noise 110 GHz monolithic two-stage amplifier. *IEEE MTT-S Int. Microwave Symp. Dig.*, p. 783 (1993).

4.116 A. K. Sharma, Solid-state control devices: State of the Art Reference. *Microwave J.*, p. 95 (1989).

4.117 D. J. Seymore, D. D. Heston, and R. E. Lehmann, Monolithic MBE-grown GaAs PIN diode limiters. *IEEE Microwave Millimeter Wave Monolithic Circuits Symp. Dig.*, 1987, p. 35 (1987).

4.118 D. J. Seymore, D. D. Heston, and R. E. Lehmann, X-band and Ka-band monolithic GaAs PIN diode variable attenuator limiters. *IEEE Microwave Millimeter Wave Monolithic Circuits Symp. Dig.*, 1988, p. 147 (1988).

4.119 L. Stark, Microwave theory of phased array antennas: A review. *Proc. IEEE* **62** (12), 1661 (1974).

4.120 W. E. Hord, Microwave and millimeter-wave ferrite phase shifters. State of the Art Reference. *Microwave J.*, p. 81 (1989).

4.121 D. E. Lawson et al., An analog X-band phase shifter. *IEEE Microwave Millimeter Wave Monolithic Circuits Symp. Dig.*, 1984, p. 6 (1984).

4.122 D. M. Krafcsik et al., A dual varactor, analog phase shifter operating at 6 to 18 GHz. *IEEE Microwave Millimeter Wave Monlithic Circuits Symp. Dig.*, 1988, p. 83 (1988).

4.123 L. C. T. Liu et al., A 30 GHz monolithic receiver. *IEEE Trans. Microwave Theory Tech.* **MTT-34** (12), 1548 (1986).

4.124 C. L. Chen et al., A low loss Ku-band monolithic analog phase shifter. *IEEE Trans. Microwave Theory Tech.* **MTT-36** (3), 315 (1987).

4.125 H. Endler, Technology advances design of microwave diode phase shifters. *The Microwave Designer's Handbook*, 4th ed., **16** (7), 321 (1986).

4.126 F. Ali et al., A single chip C-band GaAS monolithic five bit phase shifter with on chip digital decoder. *IEEE MTT-S Int. Microwave Symp. Dig.*, p. 1235 (1990).

4.127 M. C. Tsai, A new compact wideband balun. *IEEE MTT-S Int. Microwave Symp. Dig.*, p. 141 (1993).

4.128 M. Rhodes and A. A. Lane, Monolithic five-bit phase shifter for Artemis space application. *IEE Colloq. Active and Passive Components Phased Array Syst.*, April 1992, p. 10/1.

4.129 J. J. Komiak and A. K. Agrawal, Design and performance of octave S/C-band MMIC T/R module for multifunction phased arrays. *IEEE Trans. Microwave Theory Tech.* **39** (12), 1955 (1991).

4.130 G. B. Norris, A fully monolithic 4-18 GHz digital vector modulator, *IEEE MTT-S Int. Microwave Symp. Dig.*, p. 789 (1990).

4.131 C. Peignet, Y. Mancuso, and J.-C. Resneau, T/R modules for phased array antennas. *Proc. 1991 IEEE Natl. Radar Conf.*, p. 63.

4.132 A. Gupta et al., A 20 GHz 5-bit phase shift transmit module with 16 dB gain. *IEEE GaAs IC Symp.*, p. 197 (1984).

4.133 A. J. Slobodnik, Jr., R. T. Webster, and G. A. Roberts, A monolithic GaAS 36 GHz four-bit phase shifter. *Microwave J.*, 106 (1993).

4.134 S. Hara, T. Tokumitsu, and M. Aikawa, Novel unilateral circuits for MMIC circulators. *IEEE Trans. Microwave Theory Tech.* **38** (10), 1399 (1990).

4.135 P. Katzin et al., 6 to 18 GHz MMIC circulators. *Microwave J.* **35** (65), 248 (1992).

4.136 I. D. Robertson and A. H. Aghvami, Novel monolithic ultra-wideband unilateral 4-port junction using distributed amplification techniques. *IEEE MTT-S Int. Microwave Symp. Dig.*, p. 1051 (1992).

4.137 D. C. Webb, Status of ferrite technology in the United States. *IEEE MTT-S Int. Microwave Symp. Dig.*, p. 206 (1993).

4.138 E. Schloemann and R. E. Blight, Broadband/miniature circulators for use with microwave integrated circuits. *Electron. Prog.* (1), 28 (1988).

4.139 G. P. Rotrigue, Circulators from 1 to 100 GHz. State of the Art Reference. *Microwave J.*, p. 115 (1989).

4.140 L. E. Davis, Novel circulator configurations with possible applications in phased arrays. *IEE Colloq. Active Passive Component Phased Array Syst.*, April 24, p. 8.1 (1992).

4.141 Y. Murakami, Microwave ferrite technology in Japan: Current status and future expectations. *1993 IEEE MTT-S Dig.*, p. 207.

CHAPTER 5

5.1 G. A. Deschamps, Microstrip microwave antennas. *USAF Symp. Antennas, 1953* (1953).

5.2 K. R. Carver and J. W. Mink, Microstrip antenna technology. *IEEE Trans. Antennas Propag.* **AP-29** (1), 2 (1981).

5.3 R. J. Mailloux, J. F. McIlvenna, and N. P. Kernweis, Microstrip array technology. *IEEE Trans. Antennas Propag.* **AP-29** (1), 25 (1981).

5.4 I. J. Bahl and P. Bhartia, *Microstrip Antennas*. Artech House, Dedham, MA, 1980.

5.5 J. R. James, P. S. Hall, and C. Wood, *Microstrip Antenna Theory and Design*. Peter Pelegrinus, London, 1981.

5.6 J. R. James and P. S. Hall, *Handbook of Microstrip Antennas*. Peter Pelegrinus, London, 1989.

5.7 A. C. Ludwig, The definition of cross-polarization. *IEEE Trans. Antennas Propag.* **AP-21**, 116 (1973).

5.8 R. S. Elliott, The design of waveguide-fed slot arrays. In *Antenna Handbook, Theory, Applications and Design*, Chapter 12. Van Nostrand-Reinhold, New York, 1968.

5.9 R. T. Compton, Jr. and R. E. Collin, Slot antennas. In *Antenna Theory* (R. J. Collin and F. J. Zucker, eds.), Part 1, Chapter 14. McGraw-Hill, New York, 1969.

5.10 H. Y. Yee, Slot antenna arrays. In *Antenna Engineering Handbook* (R. C. Johnson and H. Jasik, eds.), Chapter 9. McGraw-Hill, New York, 1961.

5.11 R. J. Mailloux, *Phase Array Antenna Handbook*. Artech House, Norwood, MA, 1994.

5.12 B. Rama Rao, 94 GHz slotted waveguide array fabricated by photolithographic techniques. *Electron. Lett.* **20** (4), 156 (1984).

5.13   H. Y. Yee, Impedance of a narrow longitudinal shunt slot in a slotted waveguide array. *IEEE Trans. Antennas Propag.* **AP-22**, 589 (1974).

5.14   A. A. Oliner, The impedance properties of narrow radiating slots in the broadface of rectangular waveguide. *IEEE Trans. Antennas Propag.* **AP-5**, 12 (1957).

5.15   P. S. Hall and J. R. James, Survey of design techniques for flat profile microwave antennas and arrays. *Radio Electron. Eng.* **48** (11), 549 (1978).

5.15a K. F. Lee and J. S. Dahele, Characteristics of microstrip patch antennas and some methods of improving frequency agility and bandwidth. In *Handbook of Microstrip Antennas* (J. R. James and P. S. Hall, eds.). Peter Pelegrinus, London, 1989.

5.16   R. Munson, Conformal microstrip antennas. *Microwave J.*, March, p. 91 (1988).

5.17   Y. T. Lo, D. Solomon, and W. F. Richards, Theory and experiment on microstrip antennas. *IEEE Trans. Antennas Propag.* **AP-27**, 137 (1979).

5.17a L. C. Shen et al., Resonant frequency of a circular disk, printed-circuit antenna. *IEEE Trans. Antennas Propag.* **AP-25**, 595 (1977).

5.18   J. R. James, A. Henderson, and P. S. Hall, Microstrip antenna performance is determined by substrate constants. *Microwave Syst. News*, August, p. 73 (1982).

5.19   J. M. Griffin and J. F. Lowth, Broadband microstrip patch antennas. *Milit. Microwaves Conf. Proc.*, *1984*, London, p. 237 (1984).

5.20   B. Roudot, J. Mosig, and F. Gardiol, Surface wave effects on microstrip antenna radiation. *Microwave J.* **31** (3), 201 (1988).

5.21   D. M. Pozar, General relations for a phased array of printed antennas derived from infinite current sheets. *IEEE Trans. Antennas Propag.* **AP-33** (5), 498 (1985).

5.21a D. M. Pozar, *Antenna Design Using Personal Computers*. Artech House, Dedham, MA, 1985.

5.22   D. M. Pozar and D. H. Schaubert, Scan blindness in infinite phased arrays of printed dipoles. *IEEE Trans. Antennas Propag.* **AP-32**, 602 (1984).

5.23   P. D. Patel, Approximate location of scan-blindness angle in printed phased arrays. In Hal Schrank's Antenna Designer's Notebook. *IEEE Antennas Propag.* **34** (5), 1992, 53 (1992).

5.24   D. M. Pozar and D. H. Schaubert, Comparison of architectures for monolithic phased array antennas. *Microwave J.*, March, p. 93 (1986).

5.25   D. M. Pozar, Rigorous closed-form expressions for the surface wave of printed antennas. *Electron. Lett.* **26** (13), 954 (1990).

5.26   D. M. Pozar, Microstrip antennas. *Proc. IEEE* **80** (1), 79 (1992).

5.27   F. K. Schwering, Millimeter wave antennas. *Proc. IEEE* **80**, Iss. 1, 92 (1992).

5.27a P. B. Katehi, D. R. Jackson, and N. G. Alexopoulos, Microstrip dipoles. In *Handbook of Microstrip Antennas* (J. R. James and P. S. Hall, eds.). Peter Pelegrinus, London, 1989.

5.28   D. M. Pozar, Considerations for millimeter wave printed antennas. *IEEE Trans. Antennas Propag.* **AP-31**, 740 (1983).

5.29　B. Edward and D. Rees, A broadband printed dipole with integrated balun. *Microwave J.*, May, p. 339 (1987).

5.30　S. M. Rao et al., Electromagnetic radiation and scattering from finite conducting and dielectric structures: Surface/surface formulation. *IEEE Trans. Antennas Propag.* **39** (7), 1034 (1991).

5.31　R. I. Wolfson and W. G. Sterns, A high-performance, microstrip, dual-polarized radiating element. *1984 IEEE AP-S Dig.* **2**, 555 (1984).

5.32　J. P. Daniel et al., Research on planar antennas and arrays: Structures rayonnantes. *IEEE Antennas Propag. Mag.* **35** (1), 14 (1993).

5.33　F. Johansson, L. Rexberg, and N. O. Peterson, Theoretical and experimental investigation of large microstrip array antenna. *IEE Colloq. Recent Developments Microstrip Antennas*, p. 4/1 (1993).

5.33a　K. Ito, T. Teshirogi, and S. Nishimura, Circularly polarised antenna arrays. In *Handbook of Microstrip Antennas* (J. R. James and P. S. Hall, eds.). Peter Pelegrinus, London, 1989.

5.33b　T. Shiba, Y. Suzuki and N. Miyano, Suppression of higher order modes and cross polarised component for microstrip antennas. *IEEE AP-S Int. Symp. Antennas Propag. Dig.*, p. 385 (1982).

5.34　M. H. Paquay et al., A dual polarised active phased array antenna with low cross polarisation for a polarimetric SAR. *Int. Conf. Radar* 92, Oct. 12–13, 1992, p. 114.

5.35　J.-F. Zurcher et al., Dual polarized, single- and double-layer strip-slot-foam inverted patch (SSFIP) antennas. *Microwave Opt. Technol. Lett.*, **17** (9), 406 (1994).

5.36　R. M. Sorbello, Advanced satellite antenna development for the 1990s. *AIAA Int. Commun. Satellite Sys. Conf.*, *12th*, Arlington, VA, 1988, Collec. Tech. Pap., p. 652 (1988).

5.37　S. Assailly et al., Low cost stacked circular polarized microstrip antenna. *IEEE Int. Symp. AP-S Soc.*, p. 628 (1989).

5.38　C. Terrei et al., Stacked microstrip antennas, advantages and drawbacks. *PIERS '91*, Cambridge, MA (1991).

5.39　G. Kossiavas and Papienik, A circularly or linearly polarized broadband microstrip antenna operating in L-Band. *Microwave J.*, May, p. 266 (1992).

5.40　A. N. Tulintseff, S. M. Ali, and J. A. Kong, Input impedance of a probe-fed stacked circular microstrip antenna. *IEEE Trans. Antennas Propag.* **39** (3), 381 (1991).

5.41　J. T. Aberle and D. M. Pozar, Phased arrays of probe-fed stacked microstrip patches. *IEEE Trans. Antennas Propag.* **42** (7), 920 (1994).

5.42　D. M. Pozar, Characteristics of infinite arrays of subarrayed microstrip antennas. *IEEE Int. Sympo. Antennas Propag.*, p. 159 (1992).

5.43　E. Levine and S. Shtrikman, Experimental comparison between four dual-polarised microstrip antennas. *Microwave Opt. Lett.*, **3** (1), 17 (1990).

5.44　J. Huang, Dual-polarised microstrip array with high isolation and low cross-polarisation. *Microwave Opt. Technol. Lett.*, p. 99 (1991).

5.45 C. H. Chen, A. Tulintseff, and R. M. Sorbello, Broadband two-layer microstrip antenna. *IEEE Int. Symp Antennas Propag.*, p. 521 (1984).

5.46 J. Huang, A technique for an array to generate circular polarisation with linearly polarised elements. *IEEE Trans. Antennas Propag.* **AP-34**, 1113 (1986).

5.47 S. M. Duffy and D. M. Pozar, Aperture coupled microstrip subarrays. *Electron. Lett.* **30** (23), 1901 (1994).

5.47a H. Legay and L. Shafai, New stacked microstrip antenna with large bandwidth and high gain. *IEE Proc. Microw. Antennas Propag.* **141** (3), 199 (1994).

5.48 K. Itoh, Polarimetric integrated antennas composed of microstrip patches. *Direct and Inverse Method in Radar Polarimetry* (W.-M. Boerner et al., eds.). Kluwer Academic Publishers, Dordrecht/Boston/London, p. 1335 Part 2, 1992.

5.49 P. M. Haskins, P. S. Hall, and J. S. Dahele, Polarisation-agile active patch antenna. *Electron. Lett.* **30** (2), 98 (1994).

5.50 J. S. Roy et al., Circularly polarized far fields of an axially magnetized circular ferrite microstrip antenna. *Microwave Opt. Technol. Lett.* **5** (5), 228 (1992).

5.51 P. R. Haddad and D. M. Pozar, Characterisation of aperture coupled microstrip patch antenna with thick ground plane. *Electron. Lett.* **30** (14), 1106 (1994).

5.52 V. H. Rumsey, *Frequency Independent Antennas*. Academic Press, New York, 1966.

5.53 P. S. Hall, Microstrip antenna array with multi-octave bandwidth. *Microwave J.*, March, Vol. 29, No. 3, p. 133 (1986).

5.54 R. H. DuHamel, Dual polarised sinous antennas, 1986, Feb. 19. Eur. Pat. Appl. 019 8578 (1986); (U.S. Pat. 703, 042 (1985).

5.55 R. C. Hansen, ed., *Microwave Scanning Antennas*, Vol. 2. Academic Press, New York, 1966.

5.56 H. Nagano et al., A spiral antenna backed by a conducting plane reflector. *IEEE Trans. Antennas Propag.* **AP-34**, 791 (1986).

5.57 J. J. H. Wang and V. K. Tripp, Design of multioctave spiral-mode microstrip antennas. *IEEE Trans. Antennas Propag.* **39** (3), 332 (1991).

5.58 V. K. Tripp et al., A versatile, broadband, low-profile antenna. *IEEE*, p. 227 (1993).

5.59 P. J. Gibson, The Vivaldi aerial. *Proc. Eur. Microwave Conf.*, *9th*, Brighton, p. 101 (1979).

5.60 S. N. Prasad and S. Mahapatra, A novel MIC slot-line antenna. *Proc. Eur. Microwave Conf.*, *9th*, Brighton, p. 120 (1979).

5.61 E. Gazit, Improved design of the Vivaldi antenna. *IEE Proc., Part H*: *Microwaves, Opt. Antennas* **135** 89 (1988).

5.62 N. Fourikis, N. Lioutas, and N. V. Shuley, Parametric study of the co- and cross-polarisation of tapered planar and antipodal slotline antennas. *IEE Proc., H: Microwaves, Opt. Antennas* **140** (1), 17 (1993).

5.63 L. D. S. Langley, P. S. Hall, and P. Newham, Novel ultrawide-bandwidth Vivaldi antenna with low cross polarisation. *Electron. Lett.* **19** (23), 2005 (1993).

5.64 K. S. Yngvesson, J. Johansson, and E. L. Kollberg, Millimeter wave imaging system with an endfire receptor array. *Conf. Dig. Int. Conf. Infrared Millimeter Waves*. IEEE, New York, p. 189, 1985.

5.65 S. Y. Kim and K. S. Yngvesson, Characteristics of tapered slot antenna feeds and feed arrays. *IEEE Trans. Antennas Propag.* **38** (10), 1559 (1990).

5.66 M. J. Povinelli and J. A. Johnson. Design and performance of wideband dual-polarized stripline notch arrays. *IEEE AP-S Int. Symp.* **1**, p. 200 (1988).

5.67 M. J. Povinelli, A planar broad-band flared microstrip slot antenna. *IEEE Trans. Antennas Propag.* **AP-35** (8), 968 (1987).

5.68 M. J. Povinelli, Further characteristics of wideband dual-polarized microstrip flared slot antenna. *IEEE AP-S Int. Symp.* **2**, 712 (1988).

Index

Amplifier (see Tube, vacuum and solid-state devices)
AMRAAM, 80
Antenna (filled/continuous aperture)
  difference patterns, 91–100
  directivity, 24
  dual reflector systems
    Cassegrain symmetrical, 101, 106
      near field, 106–107
    Gregorian symmetrical, 106
  efficiency, ohmic, aperture, 24
    conventional measures, 101–105
      beam, 101, 158, 161
      blocking, 102, 106
      illumination, 102, 103, 158, 161–162
      scattering, 102
        random surface errors, 103
        systematic surface errors, 103–105
      spill-over, subreflector, main, 102–103
    unconventional measures, 105
    radio image, 105
  far-field, patterns, 90–100
  focal plane imaging systems, 114
    Cassegrain, Gregorian, 114
    prime focus, 114–115
    spacing of antenna elements, 116
    applications, 116
      applied science, 117–119
      radioastronomy, 116–117
      total number of antenna beams, 105, 115
  fundamental limitations, 122
  gain, 24
  HPBW, 90, 94
  illumination, functions, 94–101
  lambda, functions, 90, 93–100, 178

  Lommel, property, 94
  maximum diameters, 105
  modern approaches, 122–123
  offset reflectors, 106
    prime focus, 106–107
    Cassegrain, 106–107
    near field, 106–107
    Gregorian, 109–112
    cross-polarization, g. o., 107–109
      cancellation, Gregorian, Cassegrain, 109–114
  sum pattern, 91–100
Antenna element/s
  arrays scanned in one dimension, for, 340–342
  arrays scanned in two dimensions, for, 342
  candidates, 33
  concluding remarks, 385–386
  FMA, ferrite microwave antenna, 378
  ideal feed horns, 127
    co-polarization, patterns, 127–128
    cross-polarization, patterns, 127–129
  integration to the T/R module, 223, 302–303
  microstrip
    design guidelines, (conventional microstrip), 358–359
    dipole, 360–363
      applications, crossed-, 364–365
      comparisons with patches, 361–363
      feeding methods, 360
      printed, wideband, 363–364
    dual-polarized, quest for, 365–378
    high-quality, quest for, 365–378
    mutual coupling, 353

INDEX **421**

Antenna element/s (*Continued*)
　microstrip (*Continued*)
　　patches, 342-360
　　　array context, 366, 368
　　　bandwidth, 348-351
　　　circular, 343
　　　circular, polarization, yielding, 344
　　　circular polarization, two/four feed, 367
　　　comparisons with dipoles, 361-363
　　　conventional, dual polarization, 368-372
　　　　co-/cross-polarization, 370-372
　　　cross-polarization, worst case, 350, 353
　　　dimensions, 342-346
　　　efficiency, 350-351
　　　electric field, 346
　　　EMC, 366, 369, 372
　　　far-field, 346-348
　　　feeding, methods, 359-360
　　　limitations, 360
　　　methods of feeding, 359-360
　　　Q_T, Q_{SW}, Q_d, Q_c, 348-350
　　　rectangular, 343-344
　　　subarrays, similar, 368, 372-376
　　　　applications, 368, 374-375
　　　　reflection coefficient, 372-374
　　　　beam efficiency, 372-374
　　　subarrays, sequential, 375-377
　　　TEM, quasi, 342
　　　TM_{10}, TM_{02}, 348
　　　scan blindness, surface waves, 134, 352-358
　　　　angles, 356-357
　　　　surface waves, losses, 358-359
　　　　TE, TM, surface waves, 355
　　wideband, 223, 225
　　　flared slot, 385
　　　frequency independent, 378
　　　log-periodic, 223, 379-381
　　　spiral-mode, 223, 379, 382
　　　slotline, tapered, Vivaldi, 223
　　　planar, antipodal and symmetric antipodal, 379, 381-385
　　　　input return loss, 384
　　　　cross polarization, 384
　　　planar slots, Povinelli, 223, 385
　　work in progress, 376, 378
　　shortcomings, 338
　PCA, 223, 337
　requirements, outline, 338-339
　tunability, 224
Antenna temperature, *see* Radioastronomy
AOA, angle of arrival, 48
Archimedes, burnt the Roman ships, 1, 3

ARM, antiradiation missile, 35
ARPA (DARPA), 223, 259, 270, 315
Array, phased
　active, 201-203
　active hybrid, 187-191
　advanced, redefine radar, 84
　AESA, 85
　affordable, 255-257
　architecture, 210-212
　　applications, 220-223
　　brick, LITA, 212
　　tile, TILA, 212
　AGF, 135-137, 139-140, 146, 151, 156, 157, 165, 170, 174, 190-191
　annulus, 174
　circular, array, 175-176
　bandwidth, 143-144
　beamformers, (*see also* Beamformers), 241-254
　　operating at IF, RF, 250
　beamwidth, 143-144
　beam broadening, 149, 152, 153, 191
　beam, main, efficiency, 55, 158, 161
　binomial, 146-147
　conformal, 8
　canonical, narrow band, 131
　cost, 215-216, 274, 281-282
　　acquisition, initial, 216, 274, 281-282
　　LCC, life-cycle, 216, 274, 281-282
　cooling, 217-218
　circular, cylindrical, 173-174
　EIRP, 205-206
　electronically steerable, one dimension, 14
　electronically steerable, two dimensions, 14
　errors
　　quantizations, 189-196
　　　amplitude, lobes, 189-192
　　　delay, 195-196
　　　phase, 192-195
　　　phase, delay, applications, 196
　　random, amplitude, phase, 196-201
　　　examples, 198-199
　fault isolation/correction, 254-255
　fixed beam, 11-12
　FOV, field of view, 132-134
　　instantaneous, 132-133
　　total, 132-133
　　　horizon, 132-133
　　　surveillance volume, 132-133
　future trends, 133
　geodesic applications, 135
　G/T, 69, 205
　grating lobes, 133-134, 138-140, 142-143, 165-170, 176, 189-191, 193-195

Array, phased (*Continued*)
  inertialess beam, 7, 85
  historical developments, 18
    radar, 19, 20
    radioastronomy, 20, 21
  interconnect approaches, 212–214
    planar corporate feed, 213–214
    series-fed, 214
    space-fed, 214
      lens type, 214
      reflect-array, 214–215
    volume corporate feed, 214
  interelement spacing, 142–143
  linear, 10
    bandwidth, 143–144
    directivity, 144–145
    far-field radiation pattern, 135–142
    grating lobes, 138–140, 142–143, 189–192, 193–195
    equispaced elements, non-uniform amplitudes, 146–164
    HPBW, 143–144
    (HPBW) × (directivity) product, 145
  main lobe power, ML, 227
  MBA, 231–233
  M/NRA, 65–66, 238–240
    applications, 240–241
    ESTAR, 65–66, 240–241
  mounted on fully steerable structures, 123
    annular sythesis antennas, 123–126
    MAM, multiplebeam array model, 123
  multiplicative
    compound interferometers, 170
  offering, 7–8, 21–24
  passive, 202–203
    losses, 208–209
  performance monitoring, 254–255
  planar
    directivity, 171–172
    far-field radiation pattern, 164–169
    grating lobes, 165–169
      system considerations, 169–170
    HPBW, 171–172
    (HPBW) × (directivity) product, 172
    sidelobe
      average sidelobe level, 227
      design, 200–201
      level, SLL, 147–148, 152–155, 158–163, 181–183, 187–189
      MSSL, 197–199
      residual, 200–201
      specified, 200–201
      statistical prediction of peak, 199–201
      system considerations, 170
    sin θ space, 166–169

  power supplies, 218–220
    SMPS, 219–220
    linear, 219–220
  power addition, 7, 275
    spatial, 321–323
  remote sensing, 135
  SNR, 173
  scan blindness, and mutual coupling, *see* Antenna elements
  Schelkunoff circle, 151, 153
  scanning, 69–72
  spatial tapers
    applications, 184
    Frank–Coffman, 187–191
    Mailloux–Cohen, 187
    Willey, 182
  staring, 69–72
  subarrays, *see* Antenna elements, subarrays
  synthesis procedures, 147–164
    Bayliss, 161–164
    Dolph–Chebyshev, 147–153
      directivity, 150, 152
    Elliott, 153–156
    Hansen, circular distribution, 160–161
    Lee, J. J., 179–182
    other, 164
    Taylor, 186
      one parameter, 156–158
      $\bar{n}$ distribution, 158–160
      $\bar{n}$ circular distribution, 161
    Wild, J.P., J_n^2, 174–178
  thermal problems, 216
  wide band, 81–82, 131, 224–238
    Cohen, M. N., approach, 238
    Cottony approach, 236–238
    AEW, U. S. Navy, 234–235
    examples/applications, (ESM, ECM), 229–234
    programs, ICNIA, PAVE PILLAR, 235
ASCM, 230
ASCE, *see* Radar, air traffic control
Attenuation
  atmospheric, 26–28
  depolarization, due to rain, 39

Beam (main) efficiency, 55, 101, 158, 161
Beamformers, 241–252
  agile beam, applications, 250–251
  agile beams, several, 251–252
  beam cross-over, 243
  Blass matrix, 248–249
  Butler matrix, 248–249
  cables using, 244–245
  general considerations, 243–244

INDEX 423

Beamformers (*Continued*)
  digital, 253
  photonics-based, 226, 245–247, 252–253
    losses, 226
  non-linear, 253–254
    RBF, 254
  resistive networks, 247–248
  scanning beams, 250–251
  staring beams, 244–249
BJT, *see* solid-state devices
Black-body radiation, 53–54
BWO, *see* Tube, vacuum

CBIR, 64–65
CFA, *see* Tube, vacuum
CFT, *see* Tube, vacuum
Chebyshev, functions, 148–150
Christiansen's (Chris) cross, *see*
  Radioastronomy systems
CIM, 317
Circulators, 331–335
  ferrite, 334–335
  miniature, 335
  MMIC-based, 335
  requirements, 332–333
C/N, carrier-to-noise ratio, 68
COMINT, 47, 78
Combining efficiency, 288–291
Communications, 35
  satellite, systems, 68–69
Cooperative Engagement Capability, 1

DARPA (ARPA), 223, 259, 270, 315
Depolarization, rain induced, 39
Duty factor, 25, 274
  SSD, 290–292

ECCM, 5, 35, 47, 50–51, 82, 87, 170, 255
ECM, 48, 49–50, 82, 134, 255
EDFI, 245–246
EFIE, 363
EIKA, 261
EIKO, 261
EIRP, *see* Array, phased
ELINT, 47, 78
EMI, 252
EMP, 293
EOM, 245–246
EPC, *see* Module MPM
ERP, 49
ESM, 35, 47–48, 78, 82. *See also* Arrays, phased, wideband
ESTAR, 65–66, 240–241
EW, 35, 80, 87

advanced systems, 80
concepts, 47
ECM, *see* ECM
ECCM, 5, 35, 47, 50–51, 82, 87
ESM, *see* ESM

Flux density, *see* Radioastronomy
FMA, ferrite microwave antenna, 378
FMCW, 36, 72–73
  PILOT, 72–73
Frequency bands, designations, 6
FPA, 202, 208

GBR, ground based radar, 221
global star, 87
g. o., cross polarization, 107–109
graceful degradation, 274–275, 286–288

HBT, *see* Solid-state devices
HDMP, 315, 317
HEMT, *see* Solid-state devices
Hot maintenance, 288
Hybrid, limited scan, system, 119–120
  applications, 83, 120–122
Hybrid architecture, 203

ICBM, 184
IFF, 78
IMPATT, *see* Solid-state devices
INTELSAT, 68–69
Iridium, 87
ISAR, *see* Radar

J/S, jam-to-signal ratio, 50
Jansky, K. G., 52

KGD, 316

LBT, *see* Tube, vacuum
LCC, *see* Array, phased, cost
Leonet 1–2, 87
Limiter, protector, receiver, 324–326
light bucket mode, 7
LITA, *see* Phased, array, architecture
LPI, low probability of intercept, 35–36, 72, 82
  PILOT, 72–73

MAFET, 315, 318
MAG, 310
MAM, 123–124
MCA, 318
MCM, 221
MESFET, *see* Solid-state devices
MHDI, 318
MIMIC, 223, 259, 336

424 INDEX

Mills cross, *see* Radioastronomy, systems
MMA, millimeter array, 86
M/NRA, *see* Array, phased
Module
   double and single chip, 312–315
      MCM, recent experiences, 316–317
      COBRA, 316–317
      MPM, 260, 270, 294–298, 299
         applications, 294
         EPC, 296–297
         MMIC amplifier, 294–296
         VPB, 295
   RF wafer scale integration, 315–316
Molecules, interstellar, 52
MSAG, *see* Yield
MSSL, *see* Array, phased, sidelobe
MTBCF, 220
MTBF, 81, 85, 218, 274
   SSD, 280–281
   modules, 281
   tubes, 280
MTI-stability, 274, 286

Noise figure, 201–202, 277
   power preamplifier, 279
   TWT, 277
Non-cooperative target recognition, 83
NPD, *see* Thermal, NPD

Odyssey, 87
Oscillator, *see* Tubes, vacuum and SSD
Optical astronomy, developments, 86
OTHR, *see* Radar, systems
OVLBI, *see* Radioastronomy systems

PAE, 217, 276, 282–286, 300, 320
PAR, 83, 121
PAVE PAWS, *see* Radar, systems
PCA, 223, 337
PILOT, 72–73
POI, probability of interception, 49
   concepts, 49
Polarizations, *see* Radar, polarimetric
power addition, 7, 275
   spatial, quasi-optical, 321–323
Power supplies, *see* Array, phased
Power output improvement per decade, 291–292
Phase-shifters
   analog, 327–328
   digital, 328
   ferrite, 327
   IF/RF, 330–332
   programmable, 326–329

   requirements, 326
   solid-state, 327–329
   vector modulator, 251, 303–304, 329–330

Radar
   AAR, active array radar, 216, 282
   AESA, 85, 215
   AEW, U. S. Navy's, 234–235
   AFCR, 81
   Air traffic control
      ASR, PAR, TDWR, ASDE, 83
   bearing, 4
   bistatic, 41–46, 81
      scanning rate, 45–46
      contours of constant detection range, 44–45
   equation, 25–31, 203
   examples, illustrative, 31–34
   functions, 4, 69
      detection, 4
      identification, 5, 74–75
         unconventional approaches, 78
      other, 22–23
      surveillance, 4, 34
      tracking, 4, 34–35
         friend-or-foe, 78
   FMCW, 36, 72–73
      PILOT, 72–73
   gain, 24
   future challenges, 79–81
      LPI, low probability of interception, 35–36, 72, 82
      ISAR, inverse SAR, 46–47, 74–78
         phased array-based, 77–78
      multistatic, 41–46
      multifunction, 35, 72, 74–75, 82
         shared aperture, 81–84
         wideband, multiband, 81–84
   polarimetric, 36–41
      benefits, 41
      implementation, 38–39
      instrumental requirements, 39–41
      theoretical considerations, 36–38
      polarization, isolation, 39–41
      polarization diversity, 303–304
   pulsed, 25, 35, 72
   RCS, 85
   RWR, 48
      SNR, 48
   SAR, 41, 46, 74–75, 256
      UAV, on board, 84, 256
   scanning, 69–71
   SNR, considerations, 28–30
   staring, 69–71

INDEX   425

Radar (*Continued*)
  systems, 9–18
    advantages (over conventional systems), 9
    Aegis, on board, 84
    air-traffic control, 83
      ASR, PAR, TDWR, ASDE, 83
    AWACS, 14, 16, 170, 340–341
    B-1B, on board, 84
    BMEWS, 281
    Cobra Dane, 184, 196, 208
    ELRA, 251
    Ericsson's AEW, 10
    ERIEYE, 340–341
    F-15, -16, -18, on board, 13, 14, 15
    GBR, 221
    Martello, 71
    Mig-31, on board, 84
    NAVSPASUR, 281–283
    OTHR, 133–134
    PAR, 83, 121
    Patriot, 214–215
    PAVE PAWS, 17, 19, 89, 281
    Rafale, on board, 84
    SEASAT, 12, 13
    SIR-A, 12, 13
    URR, 218–220
  wideband, the case for, 81–84
Radioastronomy
  antenna temperature, (equivalent), 52, 55, 206–207
  basic relationships, 55
  arrays, *see* Array, phased
  continuum observations (spectral), 52
  developments, 86
  emission mechanisms, 55–56
  flux density, 51, 55
  fundamental, concepts and measurements, 51–52
  image theory, 57–58
    Fourier transform-by-Fourier transform, 58, 65
    picture point-by-picture point, 58, 60
  molecular, (spectral line) observations, 52
  objectives, 5
  radiotelescopes, 5
    resolution, 5, 7, 8
  systems, 51
    Christiansen's (Chris) cross, 21, 23, 170
    Mills cross, 21, 22, 60–61, 64, 247
    Radioheliograph, Culgoora, 60–63, 178
    OVLBI, 59, 67, 88, 135
    VLA, 17, 18, 66, 89
    VLBA, 67
    VLBI, 59, 66, 135

    future, with many FOV, 130
Radiometers, 56, 59
  radiometry, 82
  radiometric, systems, 87
  total power, 56–57
  Dicke-switched, 56–57
RAPPORT, 234
RBF, 254
Receiver protector, limiter, 324–326
RF wafer-scale integration, *see* Yield
RIN, 245
RWR, *see* Radar

SAG, *see* Yield, MSAG
SAR, *see* Radar, systems
S/C, signal to clutter ratio, 41–42
SFDR, 246
SMM, *see* Antenna elements
SMPS, *see* Arrays, phased, power supplies
SNR, 28–30, 48, 173
Solid-state
  advantages offered by, 298–299
  amplifiers
    FPA, 202, 208
    LNA, 201–203, 205, 260, 324
    baseline characteristics
      power, chips, 318
      power, 321–322, 324
      LNA, chips, 318
      LNA, 323–325
      other, 318–321
  devices, 271–273
    BJT, 276, 278
    FET, 272
    GUNN, 276
    HBT, 272, 276–277
    HEMT, 272–273, 276–277
    IMPATT, 272, 276, 278–279
    MESFET, 272, 276, 278
    MODFET, 272
    MSAG, *see* Yield
    SAG, *see* Yield
    TEG-FET, 272
    thermal NPD, 277–280
    transmitters, 271–273, 277
  power output versus frequency, 274–277
Solar Thermal Test Facility, 1–2
Specific-weight factor, 292–293
spectral purity, SSD and tubes, 283
synchrotron radiation, 55–56

Target
  bearing, 4

Target (*Continued*)
  Doppler shift, 4
  identification, 5
  radial velocity, 4
  range, 4
  range rate, 4
  RCS, 26, 79, 87
    reduction, 79–80
TDRSS, tracking and data relay satellite system, 67
TDWR, *see* Radar, air-traffic control
TEM, quasi, *see* Antenna elements
Thermal power noise density, 277–280
Thinned aperture, 182–186
TILA, *see* Array, phased, architecture
TOA, time of arrival, 48
T/R, module, 84, 85, 87, 134, 155, 179–180, 182, 186–188, 191, 202, 205, 208, 216, 223, 257–258, 259–260
  concluding remarks, 335–336
  solid-state, MMIC, 300–318
  MMIC, realization options, 301
  realization approaches, 303–306
  MMIC with or without the antenna, 302–303
  wideband, 225
TSA, tapered slotline antenna, *see* Antenna elements
Tube, vacuum
  advantages offered by, 299
  BWO, 261, 270
  CFA, 261, 269
  fast-wave devices, 269
    gyrotron, 269, 270–271, 276
  slow-wave, cross-field tube, 266–267
    magnetron, 266, 269, 270–271
  CFA, 266–267, 276
  slow-wave, linear beam tube, 261
    klystron, 261–264, 276
    amplifier, 269
    EIKA, 261
    EIKO, 261
  thermal NPD, 277–280
  TWT, 84, 85, 208, 261, 264–270
  TWTA, 261
  CCTWT, 265–266, 269

URR, 218–220
UAV, 84, 256

Vector modulators, *see* Phase-shifters
VLA, VLBA, VLBI, *see* Radioastronomy systems
VPB, *see* Module, MPM

WDM, 246

Xu, solutions proposed by, 81

Yield
  MMICs, 305–318
  deterministic approaches to increase, 306
  design guidelines for increased, 307–308
  high yield processes
    MSAG, 308–310
    RF wafer-scale integration, 315–317
    other approaches, 310–311
  intrinsic, 311
  statistical approaches to increase, 305–306